Plant ecology in the subarctic
Swedish Lapland

Ecological Bulletins No. 45

Plant ecology in the subarctic Swedish Lapland

Edited by
P. S. Karlsson and T. V. Callaghan

**This volume is dedicated to the Director
of Abisko Scientific Research Station 1974–1996,
Professor Mats Sonesson.**

Photo: Anita Sonesson

Contents

Ecological Bulletins

ECOLOGICAL BULLETINS are published in cooperation with the ecological journals Ecography and Oikos. Ecological Bulletins consists of monographs, reports and symposia proceedings on topics of international interest, often with an applied aspect, published on a non-profit making basis. Orders for volumes should be placed with the publisher. Discounts are available for standing orders.

Editor-in-Chief and Editorial Office:
Pehr H. Enckell
Joint Editorial Office
Ecology Building
S-223 62 Lund
Sweden
Fax: +46-46 222 37 90

Technical Editors:
Linus Svensson and Gunilla Andersson

Editorial Board:
Björn E. Berglund, Lund.
Tom Fenchel, Helsingør.
Erkki Leppäkoski, Turku.
Ulrik Lohm, Linköping.
Nils Malmer (Chairman), Lund.
Hans M. Seip, Oslo.

Published and distributed by:
MUNKSGAARD International Publishers Ltd.
P.O. Box 2148
DK-1016 Copenhagen K, Denmark
Fax: +45-33 12 93 87

and

MUNKSGAARD International Publishers Ltd.
238 Main Street
Cambridge, MA 02142-9740
USA
Fax: +1-617 547 7489

Suggested citation:
Karlsson, P. S. and Callaghan, T. V. (eds) 1996. Plant ecology in the subarctic Swedish Lapland. – Ecol. Bull. 45.

or

Author's name. 1996. Title of paper. – Ecol. Bull. 45: 000–000.

Cover photo: P. S. Karlsson ©

Preface

The Torneträsk region of northern Swedish Lapland is dominated by scenically beautiful and varying landscapes. It is rich in wildlife, and contains important geological and geomorphological features. The area has been settled for some hundreds of years by reindeer herding Same and is largely unspoiled by human land use. The Torneträsk region is an important part of the Fennoscandian subarctic birch forest zone which is one of the last remaining wildernesses of Europe.

Within the Torneträsk region, the Royal Swedish Academy of Sciences operates the Abisko Scientific Research Station (ANS), first established in 1912, to undertake and facilitate research into the natural sciences of the area. The excellent facilities and location of the ANS have provided an important focus for Swedish and international scientists and also for environmental monitoring together with national and international projects within large collaborative research programmes. Recently, the worrying trends and possibilities of many environmental impacts such as global change, which are likely to amplified in the North, have stimulated even greater activity at ANS which provides critical baseline and long term records of environmental history which can be used to assess recent changes. Currently, the ANS hosts c. 100–125 scientists each year.

Since 1973, the ANS has been directed by Professor Mats Sonesson, and many of the developments and achievements of the Station and the researchers working there owe their success to his encouragement, foresight and support. Mats came originally from the north of Sweden, Arnundsjö in Ångermanland, but moved to Lund to study at the University there. His Ph. D. thesis "Studies on mire vegetation in the Torneträsk area, Northern Sweden" and subsequent research addressed particularly the poor mire systems of the region and his formative reconstruction of historical vegetation development is confirmed in this Festschrift using modern techniques unavailable to Mats. His classification of modern forest structure is, again, the definitive study and is cited in many of the ecological papers originating from the Torneträsk area.

Following his Ph. D., Mats became leader of the Swedish International Biological Programme (IBP) Tundra Biome project based at Stordalen, near to Abisko. This study was a successful and important contribution to the IBP and established baseline data from the early 1970's which are currently in use. At the end of this project, Mats applied for, and was appointed to the Directorship of the Abisko Station. While encouraging all fields of research, his interest in plant ecology stimulated a rapid expansion of existing and new concepts and methodologies within this subject area. Concurrently, he expanded the station adding new laboratories, workshops and accommodation which transformed the station to a facility well suited for both field and laboratory studies. This led to new cohorts of Swedish plant ecologists and a network of close international collaboration strongly focused at Abisko. During this period, Mats has thus been involved in research ranging from vegetation science, through population ecology, to ecophysiology. However, Mats has always had a particularly close interest in mosses and lichens and has been concerned that these groups which are so ecologically important in northern latitudes, are often under-represented in research studies. He has, therefore, concentrated on these methodologically difficult groups and has made important contributions to our understanding of their ecophysiology both in terms of basic research and the implications of changes in atmospheric CO_2 and UV-B radiation. In addition to stimulating research at Abisko, Mats has been an important ambassador for the science at Abisko in particular, for the Royal Swedish Academy of Sciences, and for Sweden in general.

Now (June 1996), after 23 formative and productive years as Director of the Abisko Scientific Research Station, Mats will retire from his duties. To honour Mats, to celebrate his career, and to thank him for all that he has done to enhance our research, we have asked those scientists who have worked at Abisko within the subject areas closest to Mats' interests, to contribute to this Festschrift which overviews the ecology and environment of the Torneträsk region. Twenty three chapters written by 41 scientists from 10 countries, have been contributed, and this is a measure of his friends' and colleagues' esteem for him. We wish to thank these authors for contributing to this Festschrift and we also thank the NFR for financially supporting its publication.

On behalf of Mats Sonesson's friends and colleagues, we thank Mats for his contributions to ecological and environmental research in the North, for his participation in, and facilitation of our research, and for his great hospitality at the ANS; we wish him a happy and fulfilling retirement with his wife Anita and family, and we formally dedicate this Festschrift to him.

Terry V. Callaghan
P. Staffan Karlsson

Ecological Bulletins 45: 11–14. Copenhagen 1996

The Abisko Scientific Research Station

N. Å. Andersson, Abisko Scientific Research Station, S-981 07 Abisko, Sweden. – T. V. Callaghan, Sheffield Centre for Arctic Ecology, Univ. of Sheffield, Tapton Experimental Gardens, 26 Taptonville Rd, Sheffield, U.K. S10 5BR. – P. S. Karlsson, Dept of Ecological Botany, Univ. of Uppsala, Villavägen 14, S-752 36 Uppsala, Sweden.

The Abisko Scientific Research Station is one of the oldest and largest scientific field stations north of the Arctic circle (Sonesson 1987). The purpose of the station is "to provide Swedish and foreign visiting scientists with the opportunity of conducting scientific work based on the specific conditions of the environment surrounding the station and to conduct such research with its own personnel". The station's location at the southern shore of Lake Torneträsk, northern Sweden, at 68°21′N, at an altitude of 380 m, c. 300 m below the alpine tree line, provide convenient access to a wide range of subarctic, subalpine and alpine environments.

There are several reasons for the strong, and currently increasing (see below), interest from scientists to base their field work at the Abisko station. Sonesson (1979) summarised them as: 1) the pristine nature of the station's surroundings, i.e. the Torneträsk area; 2) accessibility; 3) the geological, climatological and biological diversity of the Torneträsk area and 4) research facilities and research traditions. In the light of the development during the last decade with an increasing concern about the effects of human induced changes of the global environment, one could add the value of a long record of scientific documentation of the area including the station's climatological database – the Abisko station has continuous climate records since 1913. Furthermore, the facilities at the station have, both in terms of personnel and laboratories, been expanded to enable the station to offer excellent working conditions, comparable to fully equipped university laboratories. Below we will briefly expand on some of these points.

The historical background of the Abisko station

The history of the Abisko station has been compiled by Bernhard (1989a, b) and the presentation below is mainly extracted from these sources. Although the history of the scientific exploration of this area goes back c. 200 yr, the completion of a railway between Kiruna and Narvik in 1902 dramatically improved the access to, and thus the scientific activity in this area. A small research station was established 1903 at Katterjokk 28 km north-west of Abisko. A few years later, in 1912, it was replaced by a new research station in Abisko. The Royal Swedish Academy of Sciences took over the responsibility for the station in 1934. The first full time Director of the station, G. Sandberg, was in charge from 1949 to 1974 when Professor Sonesson was appointed as Director.

During the last twenty years, i.e. during the leadership of Prof. Sonesson, the station has grown considerably in terms of facilities and personnel. The station now has well equipped workshops, laboratories and dormitories with a capacity for up to 70 visitors and hosts, besides research teams, student groups, field courses, excursions and conferences.

The environment

The catchment area of lake Torneträsk is c. 3290 km^2 and ranges in altitude from the lake itself at 342 m up to peaks > 1900 m (Fig. 1). Due to the proximity to the Atlantic ocean, the climate of the Torneträsk area is warm in comparison to other areas at the same latitude. The large-scale climatic conditions in the area can thus be classified as Subarctic (cf. Löve 1970, Barry and Ives 1974). At the Abisko station's climatological observatory (at 388 m altitude), the July mean temperature is 11.0°C (30-yr mean for 1961–1990, Fig. 2) and is thus 1°C warmer than the 10°C July isotherm that is often used to delimit the Arctic region (Barry and Ives 1974). The winter months are relatively warm, e.g. January has a mean temperature of −11.9°C. A considerable proportion of the Torneträsk area is located at altitudes above the alpine tree line (at c. 650 m altitude). In these areas the summers are cooler while the winter temperatures are warmer.

The geology (Lindström 1987), topography (Rapp 1961, Melander 1980) and climatic conditions vary

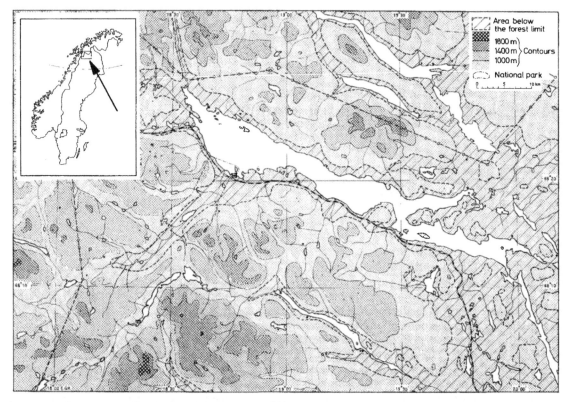

Fig. 1. Map of the Torneträsk area (from Sonesson et al. 1980).

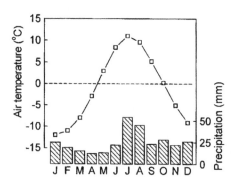

Fig. 2. Monthly mean air temperature and precipitation at the Abisko station's meteorological observatory. All values are 30-yr means for the period 1961–1990.

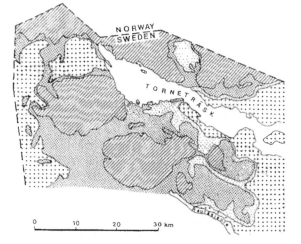

Fig. 3. Major features of regional geology according to Kulling (1964). Adopted and modified from Lindström et al. (1985). Crystalline basement and Precambrian crystalline rocks of western "windows" (crosses), autochthonous sedimentary rocks and Rautas complex (dots), Abisko Nappel (dense dots) and Seve-Köli complex (hatched).

considerably within the Torneträsk area (cf. Figs 1 and 3). For example, the western parts have an oceanic climate with an annual precipitation of c. 1000 mm while the Abisko valley receives only c. 300 mm per year. The variation in these physical factors create a number of environmental gradients which, in turn, contribute to create a high diversity in the biological characteristics.

Although this area is often perceived as a wilderness there have been human activities in the area for at least

400 yr (Sonesson 1979, Emanuelsson 1987). Furthermore, the relatively recent activities with the construction of first a railway (completed 1902) and later a road (open to Abisko 1980) has further increased the human

impact on the area. Nevertheless, the Torneträsk area probably contain some of the least disturbed natural areas in Europe.

Many features of the Torneträsk environment are documented in great detail in a number of publications (e.g. Mårtenson 1956, Ekman 1957, Sonesson 1970a, b, 1979, 1980, Sonesson and Lundberg 1974, Rosswall et al. 1975, Lindström et al. 1985, Lindström 1987 and Rapp 1961, 1986). Many other references can be found in Andersson et al. (1994).

Research activities

When Prof. Sonesson now retires we can look back on a very active period in the Abisko station's history. In the early seventies the Abisko station had c. 200 visitors each year spending a total of about 1500 days at the station. Currently the station hosts c. 6000 visitor-days per year. This include researchers and their assistants as well as student excursion groups and conference participants.

About 2200 reports and papers, in a wide range of topics, have been published based on results collected with the station as base. The number of publications is increasing at a rate of c. 40–50 papers per year. Publications related to the Torneträsk area are recorded in the station's bibliography (Andersson et al. 1994).

Mats Sonesson has put much effort in attracting both young research students as well as senior scientists from Sweden and other, mainly European, countries to locate research projects to the Torneträsk area. Presently, slightly less than half of all visitors (c. 45%) are Swedish and the reminder coming from countries from most continents. Mats Sonesson is a plant ecologist (as well as his predecessor). This has influenced the activities of the station so that plant ecology is one of its major research fields. Although this volume has a plant ecological focus, we want to stress that the station is open for a wide range of scientific activities and that many other fields contribute significantly to the activities of the station. For example, research projects in geology, geomorphology, meteorology/climatology, hydrology, cultural geography, animal ecology, limnology, environmental monitoring (cf. Bernhard 1989a, b) have been, or are, based at the Station.

During the last 20 yr the plant ecological research at the Abisko station has shifted from largely descriptive vegetation science towards more experiment approaches applied to processes at the individual or population levels. As reflected in this volume, the major research focus in plant ecology at the Abisko Station is currently physiological, reproductive and population ecology. These studies focus both on basic biological problems as well as on effects of global changes in environmental conditions such as increased CO_2, temperature and UV-B levels.

Among the large international ecological research projects based at the Abisko station, Mats Sonesson has been personally involved in several. In the first part of the seventies he lead the International Biological Programme (IBP) Tundra Biome project at Stordalen/Abisko (Sonesson 1980). During the latter years he has encouraged, and/or taken an active part in, several international teams working on the biological effects of global change phenomena such as the International Tundra Experiment (ITEX), The European Community project "Effects of enhanced CO_2 and UV-B on heath ecosystems". Mats Sonesson has also been responsible for the Sweden's participation in ecological research networks such as the Man and Biosphere (MAB), Northern Sciences Network (NSN) and International Centre for Alpine Environments (ICALPE).

Objective of this volume

This volume is dedicated to Prof. Mats Sonesson and honours his career as scientist and Director of the Abisko Scientific Research Station by presenting papers from a subset of the research carried out there. This particular subset contains papers dealing with many aspects of the interrelationships between plants and their biotic and abiotic environments in the North. This theme is represented by studies which are close to the research interest and expertise of Mats Sonesson. Indeed Mats Sonesson carried out the formative investigations for some of these studies while greatly encouraging and facilitating all of them.

This volume first presents studies on the environment surrounding the Abisko station, describing recent and longer terms changes in environment and vegetation. The following contributions present fundamental studies on plant species, populations and their interactions with the subarctic environment. Finally, the concluding studies address the implications for the vegetation of the Abisko area and arctic in general, of future changes to the environment predicted to occur in response to anthropogenic impacts on the greenhouse gases and stratospheric ozone. We end the volume by reviewing the current state of plant-environment studies represented by the preceding studies and extract implications for potential future directions of research at Abisko.

Acknowledgements – The studies presented and reviewed in this book benefitted greately from the encouragement of Prof. Mats Sonesson, the facilities of the Abisko Scientific Reserach Station, and the hospitality and help of its staff. The publication of this book was possible through a grant from the Swedish Natural Sciences Research Council.

References

Andersson, N. Å., Tjus, M. and Wanhatalo, L. 1994. Abisko bibliography 1992. – Abisko Sci. Res. Stat., The Royal Swedish Acad. Sci.

Barry, R. G. and Ives, J. D. 1974. Introduction. – In: Ives, J. D. and Barry, R. G. (eds), Arctic and alpine environments. Methunen, London, pp. 1–13.

Bernhard, C. G. 1989a. The Abisko Scientific Research Station. – The Royal Swedish Acad. Sci., Stockholm.

– 1989b. Research institutes. – In: Frängsmyr, T. (ed.), Science in Sweden, The Royal Swedish Acad. Sci. 1739–1989. Science History Publ., USA, pp. 262–267.

Ekman, S. 1957. Die Gewässer des Abisko-Gebietes und ihre Bedingungen. – Kungl. Vetenskapsakad. Handl. (Stockholm) Ser. 4 6(6).

Emanuelsson, U. 1987. Human influence on vegetation in the Torneträsk area during the last three centuries. – Ecol. Bull. 38: 95–111.

Kulling, O. 1964. Översikt över norra Norrbottenfjällens kaledonberggrund. – Sveriges Geolog. Undersökning, Ser. Ba 19.

Lindström, M. 1987. Northernmost Scandinavia in the geological perspective. – Ecol. Bull. 38: 17–37.

Lindström, M., Bax, G., Dinger, M., Dworatzek, M., Erdtmann, W., Fricke, A., Kathol, B., Klinge, H., von Pape, P. and Stumpf, U. 1985. Geology of a part of the Torneträsk section of the Caledonian front, northern Sweden. – In: Gee, D. G. and Sturt, B. A. (eds), The Caledonide Orogen – Scandinavia and related areas. John Wiley and Sons, pp. 507–513.

Löve, D. 1970. Subarctic and subalpine: Where and what? – Arct. Alp. Res. 2: 63–73.

Mårtensson, O. 1956. Bryophytes of the Torneträsk Area, Northern Swedish Lappland. III. The General Part. – Kungl. Vetenskapsakad. Avhandl. i naturskyddsärenden (Stockholm) 15: 7.

Melander, O. 1980 Inlandsisens avsmältning i nordvästra Lappland. – Stockholms Univ., Naturgeografiska Inst., Forskningsrapport 36.

Rapp, A. 1961 Recent development of mountain slopes in Kärkevagge and surroundings, Northern Scandinavia. – Geogr. Ann. 42: 65–200.

– 1986. Slope processes in high latitude mountains. – Progr. Phys. Geogr. 10: 53–68.

Rosswall, T., Flower-Ellis, J. G. K., Johansson, L. G., Jonsson, S., Rydén, B. E. and Sonesson, M. 1975. Stordalen (Abisko), Sweden. – In: Rosswall, T. and Heal, O. W. (eds), Structure and function of tundra ecosystems. Ecol. Bull. 20: 265–294.

Sonesson, M. 1970a. Studies on mire vegetation in the Torneträsk area, Northern Sweden. III. Communities of the poor mires. – Opera Bot. 26.

– 1970b. Studies on mire vegetation in the Torneträsk area, Northern Sweden. IV. Some habitat conditions of the poor mires. – Bot. Not. 123: 67–111.

– 1979. Abisko Scientific Research Station: Environment and Research. – Holarct. Ecol. 2: 279–283.

– (ed.) 1980. Ecology of a subarctic mire. – Ecol. Bull. 30.

– , Jonsson, S., Rosswall, T. and Rydén, B. E. 1980. The Swedish IBP/PT tundra biome project objectives – planning – site. – Ecol. Bull 30: 7–25.

– 1987. Preface. – Ecol. Bull. 38: 3.

– and Lundberg, B. 1974. Late Quaternary forest development of the Torneträsk area, North Sweden. 1. Structure of modern forest ecosystems – Oikos 25: 121–133.

Ecological Bulletins 45: 15–30. Copenhagen 1996

Holocene forest dynamics and climate changes in the Abisko area, northern Sweden – the Sonesson model of vegetation history reconsidered and confirmed

Björn E. Berglund, Lena Barnekow, Dan Hammarlund, Per Sandgren and Ian F. Snowball

Berglund, B. E., Barnekow, L., Hammarlund, D., Sandgren, P. and Snowball, I. F. 1996. Holocene forest dynamics and climate changes in the Abisko area, northern Sweden – the Sonesson model of vegetation history reconsidered and confirmed. – Ecol. Bull. 45: 15–30.

A new palaeoecological and palaeoclimatic project based on the study of Holocene lake sediments in the subalpine Abisko area, N Sweden (northern Scandes) is described. The palaeobotanical framework was founded upon the pollen zone system defined by Sonesson. This model of vegetation history has been confirmed, although the chronology is partly revised. Sonesson's pollen diagrams are combined with new results from a lake sediment sequence at the tree-limit, which include sedimentologic, mineral magnetic, oxygen isotope and plant macrofossil studies. Since the deglaciation at c. 9 000–8 500 ^{14}C yr BP the vegetation of the Abisko area has been dominated by a subalpine birch woodland tundra. However, between c. 5 500 and c. 3 500 BP the boreal pine–birch forest zone reached the area. During this period pine expanded to a level of c. 100–150 m higher than today. The climate has generally been subarctic-oceanic since the deglaciation except for the period c. 5 500–3 500 BP when temperate-continental conditions prevailed. This climatic development differs from the situation in the central Scandes. Our studies indicate that after c. 3 500 BP soil erosion increased and the Kårsa glacier reformed due to a climatic cooling.

B. E. Berglund, L. Barnekow, D. Hammarlund, P. Sandgren and I. F. Snowball, Dept of Quaternary Geology, Tornavägen 13, S-223 63 Lund, Sweden.

In a series of papers Sonesson published his pioneer work on the Holocene vegetation history of the Torneträsk region (Sonesson 1968, 1974) within the northern Scandes. His results have been fundamental for understanding the long-term vegetation dynamics. Sonesson's studies were based mainly on pollen-analytical investigations of lake and peat deposits.

In 1990 the Dept of Quaternary Geology in Lund initiated a multidisciplinary project in the Abisko area, with the Abisko Scientific Research Station serving as a base for fieldwork. Mountain sites like the Abisko region are key areas for studying environmental response to climate change, one objective of the Global Change Project (IGBP/PAGES). The initial part of our project involved sedimentologic, mineral magnetic and geochemical studies of Holocene lake sediments in the presently glaciated Kårsa valley, situated just above the tree-limit and west of the Abisko valley (Fig. 1). Results from these studies (presented in Snowball 1993a, b,

1994, 1995, 1996) contribute to discussions about glacial activity, soil erosion and lake development in response to climatic changes. Our investigations have later been extended to the Abisko valley to obtain a deeper insight into the relationship between vegetation dynamics and climate development. Within the project we will study vegetation changes through the analysis of regional pollen records from medium sized lakes and plant macrofossil records in smaller lake basins. We anticipated that a more precise chronology could be obtained from the ^{14}C dating of terrestrial macrofossils deposited in organic-rich lake sediments situated within the mountain forest belt, rather than from organic-poor sediments deposited above the tree-limit. Sonesson's early studies indicated that the second largest lake in the Abisko valley, Vuolep Njakajaure, contains laminated sediments in the deepest parts (Sonesson 1968) and that a complete sequence of annually laminated sediments would form an excellent chronologic tool. In

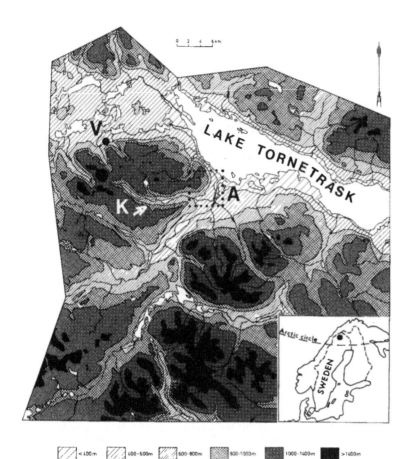

Fig. 1. Topographic map of the western parts of the Torneträsk region (from Melander 1977). The Abisko valley (area below c. 600 m a.s.l.) is marked with an A, the E-W trending Kårsa valley is marked with a K. Mire Vassijaure is shown with V and a solid dot. The position of Fig. 4 is represented by a dotted square.

Legend: `< 400m` `400–600m` `600–800m` `800–1000m` `1000–1400m` `>1400m`

addition, independent palaeoclimatic records may be obtained by using stable isotope geochemistry and fossil insect analysis of sediments from smaller lakes. Prior to the onset of our research project, a detailed dendroclimatologic record had been obtained through the study of fossil pine timber in the Torneträsk region (Schweingruber et al. 1988, Briffa et al. 1990). Megafossils of pine had also been collected and dated mainly in the central Scandes (southern Lappland and Jämtland) by Kullman (1988, 1992, 1993, 1995b) and in Lappland by Karlén (1976). Similar work had been undertaken in other parts of northern Scandes, i.e. northern Finland (Eronen and Huttunen 1993) and adjacent parts of northern Norway (Eronen and Hyvärinen 1982). Kullman's studies constitute an important part of research aimed at elucidating the relationship between long-term changes in the tree-limit in the central Scandes and climate change. However, it is questionable if the situation in northern Lappland is the same as in the central Scandes, or if it is possibly more similar to northern Finland.

Over the last 20 yr a number of new techniques and methods have developed. In Fig. 2 the strategy and techniques used by Sonesson (1968, 1974) are compared to those used in our project. The main differences

compared to Sonesson's work are: 1) the potential to combine several palaeobiological methods with geophysical/geochemical techniques; 2) access to the Accelerator Mass Spectrometry (AMS) radiocarbon dating technique which enables small fragments of terrestrial plants to be dated; 3) the ability to use a precise

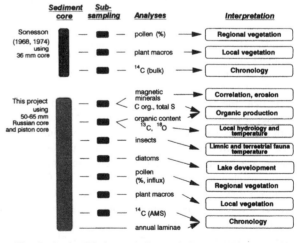

Fig. 2. A simplified comparison of the available analytical techniques used by Sonesson (1968, 1974) and the range of methods that will be applied in the present study.

chronology and volumetric sampling techniques to quantify the influx of, for example pollen and plant macrofossils, and 4) the potential to apply multivariate statistics. This multidisciplinary approach will be applied to sediment cores from Lake Vuolep Njakajaure and an adjacent small lake, Lake Badsjön, in the Abisko valley proper, and Lake Tibetanus in the Kårsa valley (Fig. 4; Barnekow unpubl.). The specific methods used in the Lake Tibetanus case study are described below.

In this paper we will present, i) a summary of Sonesson's regional pollen zones with a revised radiocarbon chronology, ii) preliminary results from a site (Lake Tibetanus) situated in the eastern part of the Kårsa valley at an altitude of c. 600 m a.s.l., and iii) an overview of the vegetation dynamics in an E-W transect from the Abisko valley bottom (c. 400 m a.s.l.) to the mountain slope at the present tree-limit (c. 600 m a.s.l.). Our synthesis is based on a combination of Sonesson's classical study of Lake Vuolep Njakajaure and our new data from Lake Tibetanus. A more detailed study on the vegetation dynamics and its relation to climate will be published in a doctoral thesis (Barnekow unpubl.).

Investigation area

Geology and topography

The Torneträsk region belongs to the Caledonian mountain range, also named the Scandes, which has a N-S extension. Thrusts of allochthonous nappes, dipping westwards, form the main range and cover the crystalline basement of the Baltic Shield (Kulling 1964). Lake Torneträsk forms a basin at an altitude of 341 m a.s.l. and mountain peaks in the south reach almost 2 000 m a.s.l. The Torneträsk basin continues to the west into the Vassijaure basin at 400–600 m elevation. The Abisko valley is also situated below 600 m whereas the tributary Kårsa valley and associated lakes in the west are located at 600–800 m a.s.l. (Fig. 1).

The soils are dominated by tills and glaciofluvial sands. Peat occurs in local depressions. The soils are poor in nutrients in the west and in the east but rich, particular in calcium, in the central part, i.e. the Abisko valley and adjacent mountains to the west (Kulling 1964, Sonesson and Lundberg 1974). At a small scale the relief of the Abisko valley is characterized by a hummocky dead-ice topography which explains the abundance of lakes (Holdar 1957).

Deglaciation age

The deglaciation chronology of the Torneträsk region has been much debated. Sonesson (1974) determined the age of the deglaciation from conventional radiocarbon datings of bulk samples, obtained from just above the contact between late glacial mineral substrate and overlying lake sediments or peat rich in organic matter. His results, based on dates obtained from lakes in the Abisko valley, point to a deglaciation age of c. 10 000 ^{14}C yr BP (individual radiocarbon dates vary from 10 590 to 8 980 ^{14}C yr BP; Sonesson 1974). Karlén (1979) based the deglaciation age on dates obtained from a number of sites in northern Lappland and concluded that the deglaciation took place 9 000–8 500 ^{14}C yr BP. Most of the dates were obtained on bulk sediment samples. One serious error with bulk sediment dates is the lake reservoir effect. Dissolution of ^{14}C deficient carbonate rocks or the influence of old groundwater may lead to depletion of ^{14}C in dissolved inorganic carbon of lake water with respect to the atmosphere (MacDonald et al. 1991). A second problem with bulk sediment dates is contamination by "old" carbon, present in soils and bedrock, which can also cause erroneously high ages (Donner and Jungner 1973). From our revision of Sonesson's pollen zone chronology (see below) we conclude that the deglaciation in the Abisko valley occurred c. 8 500 ^{14}C yr BP, corresponding to c. 9 500 calender yr BP (Table 1). However, the regional deglaciation could well be slightly earlier than the final melting of stagnant ice bodies in a dead-ice landscape like the Abisko valley. The chronology in our project will be based on AMS dates of terrestrial plant macro remains (Barnekow et al. unpubl.). In the following text all ages are expressed as uncalibrated radiocarbon years before present (1950 AD).

Climate

The present climate in the Torneträsk region is more or less oceanic (Ångström 1958), with a pronounced oceanic-continental gradient from west to east. There is also a strong orographic effect which creates a "rain-shadow" and causes the local climate of the Abisko valley to contrast with the surrounding mountains (Josefsson 1990). In Abisko the mean annual precipitation is c. 300 mm, while in Katterjåkk, 35 km west of Abisko, it increases to c. 800 mm. The average January temperature in Abisko is −12.5°C and the average July temperature is +11.0°C. Mean annual temperature is −0.8°C (1966–1988, Abisko Research Station, unpubl.).

Vegetation

The mountain areas in the Torneträsk region belong to the alpine zone and are situated above the tree-limit (Sjörs 1963). The subarctic forest zone reaches the tree-limit at 550–650 m in the western part and at 700–800 m in the eastern part of the region as a subalpine

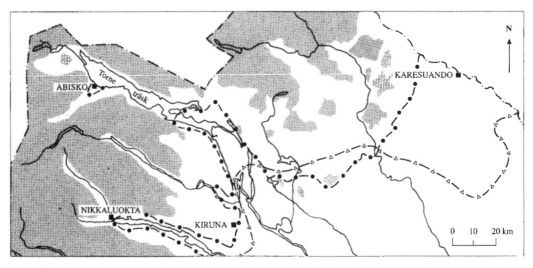

Fig. 3. The distribution of *Picea* (hatched line with triangles) and *Pinus* (hatched line with solid dots) in the northernmost parts of Sweden (based on Kullman and Engelmark 1991). *Betula* (white) can be found to the tree-limit (the area above the tree-limit is shaded). Note that *Picea* extends further to the north than *Pinus* northeast of Kiruna and that *Pinus* occurs as an exclave in the Abisko valley.

belt where mountain birch *Betula pubescens* ssp. *tortuosa* dominates (Fig. 3). Other deciduous tree species within the mountain birch forest belt are aspen *Populus tremula*, grey alder *Alnus incana* and willows *Salix* spp. Pine *Pinus sylvestris* reaches the eastern part of the Torneträsk area with a continuous distribution but occurs sporadically further to the west. In the Abisko valley pine occurs as an exclave within the subalpine belt (Figs 3–5). Further to the east of Lake Torneträsk spruce, *Picea abies*, reaches its western and northern limit (Fig. 3). This distribution is different from the situation further south in the Scandes where spruce normally ascends to higher elevations and reaches further westwards than pine (Kullman and Engelmark 1991). Different forest types (heath and meadow ecosystems) in the Torneträsk region have been defined and described by Sonesson and Lundberg (1974). Within the subalpine birch forest belt subalpine heaths also occur, determined by several factors such as local climate, soils and frost processes (Josefsson 1990). Changes in the spatial and temporal distribution of these heaths contribute to the landscape dynamics of subalpine areas such as the Abisko valley.

Pollen zones and revised chronology

Sonesson (1968, 1974) established a pollen zone system for the Torneträsk region. This zonation has been confirmed by our new studies in the Abisko valley (Holmqvist 1993, Barnekow unpubl.). Sonesson's zonation was made in a traditional, semi-objective way by visual identification of major changes in pollen composition. We have applied cluster analysis, a multivariate statistical technique, using the program CONISS (Grimm 1987) to identify pollen assemblage zones (PAZ) in an 'objective' way (cf. Birks and Gordon 1985). To illustrate this technique we present two of Sonesson's pollen diagrams (constructed from Sonesson's raw data), one from the peat sequence of the mire Vassijaure and a second from the sediment sequence of Lake Vuolep Njakajaure (Figs 1 and 7). Our zonation is almost identical to Sonesson's original four-zone-system T1 to T4. The only difference is that one additional pollen zone should be distinguished in the bottom of the Vuolep Njakajaure diagram, the local PAZ VN 1. In the revised pollen diagrams the pollen spectra have been plotted against a radiocarbon time scale. For the Vassijaure sequence we use the original radiocarbon dates of seven samples (Fig. 6). In the case of the Lake Vuolep Njakajaure sequence it is not possible to use the original chronology for the early Holocene presented by Sonesson (1968, 1974), because AMS dates of plant macrofossils in our new cores give distinctly younger ages in comparison to Sonesson's cores (Barnekow et al. unpubl.). This difference is probably caused by the lake reservoir effect or contamination by old carbon (see above). Ages of the pollen zone boundaries in the Vassijaure peat sequence have therefore been transferred to the Vuolep Njakajaure diagram by cross-correlation (Fig. 6). For the late Holocene (the last 4 000 yr) Sonesson's original dates obtained from Vuolep Njakajaure are in agreement with those from Vassijaure. Our revision of Sonesson's chronology is displayed in Table 1. The result of two independent dating methods, varve counting and AMS radiocarbon dating of bulk sediments and terrestrial macrofossils, will be presented in a future paper (Barnekow et al. unpubl.).

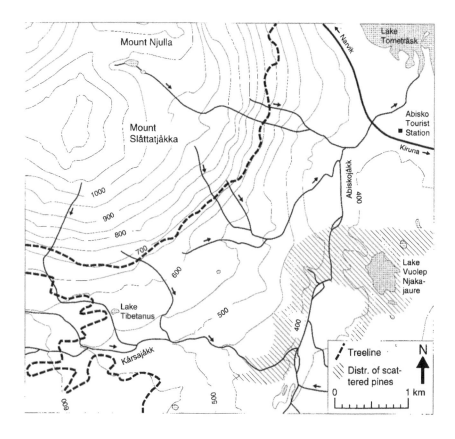

Fig. 4. Topographic map of the Abisko valley. Lake Tibetanus is situated at the mouth of the Kårsa valley close to the present day tree-limit (solid hatched line). Lake Vuolep Njakajaure lies at an altitude of 409 m a.s.l. in an area with scattered pine trees (hatched) preferentially found on south facing hill slopes. Lake Badsjön is the small lake c. 500 m NE of Lake Vuolep Njakajaure.

A case study of Lake Tibetanus

Site description

Lake Tibetanus is situated in the easternmost part of the Kårsa valley (68°20′N, 18°42′W), at an altitude of c. 560 m a.s.l. (see Figs 1, 4 and 5). The regional tree-limit, formed by mountain birch, reaches 550–600 m a.s.l. although discontinuous forest vegetation can be found to c. 700 m a.s.l. Scattered occurrences of pine exist below c. 450 m a.s.l. Today the lake is surrounded by heath vegetation with sparse birch vegetation. The lake is <4 m deep and between 85 and 100 m across. Along the slope above the lake is an outcrop of calcite marble (Kulling 1964) that constitutes a substantial supply of dissolved inorganic carbon to ground water and surface runoff. Thus, the lake water exhibits high concentrations of bicarbonate and high values of conductivity and pH (Ekman 1957).

Methods

Sampling

Multiple cores along a W-E trending transect were collected in March 1995 with a Russian peat-corer (1 m long, ⌀ 65 or 100 mm). The results presented in this paper are based on a number of overlapping cores from the deepest, western part of the ice-covered lake at a water depth of 3.9 m. The individual cores were described in detail immediately after collection and correlated in the laboratory in Lund. The sequence was then contiguously sampled in 20–68 mm slices taking into account sediment stratigraphic boundaries, resulting in 61 main samples for oxygen isotope analyses and determination of carbon content. For dating, magnetic analyses and terrestrial macrofossil analyses most of the main samples were divided into two or three subsamples ranging in thickness between 17 and 26 mm.

Oxygen isotopes

A minor part of each of the 61 main sediment samples was freeze-dried and gently passed through a 125 μm sieve to avoid contamination by mollusc shells. The sedimentary carbonate samples obtained were analysed for stable oxygen isotope composition on the <125 μm fraction on a Finnigan-Mat 250 mass-spectrometer at the Geological Inst., Univ. of Copenhagen. The preparation followed standard procedures (McCrea 1950), and the results were expressed as per mille deviations (δ-values) from the international PDB standard (Craig 1957); $\delta^{18}O_{sample} = [(^{18}O/^{16}O)_{sample}/(^{18}O/^{16}O)_{standard} - 1]10^3$ (‰). The analytical reproducibility is within ±0.07‰ on the δ-scale.

Carbon content

The carbon content of the sediments was determined by a Leco RC 412 Multiphase Carbon Determinator. The results are expressed as elemental organic and carbonate contents in percentages of total dry weight.

Magnetic minerals

All samples were magnetized in a field of 1 Tesla and the induced remanent magnetisation measured with a Molspin "Minispin" fluxgate magnetometer. This parameter, generally called saturated isothermal remanent magnetisation (SIRM), is a measure of the concentration of magnetic minerals in the sample (Thompson and Oldfield 1986).

Fig. 5. Upper: Lake Vuolep Njakajaure during the fieldwork in March 1995. View to the SE. Middle: Pine stand west of the Abisko river (Abiskojokk) at c. 400 m a.s.l. Note the birch tree-limit at c. 650 m a.s.l. in the background (SE slope of Mt. Slåttatjåkka). Lower: Lake Tibetanus during fieldwork in March 1995. Note the sparse birch vegetation near the tree-limit. (Photos: Björn E. Berglund).

Plant macro remains

Subsamples were washed with a fine jet of water over a 250 μm sieve. Plant macro remains for AMS dating were picked out from 11 levels, evenly distributed along the core, and rinsed with super-pure water before determination. All subsamples were examined under a binocular microscope. Pine needles were then removed and dried at 40°C to allow the calculation of dry weight of pine needles per litre of sediment.

Sediment stratigraphy

A c. 3 m thick sediment sequence was recovered from the deepest part of Lake Tibetanus. Based on the visual stratigraphy the sequence was divided into five units. Transitions between the individual units are gradual and indicate that sedimentation has been continuous. On top of a dark grey silt with sand layers, that could not be completely penetrated, follows a 2.8 m thick sequence of highly calcareous, organic sediments.

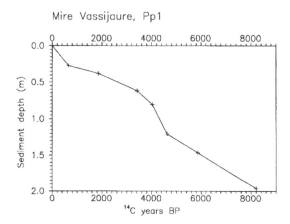

Fig. 6. Upper: Time-depth diagram from Vassijaure based on ^{14}C dates published by Sonesson (1974). Lower: Time-depth diagram from Vuolep Njakajaure based on the ^{14}C dates published by Sonesson (1974). Dates indicated Va are transferred from the Vassijaure profile, dates indicated VN are derived from the original Vuolep Njakajaure profile.

Table 1. Comparison of the Torneträsk pollen zones as identified by Sonesson (1974) and the present revision. ^{14}C ages have been calibrated and expressed in calendar years (cal. ^{14}C BP) according to Stuiver and Reimer (1993).

Sonesson (1974) ^{14}C BP	Torneträsk pollen zones	Abisko PAZ	Revised (this paper) ^{14}C BP	cal ^{14}C BP
	T4 Late Birch period	A5 Betula-Salix Juniperus- Ericales-herbs		
3 500			3 400	3 600
	T3 Pine period	A4 Pinus-Betula		
6 500– 5 800			5 500	6 100
	T2 Birch–Alder period	A3 Betula-Alnus- Juniperus- herbs		
9 000			7 700	8 400
	T1 Early Birch period	A2 Betula-herbs- Lycopodium		
			8 300	9 300
		A1 Hippophaë		
10 000 ◁———— deglaciation ————▷			c. 8 500	c. 9 500

Magnetic analysis

The concentration of magnetic minerals in Lake Tibetanus, as reflected by SIRM, is presented in Fig. 8. The minerogenic sediment of unit 1 displays values >2 mAm^2 kg^{-1} (A) with a significant decrease (<0.6 mAm^2 kg^{-1}) in the overlying calcareous sediments. Due to the very low concentration of magnetic minerals in the calcareous sediments the values are somewhat scattered and a three-point moving average has been applied to smooth the data. Two periods of low values (B$_1$ and B$_3$), 2.80–2.00 m and 1.50–1.25 m respectively, sandwich a period of higher values (B$_2$) between 2.00 m and 1.50 m. From c. 1.25 m values gradually increase (unit C) and culminate in maximum values in unit D.

Also shown in Fig. 8 is the SIRM record of the longest sediment core recovered from Lake Gaskkamus Gorsajávri (adapted from Snowball 1993a). This lake is one of three lakes studied by Snowball in the Kårsa valley (Fig. 1) only 7 km to the west of Lake Tibetanus. Although the postglacial sediments in Gaskkamus Gorsajávri have a much lower organic content (0.1–3% organic carbon) than the Lake Tibetanus sequence, the two SIRM curves show very similar trends. Correlatable features are labelled A–D.

SIRM is a proxy record of both glacial activity and catchment conditions which are both ultimately controlled by climate. During periods of cold climate glacial scour is significant and soil erosion is more intense which results in high sedimentation rates and hence high SIRM values (A, C and D). In contrast during periods of climatic warming glacial activity is reduced or even non-existent and soils are stabilized by vegetation (B). As a result sedimentation rates are low. Snowball (1993a, 1995) demonstrated that a reduced supply of minerogenic sediment, combined with increased organic carbon content led to magnetite dissolution and thus low SIRM values. This pattern was found in sediment cores recovered from all three of the Kårsa valley lakes (Snowball 1993b). Unit B$_2$, found only in Gaskkamus Gorsajávri, was not caused by glacial activity but probably by a water level lowering which caused the deposition of coarse sediment and moss remains (Snowball 1996, Snowball and Sandgren in press).

Although Lake Tibetanus has been isolated from a direct glacial influence since the regional deglaciation (A) the SIRM record can, based on the argument above, also be considered to reflect climatic changes. During cold periods erosion, mainly as slope wash of exposed soils, was more severe and the lake experienced a higher input of minerogenic matter, reflected by increased SIRM values (C and D).

The low SIRM values of unit B in the sediments of both lakes, deposited between the time of the regional deglaciation and c. 3 000 BP, reflect a period of generally warm climate and the complete melting and disappearance of the Kårsa glacier. Unit B$_2$, characterised by higher SIRM values and the deposition of plant macrofossils in both lakes, is also interpreted as a period of low water level.

A successive environmental change at c. 3 000 BP is reflected by the return to high SIRM values in both sequences. Soil erosion around Lake Tibetanus increased at the same time as the Kårsa glacier reformed (Snowball 1996).

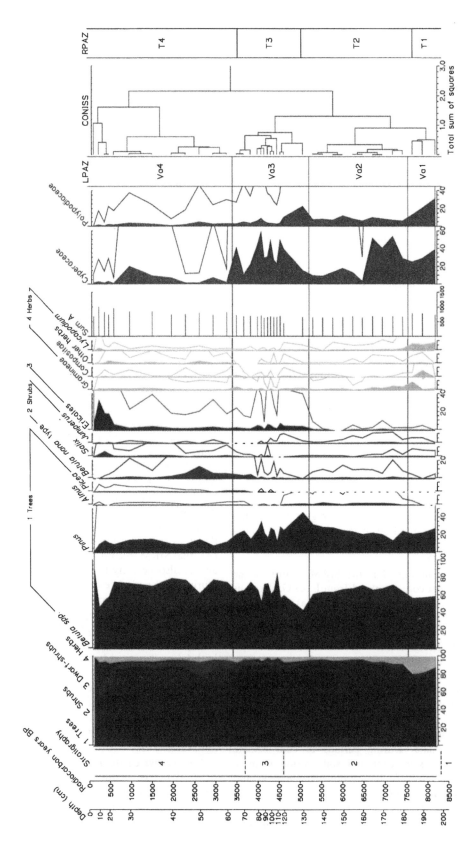

VASSIJAURE Pp 1 (460 m)

Analysts: M. Sonesson, M. Varga 1964

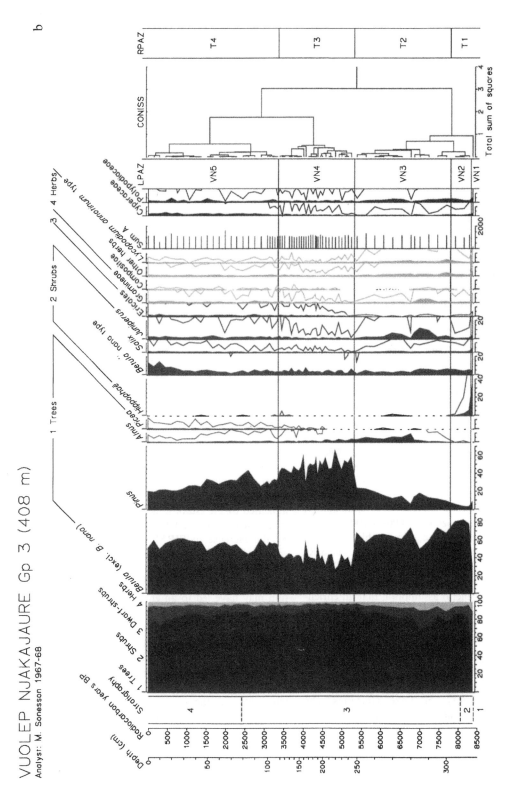

VUOLEP NJAKAJAURE Gp 3 (408 m)
Analyst: M. Sonesson 1967-68

Fig. 7. a: Pollen percentage diagram from the mire Vassijaure (Sonesson 1974). Simplified stratigraphy: Layer 1 = peat, layer 2 = *Carex*-moss peat, layer 3 = Ericales-*Carex* peat, layer 4 = Ericales-*Sphagnum* peat. Local pollen assemblage zones (LPAZ) defined according to cluster analysis (CONISS) and compared with Sonesson's original regional pollen zones (RPAZ). b: Pollen percentage diagram from Vuolep Njakajaure (Sonesson 1968). Simplified stratigraphy: Layer 1 = sand, layer 2 = gyttja clay, layer 3 = slightly clayey laminated gyttja, layer 4 = clayey gyttja. The pollen zonation corresponds to Fig. 7a.

Oxygen isotype analysis

The oxygen isotope record is based on the 61 sedimentary carbonate sampels. $\delta^{18}O_{Sed.}$ varies between $-13.9‰$ and $-11.0‰$. A five-point moving average (Fig. 9) exhibits a trend from relatively enriched (less negative) to more depleted (more negative) values with time in unit 1–3. In unit 4a and the lower part of unit 4b minor fluctuations around $-13.4‰$ were recorded, whereas rather stable values close to $-13.2‰$ prevail in the upper part after a slight enrichment at c. 1 m. The carbonate mineralogy has been determined by X-ray diffraction analysis as being exclusively low-magnesian calcite. A highly covariant $\delta^{18}O$ record obtained on mollusc shells (*Pisidium* spp.) from the core (Hammarlund et al. unpubl.) indicates that the calcite in the sediments above c. 2.5 m was endogenically precipitated in isotopic equilibrium with the lake water. This means that the $\delta^{18}O_{Sed.}$ record may contain useful palaeoclimatic information.

The oxygen isotope composition of limnic carbonates is related to climate in a complex way by different temperature-dependent fractionation processes. $\delta^{18}O$ of precipitation water exhibits a positive correlation with air temperature at the condensation of meteoric water vapour (Dansgaard 1964). On the other hand, $\delta^{18}O$ of limnic carbonates reflects both $\delta^{18}O$ composition (positive correlation) and temperature (negative correlation) of the surrounding lake water that ultimately originates from catchment precipitation (Craig 1965). The correlation coefficient of the condensation process is numerically dominant, and causes a net positive temperature effect (e.g. Siegenthaler and Eicher 1986). Therefore an enrichment of ^{18}O with time in a carbonate sequence is often attributed to an increase in temperature during the summer (the main season of carbonate precipitation). The only comparable record in northern Fennoscandia is from Lake Vanhalampi in northeastern Finland (Hyvärinen et al. 1990) which exhibits a $\delta^{18}O$ maximum during the Boreal Chronozone (9 000–8 000 BP) followed by a successive depletion, comparable in magnitude to the recorded change in Lake Tibetanus. The $\delta^{18}O$ maximum observed by Hyvärinen et al. (1990) was interpreted to reflect an early Holocene climatic optimum. However, the general assumption that an enrichment of ^{18}O reflects a climatic warming does not account for changes in atmospheric circulation (e.g. moisture source), temporal changes in the seasonal distribution of precipitation, or hydrological conditions (inflow/evaporation ratio) that affect $\delta^{18}O$ of the lake water.

There may thus be alternative interpretations of the observed trends in $\delta^{18}O_{Sed.}$ in Lake Tibetanus. The pronounced and persistent depletion of ^{18}O in units 1–3 coincides with the terrestrial vegetation succession from open heaths or meadows to a more or less closed forest (cf. vegetation dynamics below). During the early Holocene the lake water may have been subject to

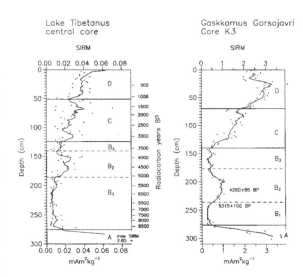

Fig. 8. Comparison of SIRM records from Lake Tibetanus and Gaskkamus Gorsajávri, the middle basin of the three lakes in Kårsa valley. The solid line corresponds to a five-point moving average. Comparable units are labelled A–D. Note how well the two datings, based on terrestrial macrofossils, in Gaskkamus Gorsajávri compares with the high resolution chronology of Lake Tibetanus.

enrichment of ^{18}O, whereas the sheltering effect of closed forest at c. 5 500–3 500 BP, as indicted by the massive occurrence of pine needle in unit 4a (Fig. 9), possibly diminished wind-induced evaporative loss of ^{16}O from the water surface, and caused a relative depletion of ^{18}O in the lake water and precipitated carbonates. Alternatively, the part of the record with the most depleted $\delta^{18}O_{Sed.}$ values (c. 2.1–1.0 m) may correspond to a period when a decreased proportion of the annual precipitation occurred during the summer, giving rise to a relative depletion of ^{18}O in groundwater, the main source of recharge to the lake. This is consistent with relatively dry summer conditions as indicated by the establishment of pine forest around the lake, and possibly a lake-water lowering as suggested by the increase in SIRM values (Fig. 8). The slight enrichment of ^{18}O at c. 1 m does not seem to be directly related to the disappearance of the closed pine–birch forest. Instead a relative increase in summer precipitation may have led to an enrichment of ^{18}O in groundwater. A decrease in lake water temperature during the summer may have contributed to the observed enrichment of ^{18}O in limnic carbonates. Isotopic studies of the present day catchment (Hammarlund et al. unpubl.) reveal that summer evaporation is insignificant and that $\delta^{18}O$ of the lake water closely reflects mean annual air temperature in the area (i.e. a rapid renewal of ground water with $\delta^{18}O$ values representative of average yearly precipitation). Furthermore, an additional explanation of the relatively enriched $\delta^{18}O_{Sed.}$ values in the lower part of the sequence may be a greater importance of summer precipitation and/or increased zonal circula-

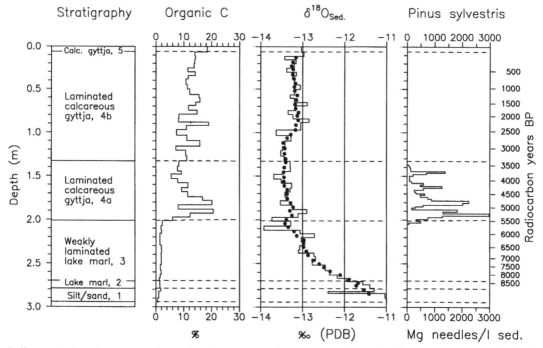

Fig. 9. Sediment stratigraphy, content of organic carbon, oxygene isotope composition of sedimentary carbonates (black dots represents a five-point moving average) and occurrence of pine needles. The chronology is based on AMS radiocarbon datings of terrestrial macrofossils.

tion. A northward displacement of the jet stream during the early Holocene associated with increased influence of Atlantic air-masses (Yu and Harrison 1995) may have led to a relative enrichment of ^{18}O in precipitation in the Abisko area due to enhanced moisture transport across the narrow mountain chain (Fig. 1) and decreased rain-out of ^{18}O before the moisture reaches the Abisko area.

Studies of additional sediment components by means of isotope geochemistry are currently in progress. Analysis of $\delta^{18}O$ in aquatic cellulose extracted from the organic material is expected to generate a Holocene record of lake water $\delta^{18}O$ according to a method described by Edwards (1993). These data can be used in combination with $\delta^{18}O_{Sed.}$ for palaeotemperature reconstructions and to restrict the number of possible processes influencing the data.

Vegetation history

Plant macrofossil analysis of the Lake Tibetanus sediment sequence shows that unit 4a has a high concentration of pine needles. Preliminary analyses also show that birch Betula pubescens seeds are very frequent in this unit. This composition most probably reflects the vegetation in the catchment area, although the postulated lake-water lowering (see above) may have caused some enrichment of plant macro remains in the deepest part of the basin. The pine–birch assemblage indicates

that between c. 5 500 and c. 3 500 PB the lake was surrounded by a pine–birch wood of boreal character. During this period the conditions around Lake Tibetanus (at 560 m a.s.l.) appear to have been as favourable or even more favourable for pine growth than the present day conditions in the Abisko valley. A modern analogue is possibly the pine–birch stand along the western shore of the river Abiskojåkk at an elevation of c. 400–420 m (Figs 3 and 5). This means that the pine-limit may have been c. 100–150 m higher than today. The general tree-limit, today situated between 650–700 m a.s.l. north of Lake Tibetanus, was possibly also higher during this period but that is not possible to judge from our present information. We may however conclude that birch (probably mountain birch) has been the dominant tree species at the tree-limit since c. 8 300 BP except for the period c. 5 500–3 500 BP when birch was mixed with pine or possibly occurred as a narrow belt just below the tree-limit.

Vegetation dynamics as a response to climate change

The following interpretation of the vegetation dynamics since the regional deglaciation is based mainly on Sonessons's pollen records from the Abisko valley and partly on our new results from Lake Tibetanus. We also make comparisons to studies undertaken on lakes in the Kårsa valley by Snowball (1995). The studied period is divided into chronologic zones according to the regional pollen zones A 1 to A 5 (Table 1). For each zone the pollen characteristics recorded in the Vuolep Njakajaure sequence (Sonesson 1968, Fig. 7b) are presented, followed by an interpretation of vegetation, landscape and climate. Comparisons are made to Finnish Lappland, neighbouring areas of northern Norway and also Jämtland in the central Scandes. The terminology for characterization of the biotic zones follows Sjörs (1963).

Subarctic shrub tundra, c. 8 500–8 300 BP

A 1, Hippophaë PAZ
This zone is characterized by high values of *Hippophaë*, *Betula nana*, *Salix*, Cyperaceae, Gramineae, Polypodiaceae and various herbs. Values of tree pollen are low and the total pollen concentration is also very low.

Vegetation, landscape and climate
This short tundra phase with sea buckthorn seems to be characteristic for at least the northern part of the Scandes, deglaciated during the early Holocene (Huntley and Birks 1983). A subarctic/subalpine vegetation dominated by sea buckthorn has no modern analogue in the Torneträsk area today although such shrub vegetation can be found in the mountains of southern Norway (Lid 1942). In southern Scandinavia sea buckthorn is regarded as a low-competitive, warm demanding, pioneer shrub characteristic of the Late Weichselian revegetation phase prior to the Allerød Chronozone, c. 12 500–12 000 BP (Iversen 1954). As can be expected in a low-alpin environment, the vegetation was also characterized by communities with grasses, sedges, herbs and ferns. Heath indicators have low frequencies which may indicate unstable and mineral-rich soil conditions. The corresponding tundra phase in Finnish Lappland and neighbouring parts of northern Norway is dominated by Ericales (probably mainly *Empetrum*) which indicates an open heath phase before the birch expansion (Hyvärinen 1975, Eronen and Hyvärinen 1982).

The rapid plant colonization of the landscape during this shrub tundra phase may represent a response to a favourable climatic situation. Based on data from the central Scandes c. 600 km further south it has been assumed that the Holocene thermal optimum occurred during the millennium prior to 8 000 BP (Kullman 1995b). The regional deglaciation occurred c. 9 000 BP and mountain forests with pine developed rapidly, so that the highest Holocene tree-limit was reached before 8 000 BP. However, there is no evidence from the Abisko area that the Holocene climatic optimum occurred at the same time as in the central Scandes.

Subarctic birch woodland tundra, c. 8 300–7 700 BP

A 2, Betula-herbs-Lycopodium PAZ
This zone is characterized by particularly high values of *Betula*, probably mountain birch *B. pubescens* as well as dwarf-birch *B. nana*. There are also rather high values of *Salix*, Gramineae, *Lycopodium annotinum* type and Polypodiaceae. The total pollen concentration is higher than in pollen zone A 1.

Vegetation, landscape and climate
The pollen diagram indicates that this period is characterized by a rather open woodland tundra with mountain birch and dwarf-shrubs like dwarf-birch and dwarf-willows. Communities with grasses, herbs and sedges were common and lycopods were still frequent. Acidophilous dwarf-shrubs (e.g. Ericales) seem to have been rather rare. The vegetation cover was denser than before and hence soils were slightly more stable. This landscape shows some similarities with the present day subalpine zone but the presently widely distributed heath communities were absent. Similar birch woodlands occurred in Finnish Lapland and neighbouring parts of northern Norway during this period (Hyvärinen 1975, Eronen and Hyvärinen 1982).

The composition of the vegetation indicates a climatic situation similar to the preceding period. The immigration of pine was delayed in comparison to the central Scandes (Kullman 1995b). This delay may have been due to climatic as well as pedological conditions. In general the composition of the vegetation indicates that an oceanic climate prevailed (cf. Yu and Harrison 1995), a distinct contrast to the central Scandes where the forest composition at the tree-limit indicates a continental climate.

Subarctic birch woodland tundra, c. 7 700–5 500 BP

A 3, Betula-Alnus-Juniperus-herbs PAZ
This zone is characterized by high *Betula* values, increasing *Pinus* values, relatively high *Alnus* values (particularly after c. 7 000 BP), high *Juniperus* values, decreasing *Betula nana* and *Salix* values, and rather high values of Gramineae, herbs, *Lycopodium* and Polypodiaceae.

Vegetation, landscape and climate
During this period the birch woods became slightly denser and open areas were less common. Pine colonized the area but was not very frequent. Stream banks and lake shores were colonized by grey-alder which reached its Holocene peak between c. 7 000 and c. 5 500 BP. Juniper was also much more common than today. Open herb- and fern-rich communities were still com-

mon which may indicate that some birch woodlands were of a meadow type (Sonesson and Lundberg 1974). During the same period pine expanded into the subarctic woodlands of Finnish Lappland and neighbouring parts of northern Norway (Hyvärinen 1975, Eronen and Hyvärinen 1982). In Jämtland the mountain forest belt was occupied by mixed woodlands of pine, birch and alder, with a gradually descending pine-limit (Kullman 1995a).

During this period erosion was minimal and the soils were well stabilized, both in the Abisko and Kårsa valleys (Snowball 1995, 1996). The Kårsa glacier and probably other minor glaciers disappeared (cf. Jonasson 1991). The vegetation composition indicates more favourable climatic conditions after c. 7 000 BP than today. The climate seems to have been more oceanic than further south and east, an assumption based mainly on the low pine pollen frequency in the Torneträsk area as shown in Sonesson's pollen diagram.

Boreal pine–birch forests, c. 5 500–3 400 BP

A 4, Pinus-Betula PAZ
Pinus dominates the pollen spectra and depresses the *Betula* values. *Alnus* decreases throughout the zone. *Picea* begins to occur after c. 4 500 BP. *Salix* and *Juniperus* exhibit low values, as also Gramineae, herbs, *Lycopodium*, sedges and ferns do, while Ericales gradually increases throughout the zone.

Vegetation, landscape and climate
This is the period of the most extensive Holocene forest cover in the Abisko area and pine reached its highest altitude, probably 100–150 m higher than today. The tree-limit was probably also higher than today. During this period woodlands with pine and birch probably also occurred in the Vassijaure area, west of Lake Torneträsk. The vegetation was boreal in character with pine woods or birch–pine woods on dry soils and birch–alder–willow woods on damp soils. Meadow communities decreased in abundance. Instead heath communities, possibly pine and birch woodlands with dwarf shrubs became more common. Spruce never colonized the Torneträsk area but probably reached its present distribution around Kiruna at c. 4 000 BP (cf. Kullman 1995a).

These types of woodlands were present in Finnish Lappland for a much longer time, between 7 000 and 3 500 BP with a maximum distribution of pine between 6 000 and 4 000 BP (Hyvärinen 1975, Eronen and Huttunen 1993, Zetterberg et al. 1994). In neighbouring areas of northern Norway pine seems to have had its optimum during the same period as in the Torneträsk area (Eronen and Hyvärinen 1982). However evidence in the south (Kullman 1995b) is different and points to a gradual lowering after c. 8 000 BP with a distinct decline c. 5 500–4 000 BP.

The forest composition in the Abisko area seems to reflect a climate with a continental character, i.e. dry and warm summers combined with cold winters. This assumption is based on pollen records (Sonesson 1968), our plant macrofossil data and possibly a water level lowering as indicated by the magnetic data from Lake Tibetanus. Similar conditions have been documented in Finnish Lapland (Hyvärinen and Alhonen 1994). The delayed expansion of pine in comparison to the central Scandes and northern Finland is difficult to explain and this problem needs to be further addressed. It is of interest to note that in southern Scandinavia one of the driest, and probably most continental, periods during the Holocene occurred c. 4 500 BP (Digerfeldt 1988). It is also worth mentioning that dendroclimatic studies in Finnish Lapland demonstrate unstable growth conditions after c. 5 000 BP, probably caused by increased climatic variability (Zetterberg et al. in press).

Subarctic birch woodland tundra, c. 3 400 BP to present

A 5, Betula-Salix-Juniperus-Ericales-herbs PAZ
Betula dominates the pollen spectra throughout this zone while *Pinus* distinctly decreases, excpet for a peak c. 2 800–2 500 BP. *Alnus* exhibits low values, *Picea* increases slightly. *Betula nana*, *Salix*, *Juniperus*, Ericales, Gramineae, herbs, *Lycopodium*, Cyperaceae and Polypodiaceae all display increasing values.

Vegetation, landscape and climate
This is the regression period for the forest ecosystems which prevailed during the mid-Holocene. The main retreat occured shortly before 3 000 BP, when the boreal forests were replaced by a subarctic woodland tundra. Pine retreated on the mountain slopes as well as in favourable valley sites. The present occurrence of pine in the Abisko valley can therefore be regarded as an isolated relic from the mid-Holocene. Mountain birch became dominant in the landscape and formed a belt where a shrub vegetation with willows and junipers was also common. Dwarf-shrub heaths (Ericales) also expanded during this period. The birch ecosystem seems mainly to represent heath communities in contrast to the birch period of the early Holocene (cf. Sonesson and Lundberg 1974). Subalpine heaths with unstable soils probably became common after c. 3 000 BP, accompanied by increased soil erosion as indicated by the magnetic stratigraphy in Lake Tibetanus.

A general decline of the tree-limit around 3 000 BP is well known from the Scandes (Hyvärinen 1975, Kullman 1988, 1995b, Eronen and Huttunen 1993, Vorren et al. 1993, Zetterberg et al. 1994). Since then, mountain birch has been more dominant in the subalpine area and today forms a rather homogeneous belt along the mountain chain. At the same time many

Table 2. Stratigraphy of Lake Tibetanus.

Sediment depth (m)	Unit	Sediment stratigraphy
0–0.065	5	Dark, greyish-green laminated calcareous gyttja
0.065–1.33	4b	Laminated grey, beige or brown calcareous gyttja. Low frequency of plant macro remains
1.33–2.01	4a	Laminated of grey, beige or brown calcareous gyttja. Plant macro remains abundant, gradually decreasing towards upper boundary
2.01–2.71	3	Weekly laminated whitish lake marl
2.71–2.785	2	Alternating layers of grey or yellow/beige calcareous lake marl
2.785–2.95+	1	Dark grey calcareous silt with sand layers

mountain glaciers in Scandinavia reformed and advanced to their greatest Holocene extent (Karlén 1973, 1988, Dahl and Nesje 1994, Snowball 1995, 1996). Slope erosion has been more extensive (Nyberg 1985, Jonasson 1991). All this evidence points to a change towards a more oceanic climate. There may have been minor climatic oscillations (Karlén 1988, Karlén et al. 1995) but these have not been identified neither in our nor in Sonesson's (1968, 1974) records.

Conclusions

1. The regional pollen zone system of the Torneträsk region, proposed by Sonesson (1968, 1974), has been confirmed by defining the boundaries with statistical analysis of palynological data. The chronology has been revised for the early and middle parts of the Holocene (Table 1).

2. The pollen spectra have been interpreted in terms of five biotic zones. This zonation shows that the Abisko area has been situated within the subalpine birch woodland tundra from c. 8 300 BP until today, with the exception of c. 5 500–3 500 BP when the boreal pine–birch forest zone reached the area. During the same period there seems to have been an altitudinal expansion of pine to an elevation of c. 100–150 m higher than today. Birch (probably mountain birch) has been the dominant tree species at the tree-limit since c. 8 300 BP except for the pine–birch period (c. 5 500– 3 500 BP). In this respect the area differs from the southern Scandes, where pine seems to have been more frequent during early Holocene.

3. The climate has been subarctic-oceanic since the regional deglaciation apart from the period c. 5 500– 3 500 BP which seems to have been temperate-continental with warm and dry summers and cold winters. This is also associated with low lake levels which may indicate that the Holocene summer temperature optimun occurred during this period. No small-scale fluctuations of the climate have been recorded.

4. There was minimal erosion of catchment soils from c. 8 000–3 000 BP. After c. 3 000 BP the erosion increased due to a change towards a more cold and humid climate. This climatic cooling caused the Kårsa glacier to reform and also initiated soil erosion in catchments isolated from a glacial influence.

5. The chronology will be improved by utilizing annually laminated lake sediments and AMS radiocarbon technique. This makes it possible to quantify biotic changes, terrestrial vegetation dynamics and limnic development.

6. The Abisko area is a key area for studies of biotic and climatic changes in the long-term perspective because of its geographical position at the continental margin where diverse palaeoecological records may reflect the changing influence of Atlantic air-masses during the Holocene. The results here will be correlated with related studies along the Scandes Mountains and in a N-S transect through Europe within the research program of Global Changes (IGBP).

Acknowledgements – We are very grateful for generous hospitality and assistance at Abisko Scientific Research Station. We express our thanks to the director M. Sonesson and to N.-Å. Andersson. Assistance in the field was provided by A. Eriksson and B. Wanhatalo. G. Possnert at the Tandem laboratory, Uppsala Univ., carried out the AMS measurements. B. Buchardt, Univ. of Copenhagen, provided access to stable isotope laboratory equipment, and L. Madsen and B. Warming performed the analysis. C. Augustsson made the carbon analyses. To all these people we are very grateful. The study was financed by the Crafoord Foundation at the Royal Swedish Academy of Sciences (KVA), and by the Swedish Natural Science Research Council (NFR).

References

Ångström, A. 1958. Sveriges klimat. 2nd ed. – Stockholm.

Birks, H. J. B. and Gordon, A. D. 1985. Numerical methods in Quaternary pollen analysis. – Academic Press, London.

Briffa, K. R., Jones, P. D., Pilcher, J. R., Karlén, W., Schweingruber, F. H. and Zetterberg, P. 1990. A 1,400 year tree-ring record of summer temperatures in Fennoscandia. – Nature 346: 434–439.

Craig, H. 1957. Isotopic standards for carbon and oxygen correction factors for mass-spectrometric analysis of carbon dioxide. – Geochim. Cosmochim. Acta 12: 133–149.

– 1965. The measurement of oxygen isotope palaeotemperatures. – In: Tongiorgi, E. (ed.), Stable isotopes in oceanographic studies and palaeotemperatures. Consiglio Nazionsale delle Ricercha, Pisa, pp. 161–182.

Dahl, S. O. and Nesje, A. 1994. Holocene glacier fluctuations at Hardangerjøkulen, central-southern Norway: a high-resolution composite chronology from lacustrine and terrestrial deposits. – The Holocene 4: 269–277.

Dansgaard, W. 1964. Stable isotopes in precipitation. – Tellus 16: 436–468.

Digerfeldt, G. 1988. Reconstructions and regional correlation of Holocene lake-level fluctuations in Lake Bysjön, south Sweden. – Boreas 17: 165–182.

Donner, J. J. and Jungner, H. 1973. The effect of re-deposited organic material on radiocarbon measurements of clay samples from Somero, southwest Finland. – GFF 95: 267–268.

Edwards, T. W. S. 1993. Interpreting past climate from stable isotopes in continental organic matter. – In: Swart, P. K., Lohmann, K. C., McKenzie, J. and Savin, S. (eds), Climate change in continental isotopic records. Geophysical monograph, American Geophysical Union 78: 333–341.

Ekman, S. 1957. Die Gewässer des Abisko-Gebietes und Ihre Bedingungen. – Proc. Royal Swed. Acad. Sci. 4(6).

Eronen, M. and Hyvärinen, H. 1982. Subfossil pine dates and pollen diagrams from northern Fennoscandia. – GFF 103: 437–445.

– and Huttunen, P. 1993. Pine megafossils as indicators of Holocene climatic changes in Fennoscandia. – In: Frenzel, B. (ed.), Oscillations of the alpine and polar tree limits in the Holocene. Paleoclimate Research 9, special issue "European Palaeoclimate and Man", Gustav Fischer, Stuttgart, pp. 29–40.

Grimm, E. C. 1987. CONISS: a FORTRAN 77 program for stratigraphically constrained cluster analysis by the method of incremental sum of squares. – Comput. Geosci. 13: 13–35.

Holdar, C.-G. 1957. Deglaciationsförloppet i Torneträsk-området efter senaste nedisningsperioden, med vissa tillbakablickar och regionala jämförelser. – GFF 79: 291–528.

Holmqvist, B. 1993. Stratigrafiska undersökningar i sjön Vuolep Njakajaure, Abisko. – Honorary thesis no. 52, Dept of Quaternary Geology, Lund Univ.

Huntley, B. and Birks, H. J. B. 1983. An atlas of past and present pollen maps for Europe: 0–13 000 years ago. – Cambridge Univ. Press, Cambridge.

Hyvärinen, H. 1975. Absolute and relative pollen diagrams from northernmost Fennoscandia. – Fennia 142: 1–23.

– and Alhonen, P. 1994. Holocene lake-level changes in the Fennoscandian tree-line region, western Finnish Lapland: diatom and cladoceran evidence. – The Holocene 4: 251–258.

– , Martma, T. and Punning, J.-M. 1990. Stable isotope and pollen stratigraphy of a Holocene lake marl section from NE Finland. – Boreas 19: 17–24.

Iversen, J. 1954. The late-glacial flora of Denmark and its relation to climate and soil. – In: Iversen, J. (ed.), Studies in vegetational history. Geol. Surv. Denmark 2,80: 87–119.

Jonasson, C. 1991. Holocene slope processes of periglacial mountain areas in Scandinavia and Poland. – UNGI report 79.

Josefsson, M. 1990. The geoecology of subalpine heaths in the Abisko valley, northern Sweden. – UNGI report 78.

Karlén, W. 1973. Holocene glacier and climatic fluctuations, Kebnekaise mountains, Swedish Lapland. – Geogr. Annal. 55: 417–424.

– 1976. Lacustrine sediments and tree-limit variations as indicators of holocene climatic fluctuations in Lapland, northern Sweden. – Geogr. Annal. 58A: 1–34.

– 1979. Deglaciation dates from northern Swedish Lapland. – Geogr. Annal. 61A: 203–210.

– 1988. Scandinavian glacial and climatic fluctuations during the Holocene. – Quat. Sci. Rev. 7: 199–209.

– , Bodin, A., Kuylenstierna, J. and Näslund, J.-O. 1995. Climate of northern Sweden during the Holocene. – J. Coastal Res. Spec. Iss. Holocene cycles: Climate, sea levels, and sedimentation 17: 49–54.

Kulling, O. 1964. Översikt över norra Norrbottensfjällens kaledonberggrund (The geology of the caledonian rocks of the northern Norrbotten mountains). – Sv. Geol. Undersökning Ba 19: 1–166.

Kullman, L. 1988. Holocene history of the forest-alpine tundra ecotone in the Scandes Mountains (central Sweden). – New Phytol. 10: 101–110.

– 1992. Orbital forcing and tree-limit history: hypothesis and preliminary interpretation of evidence from Swedish Lapland. – The Holocene 2: 131–137.

– 1993. Holocene thermal trend inferred from tree-limit history in the Scandes Mountains. – Global Ecol. Biogeogr. Lett. 2: 181–188.

– 1995a. New and firm evidence for Mid-Holocene appearance of *Picea abies* in the Scandes Mountains, Sweden. – J. Ecol. 83: 439–447.

– 1995b. Holocene tree-limit and climate history from the Scandes Mountains Sweden. – Ecology 76: 48–60.

– and Engelmark. O. 1991. Historical geography of *Picea abies* (L.) Karst. at its subarctic limit in northern Sweden. – J. Biogeogr. 18: 63–70.

Lid. J. 1942. *Hippophaë rhamnoides* i Lom. – Nytt Mag. Naturvid. 83: 67–70.

MacDonald, G. M., Beukens, R. P. and Kieser, W. E. 1991. Radiocarbon dating of limnic sediments: a comparative analysis and discussion. – Ecology 72: 1150–1155.

McCrea, J. M. 1950. On the isotopic chemistry of carbonates and a paleotemperture scale. – J. Chem. Phys. 18: 849–857.

Melander, O. 1977. Geomorfologiska kartbladet 30H Riksgränsen (öst), 30I Abisko, 31H Reurivare och 31I Vadvetjåkka. – Swedish Environmental Protection Agency, Solna, Sverige, 1–56.

Nyberg, R. 1985. Debris flows and slush avalanches in northern Swedish Lapland: distribution and geomorphological evidence. – Ph.D. thesis, Lund Univ., Dept of Physical Geography.

Schweingruber, F. H., Bartholin, T., Schär, E. and Briffa, K. R. 1988. Radiodensitometric-dendroclimatological conifer chronologies from Lapland (Scandinavia) and the Alps (Switzerland). – Boreas 17: 559–566.

Siegenthaler, U. and Eicher, U. 1986. Stable isotope and carbon isotope anlayses. – In: Berglund, B. E. (ed.), Handbook of Holocene palaeoecology and palaeohydrology. Wiley, Chichester, pp. 407–422.

Sjörs, H. 1963. Nordisk växtgeografi. – Scandinavian Univ. Books, Stockholm.

Snowball, I. F. 1993a. Mineral magnetic properties of Holocene lake sediments and soils from the Kårsa valley, Lapland, Sweden, and their relevance to palaeoenvironmental reconstruction. – Terra Nova 5: 258–270.

– 1993b. Geochemical control of magnetite dissolution in sub-arctic lake sediments and the implications for environmental magnetism. – J. Quat. Sci. 8: 339–346.

– 1994. Bacterial magnetite and the magnetic properties of sediments in a Swedish lake. – Earth Planet. Sci. Lett. 126: 129–142.

– 1995. Mineral magnetic and geochemical properties of Holocene sediments and soils in the Abisko region of northern Sweden. – Lundqua Ph.D. thesis 34, Lund Univ., Dept of Quaternary Geology.

– 1996. Holocone environmental change in the Abisko region of northern Sweden recorded by the mineral magnetic stratigraphy of lake sediments. – GFF 118: 9–17.

Sonesson, M. 1968. Pollen zones at Abisko, Torne Lappmark, Sweden. – Bot. Not. 121: 491–500.

– 1974. Late Quaternary development of the Torneträsk area, north Sweden: 2. Pollen analytical evidence. – Oikos 25: 288–307.

– and Lundberg, B. 1974. Late Quaternary development of the Torneträsk area, north Sweden: 1. Structure of modern forest ecosystems. – Oikos 25: 121–133.

Stuiver, M. and Reimer, P. J. 1993. Extended ^{14}C data base and revised calibration program. – Radiocarbon 35: 215–230.

Thompson, R. and Oldfield, F. 1986. Enviromental magnetism. – Unwin and Allen, London.

Vorren, K. D., Jensen, C., Mook, R., Mørkved, B. and Thun, T. 1993. Holocene timberlines and climate in northern Norway – an interdisciplinary approach – In: Frenzel, B. (ed.), Oscillations of the alpine and polar tree limits in the Holocene. Paleoclimate Research, 9, special issue "European Palaeoclimate and Man", 29–40, Gustav Fischer, Stuttgart, pp. 115–126.

Yu, G. and Harrison, S. P. 1995. Holocene changes in atmospheric circulation patterns as shown by lake status changes in northern Europe. – Boreas 24: 260–268.

Zetterberg, P., Eronen, M. and Briffa, K. R. 1994. Evidence on climatic variability and prehistoric human activities between 165 B.C. and A.D. 1 400 derived from subfossil Scots pine (*Pinus sylvestris* L.) found in a lake in Utsjoki, northernmost Finland. – Bull. Geol. Soc. Finland, 66, Part 2: 107–124.

– , Eronen, M. and Lindholm, H. in press. The mid-Holocene climatic change around 3 800 BC: tree-ring evidence from northern Fennoscandia. – Paläoklimaforschung-Paleoclimate Research 17.

Ecological Bulletins 45: 31–44. Copenhagen 1996

Modelling and measuring evapotranspiration in a mountain birch forest

M. Ovhed and B. Holmgren

Ovhed, M. and Holmgren, B. 1996. Modelling and measuring evapotranspiration in a mountain birch forest. – Ecol. Bull. 45: 31–44.

A combination of measurements, empirical relations and a four-layer model based on the Penman-Monteith equation has been used to determine the evapotranspiration, on a continuous basis, from a mountain birch forest, of medium density, in the Abisko area (68°21′N, 18°49′E), during the 1988 and 1989 vegetation seasons.
The total latent and sensible heat fluxes from the birch forest are 49% and 45%, respectively, of the net radiation, which, summed over the period with birch foliage, amounts to c. 600 MJ m^{-2}. The ground heat flux accounts for c. 5% and the photosynthesis for 1% of the net radiation.
Of the net radiation, the dominating component of the energy balance, 59% is absorbed by the birch canopy and 41% by the undergrowth. The woody parts of the birches absorb 12% of the net radiation. Thus, 47% is absorbed for the birch leaves. The study shows that the undergrowth evapotranspiration is an important part of the water balance of the birch forest. It contributes 37% of the total evapotranspiration when the birches are fully foliated.
Using climate records 1913–1994, summer (June–August) evapotranspiration is estimated to be somewhat higher than the summer precipitation (130 and 126 mm, respectively), assuming that the leaf area index is the same in 1988 and 1989.
In spite of the fact that the site is in an area that has the lowest annual precipitation in Sweden, its prevailing soil moisture is, on the average, high. The reasons for this are mainly the low summer temperatures and the high water holding capacity of the soils.

M. Ovhed and B. Holmgren (correspondence), Dept of Meteorology, Uppsala Univ., Box 516, S-751 20 Uppsala, Sweden and Abisko Scientific Research Station, S-981 07 Abisko, Sweden.

The mountain birch *Betula pubescens* ssp. *tortuosa* forest dominates large areas of the subarctic/subalpine regions and it often forms the forest limit in northern Scandinavia (Sonesson and Lundberg 1974). The forest gives rise to a microclimate different from that of the nearby tundra particularly as far as the snow distribution, the net radiation, soil temperatures and soil moisture are concerned (Rouse 1984b, Josefsson 1990).

Evapotranspiration is an important component of the energy and water balances in the subalpine/subarctic mountain birch forest. The aim of this study is to determine the energy and water balances on an hourly basis separately for the birch canopy and the undergrowth during the vegetation season, at a site characterised by a forest of approximately medium tree density.

Models calculating the evapotranspiration are often developed for agricultural and forestry purposes with relatively homogenous vegetation covers. Natural mountain birch forest showed large variations of density, from closed canopies to open canopies of scattered trees. The one-dimensional evapotranspiration model presented here, is based on the Penman-Monteith's equation (Monteith 1965) using a multilayer approach (Waggoner and Reifsnyder 1968). In our model the birch canopy is divided into three layers. A fourth layer, the undergrowth, is further divided into two sublayers, one consisting of shrubs and grass and the other of cryptogams.

There appear to be no experimental studies of evapotranspiration in subarctic deciduous forests in northern Scandinavia to date. In Canada, there are some studies made at the tundra/forest interface dealing with the energy balance and evapotranspiration in a willow-birch forest (Blanken and Rouse 1994), and also in coniferous stands (Rouse 1984a–c, Lafleur and Adams 1986, Lafleur and Rouse 1988).

Site

This study was performed near Abisko Scientific Research Station, ANS (68°21'N, 18°49'E, 385 m a.s.l.). This area is in the middle of the range of mountains with summits up to 1800 m a.s.l. The mountain birch forests in the area vary from dry heath types with widely spaced trees to moist and dense meadow types. The altitude of the timber line varies between 550 and 700 m a.s.l. (Sonesson and Lundberg 1974).

The results presented here were obtained during the summer seasons of 1988 and 1989. In both these years bud burst occurred in the beginning of June. The leaf growth was completed at the turn of June/July. Autumn colours appeared in the end of August and leaf abscission was completed during September.

The measuring site (20×20 m) is on flat ground inside a mountain birch forest of about medium density at the mouth of the Abisko valley. The forests of the Abisko valley are broken up by open treeless heaths and mires with typical sizes of a few hectares. Small hills of 5–10 m height are characteristic for the topography of the valley floor (Josefsson 1990).

The trees show a large diversity regarding age and size distribution. The trees at the site are multi-stemmed as well as single-stemmed. The forest density is c. 750 individual birches ha^{-1} comprising a total of c. 3400 stems ha^{-1} (>2.5 m height). Slightly $<50\%$ of the stems are within the height range of 2–4 m. About 30% are <2 m and 20% >4 m. The height of the tallest trees generally ranges between 5 and 7 m. The maximum height was c. 9 m. The total trunk basal area is 18 m^2 ha^{-1}. The projected crown diameters of the individual birches vary between 1 and 4 m. The total projected crown coverage is 26%. The forest may be described as fairly low and open which is typical for the mountain birch forest in, especially, the drier areas in northern Scandinavia.

Using the theory for radiation penetration of diffuse radiation into a community of random spatial distribution of vegetation elements (Ross 1975), the silhouette woody biomass area index (BAI, Baldocchi et al. 1986) was determined to $0.36 \pm 10\%$ by direct area measurements of individual trees combined with measurements of radiation transmission in leafless conditions under overcast skies. Leaf Area Index (LAI) was determined to 2.0, with a standard deviation of 5%, during the fully leafed seasons in 1988 and 1989. Two methods, giving about the same results, were used: number counts of leaves on representative trees and the weighing of autumn leaves that had fallen within randomly chosen squares on the ground. The leaf number was then used together with the average size of c. 1500 leaves collected at different heights in different trees was determined to 8.9 cm^2 (STD = 1.7) using a LI-COR leaf area meter.

The undergrowth below the birch canopy consists of an understorey layer (maximum height 30 cm) with unevenly distributed shrubs (*Empetrum hermaphroditum*, *Vaccinium myrtillus*, *Vaccinium uliginosum* and *Vaccinium vitis-idaea*) and grasses (e.g. *Deschampsia caespitosa* and *Deschampsia flexuosa*). The bottom layer consists of green mosses (e.g. *Hylocomium splendens* and *Polytrichum commune*) and lichens (e.g. *Peltigera aphthosa* and different *Cladonia* species). The cover of phanerogams and cryptogams was about equal (50% each) at the site. From the lysimeters, a LAI of vascular plants of 0.5 (projected) and a dry weight of cryptogams of 0.32 kg m^{-2}, were estimated to be representative for the site.

Judging from the undergrowth species, the mountain birch forest of the site may be classified as an *Empetrum-Vaccinium myrtillus* type (Sonesson and Lundberg 1974) indicating that the soil moisture conditions are intermediate between wet and dry.

A description of the soils in the birch forest and on treeless heaths has been made by Josefsson (1990). The soils of the birch forest are podzolized. The organic horizon varies between 10 and 15 cm at the site. Although some roots may extend as deep as 50 cm, the majority of the roots are found in the organic layer. The mineral soil may be defined as fine loamy sand-fine sand with frequent presence of stones. The content of silt and clay amounts to 20–50%. The water content of the soils, given as water weight percentage in relation to the wet substrate, is 45–88% for the organic soils and 14–34% for the mineral soils. Because of the relatively high amount of silt in the soils, the water holding capacity is high.

Instruments

Above the canopy, the incoming longwave radiation was measured by an Eppley PIR pyrgeometer, the incoming shortwave radiation by one pyranometer (Kipp and Zonen, model CM-11) and the photosynthetically active radiation by a PAR quantum sensor (Li-Cor, model LI-190sb, range 0.4–0.7 µm) mounted on a roof unobstructed by trees c. 100 m east of the forest site. At the forest site, two pyranometers (Kipp and Zonen, model CM-5) were mounted, upright and inverted, on a 12 m-tower, to give the temporal albedo variations of the birch forest. At the forest floor beneath the canopy one pyranometer (Kipp and Zonen, CM-5), one Eppley PIR-pyrgeometer and one PAR pyranometer (Li-Cor, LI-190sb) were installed on a carriage that moved back and forth continuously with a speed of c. 8 m min^{-1}, along an 8 m long track, oriented 100° E of north. Moving sensors were used since they give more accurate estimations of the area-averaged radiation fluxes beneath a forest canopy than fixed sensors do (Péch 1986).

Net radiation was measured at 12 m height, c. 5 m above the birch canopy, by a Siemen-Ersking net radiometer. The calibration constants for the net radiometer were obtained for shortwave as well as longwave radiation as described by (Rodskjer 1978). According to Halldin and Lindroth (1992), the net radiometer used may, under conditions with high incoming solar irradiance, give to low output (related to high sensor body temperature). No correction for this deviation is generally applied here, since periods of high incoming solar irradiance and excess temperatures are relatively few and of too short a duration to significantly modify the results in the present study. The output of the net radiometer was checked with the independent radiation measurements of the incoming and reflected solar radiation and the incoming longwave radiation above the canopy. The outgoing longwave radiation was obtained from estimated birch leaf temperature (using the Penman-Monteith equation) and the surface temperature measurements at the forest floor. The latter estimates of the net radiation were used to supplement suspect or missing data from the net radiometer.

Air temperature, inside the canopy, was measured with ventilated copper-constantan thermocouples at six levels (0.2, 0.4, 1.1, 2, 3.5 and 5 m above the ground surface). Relative humidity and temperature of the air was measured with ventilated Rotronic Ya-100 sensors at three levels (0.2, 2 and 5 m).

Wind speed was measured inside the canopy at three levels (0.2, 2 and 5 m) and at one level above the canopy (12 m), by lightweight three-cup anemometers with magnetic-contact switches, constructed by K. Lundin, at the Dept of Meteorology, Uppsala Univ. Wind direction was measured at 2 m.

Ground temperature was measured, at two levels in the upper organic layer (0.08 and 0.15 m depth) and at three levels in the mineral soil (0.22, 0.53, and 1.03 m depth).

Surface temperature of the bottom layer, consisting of lichens and mosses, was measured with thermocouples at three points, in a small clearing, in the shelter of trees and at a point of intermediate exposure. Temperatures inside and on the surface of a birch stem were measured at three levels (0.4, 2 and 3.5 m).

At the forest site, all sensor outputs were sampled by two Campbell 21x data loggers at 5 s intervals. Average, maximum and minimum values were stored on an hourly basis.

Radiation instruments above and below the canopy, were calibrated relative to each other twice within each summer. The Kipp and Zonen CM-11 was also calibrated relative to a Linke-Feussner actinometer, both in the 1988 and 1989 summers. The air temperature and the humidity measured at 2 m were calibrated relative to an Assmann psychrometer once a week. The three humidity sensors were also calibrated relative to each other, on several days every summer. All thermocouple temperature sensors were calibrated before and after the project period.

An ADC flow-through steady-state porometer model LCA3 was used for the measurements of stomatal conductance of the birch leaves, as described in Holmgren et al. (1996).

Evapotranspiration from the ground vegetation was measured with simple lysimeters, consisting of a container with a circular orifice of $0.049 \, m^2$ and a depth of 0.25 m. Roots from the vascular plants were almost completely confined to the top 15 cm of organic soil. The containers were manually weighed 1–2 times per day. After periods of rain, excess water was siphoned out of the container.

Evapotranspiration from the birches was measured with a potometric technique (Weyers and Meidner 1990), whereby medium sized birches were cut and put into small cylindrical water containers ($1.5 \, dm^3$) where the evapotranspiration rate was measured through the changes of water level in the container. The leaf conductance was measured before cutting the birches and at regular intervals (up to six times a day) after the cutting, until the end of the experiment, which usually lasted 1–3 days. Leaf conductances were simultaneously measured on some "reference"-birches. This was made in order to transfer the rate of evapotranspiration of the cut birches to that of undisturbed birches (Holmgren et al. 1996). For most birches the stomatal conductance increased after cutting and installation into the water containers.

To measure the soil moisture potential, two Delmhorst gypsum blocks were installed at 15 cm depth, in the humus layer, just above the inorganic soil, at two spots, the one relatively dry and the other relatively wet compared with average conditions.

Measurements of birch canopy interception (total rain minus throughfall) were made by 10 small rain gauges distributed at various distances from the stems of three birches.

Theory and model description

Neglecting the effects of horizontal advection, the energy balance for a vegetation column may be written as:

$$A = E\lambda + H = R_n - G - S - P \qquad (1)$$

where A is available energy for the vertical turbulent exchange of latent ($E\lambda$) and sensible (H) heat, respectively, between the vegetation and the air. R_n is the net radiation above the vegetation. G is the ground heat flux, S is the energy storage within the canopy and finally P is the biochemically stored energy. Our study includes measurements or estimates of all terms, except the sensible heat flux, which was obtained as a residual.

The net radiation

The net radiation, above the forest, R_n, is split into two components, $R_{n,u}$ and $R_{n,can}$, for the undergrowth and the birch canopy, respectively. $R_{n,u}$ is calculated from measurements of incoming shortwave and longwave radiation below the birch canopy and temperature measurements at the top of the bottom layer with an emissivity of 0.97 (Gates 1980). The absorption of the shortwave radiation is determined with due considerations to the spectral changes of the solar radiation and the spectral reflectivity of the undergrowth layer (Ovhed and Holmgren 1995). $R_{n,can}$ is calculated as the difference between R_n, and $R_{n,u}$. The distribution of $R_{n,can}$ inside the canopy is calculated for the three layers (see above), with an exponential decrease (Lindroth and Halldin 1986) in relation to the accumulated sum of LAI and BAI.

$R_{n,can}$ is further divided into two parts: $R_{n,stem}$ the part absorbed by the woody elements of the canopy and $R_{n,leaf}$, that absorbed by the birch leaves. The basis of this division originated from the results of spectroradiometric measurements (Ovhed and Holmgren 1995).

G, S and P

The ground heat flux, G, is calculated from the profile measurements of soil temperatures. Soil densities and soil moisture contents were obtained from measurements at nearby sites (Josefsson 1990). Specific heats and conductivities of organic and inorganic soils were calculated using thermal properties given by van Wijk and Vries (1963). The short term variations of G are based on soil temperature differences at 1-h intervals, between the surface and the depth of 1 m. The small heat transfers at 1 m were obtained from the measured soil temperature variations over the year at that depth. These calculations were based on the conventional heat transfer theory (Carslaw and Jaeger 1976), following the approximation of the monthly soil temperatures by six Fourier components.

The birch canopy storage S is divided into three major parts (Stewart and Thom 1973): energy stored as 1) thermal and 2) latent heat in the air volume of the canopy, and as 3) thermal heat in vegetation parts, respectively. The changes of S were obtained from profiles of air temperature and humidity and from temperature measurements in the birch stems and branches.

The storage contribution by photosynthesis P was calculated from all measurements (1300) of CO_2 changes by the ADC porometer. On the average P was 2 W m^{-2} (ranging between -1 and $+6$ W m^{-2}). These results are of the same magnitude as could be derived from Karlsson (1991). For the whole vegetation season, total P, for the birch canopy and the undergrowth, is

estimated to be 1% of the total net radiation. This contribution is small but differs from the other storage terms as it mainly acts in one direction (i.e. converting solar energy to chemical energy).

Evapotranspiration

The evapotranspiration of the birch forest is modelled according to the principles of Waggoner and Reifsnyder (1968), further discussed by e.g. Lhomme (1988). In the present study, modified versions of the Penman-Monteith equation are applied to three layers in the birch canopy and to the undergrowth layer.

Canopy layer

With the assumption that aerodynamic conductances for latent and sensible heat are equal, the evapotranspiration of the birch canopy layers is, in our case, written as (cf. Raupach and Finnigan 1988):

$$(E\lambda)_{canopy} = \sum_{i=1-3} LAI_i \frac{[\Delta_i A_i + \rho c_p(e_{si}(T_i) - e_i)g_{ai}]}{\Delta_i + \gamma\left(1 + \dfrac{g_{ai}}{g_{si}}\right)} \quad (2)$$

where LAI is the leaf area index, Δ is the slope of the saturated vapour pressure curve, A is the available energy per unit LAI, ρ is the air density, c_p is the specific heat of the air at constant pressure, $e_s(T)$ is the saturated vapour pressure at temperature T, e is the water vapour pressure, g_a is the aerodynamic conductance from the leaf to the ambient air, γ is the psychrometric constant and, finally, g_s is the stomatal conductance. All conductances are expressed per unit LAI. The subscript i indicates a value in the middle of the layer, except for LAI and A which are integrated for the layer i. The birch canopy was divided into three layers; one top layer of height > 3.5 m with LAI = 0.66, one layer in the height interval 1–3.5 m with LAI = 1.16 and a last layer of height < 1 m with LAI = 0.18.

The stomatal conductance, g_s is modelled using a modified approach of Stewart (1988). The results are validated by approximately 1300 individual measurements by the ADC porometer (Holmgren et al. 1996). For the birch leaves, g_s is given by:

$$g_s = g_{s,max} \cdot f_1(D) \cdot f_2(t) \cdot f_3(PAR) \cdot f_4(\Psi_s) \quad (3)$$

where $g_{s,max}$ is the maximum stomatal conductance. The f-functions (≤ 1) denote the effects by the water vapour pressure deficit (D), the temperature in the ambient air (t), photosynthetically active fluence rate above the canopy (PAR) and soil water potential in the humus layer 15 cm depth (Ψ_s) (Holmgren et al. 1996).

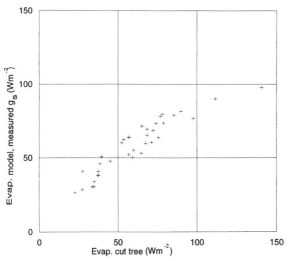

Fig. 1. Comparison between measured evapotranspiration from cut birches and modelled evapotranspiration using Eq. 2 and measurements of g_s.

Aerodynamic conductance, g_a is expressed by (Monteith 1973, Raupach and Finnigan 1988) as:

$$g_{ai} = \frac{\sqrt{u_i}}{C_1} \tag{4}$$

where u is the wind speed inside the canopy and C_1 is an empirically determined constant.

Using the model (Eq. 2) the calculated evapotranspiration was compared with the evapotranspiration of the cut birches. The value of C_1 (58 $s^{0.5}$ $m^{-0.5}$) was chosen to give the best fit (the least variance around the 1:1 line) when comparing model and measurements (Fig. 1). For high values of the measured evapotranspiration there appears to be a systematic underestimation by the model. Possible causes of the deviation are discussed in Holmgren et al. (1996).

Undergrowth layer

The undergrowth consisted of vascular plants (mainly shrubs) and cryptogams (mainly mosses and lichens). Only the former is supposed to have a stomatal conductance, the latter only evaporates after interception of precipitation.

The evapotranspiration of the undergrowth is written as;

$$E_g\lambda = R_v(LAI)_v \frac{[\Delta A_g + \rho c_p(e_s(T_a) - e_a)g_{ag}]}{\Delta + \gamma\left(1 + \frac{g_{ag}}{g_{sv}}\right)}$$
$$+ E_{crypt}\lambda \tag{5}$$

where the first term on the right refers to vascular plants and the second term to cryptogams. R_v is the relative surface coverage of vascular plants and LAI_v is their leaf area index of a homogenous cover of vascular plants. The subscript g refers to the undergrowth layer, T_a and e_a refer to measurements at the lowest level (20 cm). g_{sv} and g_{ag} are the stomatal conductance and aerodynamic conductance per unit LAI of the vascular plants, respectively.

g_{sv} was calculated from a simplified version of Eq. 3, with dependence only on PAR and D:

$$g_{sv} = g_{sv,max} \cdot f_1(d) \cdot f_3(PAR) \tag{6}$$

$g_{sv,max}$ was obtained from Körner et al. (1979), for shrubs, and from Johnson and Caldwell (1976), for grasses. The combined value used was 0.01 m s^{-1} for projected leaf area. The same function $f_1(D)$ as for the birch stomatal response, was used, while $f_3(PAR)$ was somewhat modified. A linear increase of f_3 from 0 to 1 was assumed between 0 and 250 μmol m^{-2} s^{-1}, derived from measurements of photosynthetic rate for shrubs by Karlsson (1987a). For PAR, the measurements below the birch canopy were used.

The aerodynamic conductance for the undergrowth was assumed to be the maximum of two semi-empirical relations, one for forced and the other for free convection of sensible heat:

$$g_{ag} = Max\left\{\frac{u_{20cm}}{C_2}, \frac{\sqrt{\Delta T}}{C_3}\right\} \tag{7}$$

where u_{20cm} is the wind speed at 20 cm and ΔT is the temperature difference between the bottom layer and the air temperature at 20 cm. C_2 and C_3 are determined from lysimeter measurements to give the best fit during dry conditions (i.e. when cryptogams account for < 5% of E_g, according to the model). The values of C_2 and C_3 were determined to be 88 and 34, with a standard deviation of 28% and 34% (n = 20), respectively.

In the model, the cryptogams only evaporate stored water. The evaporation of the cryptogams was calculated using the approach:

$$E_{crypt}\lambda = R_cC_4G(w)\frac{[\Delta A_g + \rho c_p(e_s(T_a) - e_a)g_{ag}]}{\Delta + \gamma} \tag{8}$$

where R_c is the relative surface coverage by the cryptogams, C_4 is an empirical constant, G(w) is a function, which depends on the stored water content, w (in relation to the dry weight of the cryptogam). For w between 0 and w_c (a critical species-related value below which the surface water pressure of the cryptogam starts to decrease below the saturation water pressure of a free water surface), G(w) is assumed to increase linearly with w from 0 to 1 (as may be derived from measurements by Larson 1977). Between w_c and w_{max},

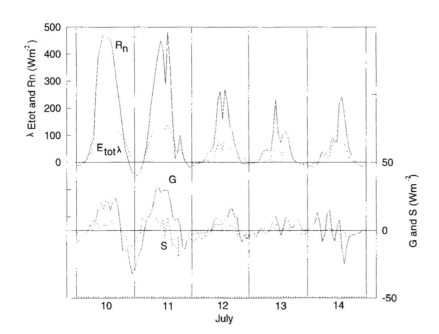

Fig. 2a. Net radiation and evapotranspiration of the birch forest (top). Ground heat flux and energy storage inside the birch canopy (bottom).

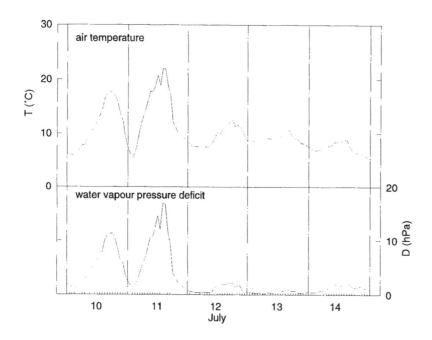

Fig. 2b. Air temperature and water vapour pressure deficit at 2 m level.

(the maximum water storage) G(w) remains constant (=1).

For w_{max}, we adopt a value of 600% as an average for lichens (Larson 1977) and mosses (Sonesson et al. 1992). w_c is assumed to be 400% which is the same relation between w_{max} and w_c as derived for lichens by Larson (1977).

The constant C_4 was included to account for the fact that the cryptogams from the bottom layer of the undergrowth with reduced incoming shortwave radiation (Ovhed and Holmgren 1995) and smaller aerodynamic conductance. About 80 daily measurements obtained within 72 h after rains, have been used in the comparison. A value of 0.25 was found by least square

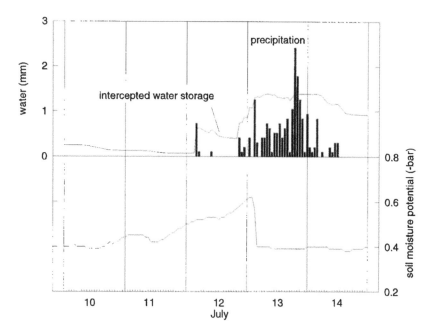

Fig. 2c. Precipitation and total storage of water by the vegetation (top) and soil moisture potential (bottom).

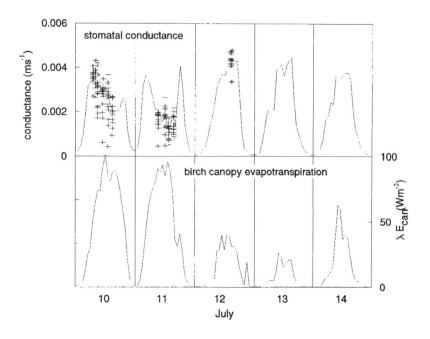

Fig. 2d. Modelled (line) and measured (+) stomatal conductance of the birches (top), and modelled evapotranspiration from the birch canopy (bottom).

fit of the difference between model and lysimeter evaporation.

It may be noted that our model for the cryptogam evapotranspiration stresses the role of water storage of the cryptogam layer and that no water flow is assumed to occur between the cryptogams and the substrate. Another formulation of the water budget in a wet meadow tundra, where the water flow between the moss surface and the substrate are of major importance, is given by Miller et al. (1978).

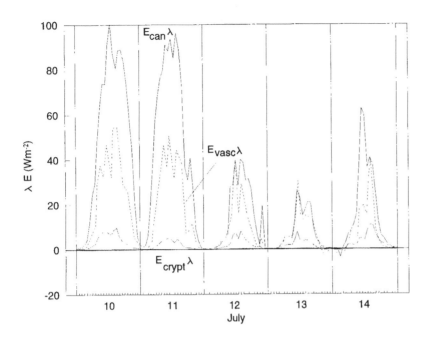

Fig. 2e. Evapotranspiration from the birch canopy, the vascular plants of the undergrowth and the cryptogams, respectively.

Interception of rain

Using the interception measurements, the maximum intercepted water by the foliated birch crowns was 1 mm per unit crown area, if the total precipitation was >2 mm. This magnitude appears to be in accordance with results obtained for hardwood forests (Zinke 1967). For precipitation <2 mm, approximately half of the rain passes through the tree crowns. In the model, the intercepted water evaporates at the rate given by Eq. 2. with $g_s = \infty$.

The interception of the undergrowth is treated differently for the vascular plants and for the cryptogams. The vascular plants are assumed to have maximum interception of 0.1 mm per unit area covered, which is the same, relative to LAI, as for the birches. Based on the results of Larson (1977) and Sonesson et al. (1992) the maximum water storage of the cryptogams was calculated to 2 mm water per unit cryptogam area.

After adopting the empirical values (Eqs 5–9) discussed above, the modelled ground evapotranspiration, accumulated over the whole observation period, was found to agree with evapotranspiration measured by the lysimeters to within 10%.

Results

Diurnal variations

A 5-day period with two dry, almost clear days and three cloudy days with rain have been selected to show characteristic hourly variations (Figs 2a–e). During the first two days, the maximum R_n exceeds 450 W m^{-2} (Fig. 2a). The following three days include some showers, a period with continuous rain, but also some short periods with sunshine. The maximum R_n is then <250 W m^{-2}. E_{tot} decreases significantly during the three cloudy days. The diurnal variations of G and S are most pronounced on the clear days with amplitudes of ± 30 and ± 15 W m^{-2}, respectively. On the cloudy days the variations of G and S are small and irregular.

Obviously related to the variations of R_n, the first two days show high amplitudes of air temperature and water vapour saturation deficit, D. On the following cloudy and rainy days D is low throughout (Fig. 2b). The soil moisture potential in the organic soil decreases until shortly after midnight on 13 July (Fig. 2c), indicating a drying out of the soil. About 24 h after the first rain, the water storage capacity of the undergrowth is filled. The water then penetrates into the soil, and the drying process is abruptly interrupted.

On the first two days the stomatal conductance is depressed around noon daytime according to our model (Fig. 2d), due to the increase of D. Similar patterns for deciduous trees are observed by Körner (1985). The crosses, referring to measurements of stomatal conductance of individual leaves, show a typical scatter for these types of measurements. The rise in the morning and the decline in the evening are due to the variations of PAR. The maximum conductances in the mornings and evenings result from a combination of low D and PAR values. At around a threshold of

300 μmol m^{-2} s^{-1}, the stomata start to close (Dolman and van den Burg 1988, Holmgren et al. 1996). According to the model, the variations of g_s on the last three days are mainly regulated by PAR. Variations of air temperature and soil moisture potential have only moderate effect during these days.

In Fig. 2e E_{tot} is split into its three components: evapotranspiration from the birch canopy, E_{can}, evapotranspiration from the vascular plants in the undergrowth, E_{vasc}, and evapotranspiration from the cryptogams E_{crypt}. E_{can} and E_{vasc} show similar variations, although with different amplitudes. The decreases of E_{can} and E_{vasc} on the last three days are mainly due to low R_n and D. On the first two days, when the moisture content in the cryptogams is low, E_{crypt} is small whereas on the following three days, E_{crypt} increases, in spite of low R_n and D.

Seasonal variations

The accumulated evapotranspiration and precipitation, show a small positive water balance, for both years 1988 and 1989, for the period 1 June to end of September (Fig. 3). The measuring periods began 10 and 18 days after snowmelt, in 1988 and 1989, respectively. Immediately after snowmelt, the soils are saturated with moisture. During the first five days of June in 1988, a frost horizon was still present below 0.5 m depth. In 1989 all ground frost was gone before 1 June.

As indicated by the Ψ_s curve (Fig. 3), the soil moisture was high in both summers, except for some prolonged dry periods. The maximum deficit of the accumulated total evapotranspiration and the precipitation is 43 mm in 1988 and 48 mm in 1989. These maxima induced simultaneous minima in the measured soil water potentials.

The start of the birch canopy evapotranspiration lags 2–4 weeks behind the undergrowth, due to the late foliage development of the birches. During the period with foliage, at the end of June to end of August, the evapotranspiration rate of the birch canopy is on the other hand higher than that of the undergrowth. Towards the end of August, the evapotranspiration rates are very small for both the undergrowth and the birch canopy, due to the low net radiation. The accumulated evapotranspiration is then higher for the birch canopy.

The total evapotranspiration by the cryptogams is small and is qualitatively in agreement with the variations shown in Fig. 2e. The total amount of water stored by the cryptogams is c. 9% of the total summer precipitation. Corresponding values for the birches and for the vascular plants of the undergrowth are 7 and 2%, respectively, of the summer precipitation.

Table 1 summarises the water and energy budgets for periods with full birch foliage in 1988 and 1989. The higher values in 1989 of R_n, A_{tot} and E_{tot} are mainly due to the fact that the development of the birch foliage was one week earlier in 1989. This week was, furthermore, a sunny period with very high values of R_n. The higher values of G in 1988 are caused by lower ground temperatures in the beginning of the summer and higher in the end. G + P was 7 and 5% of $R_{n,tot}$ in the two years, respectively.

It may be noted that the evapotranspiration in 1989 is higher than in 1988, in spite of the lower mean temperature in 1989 (11.0°C in 1988 compared with 9.2°C in 1989). However, an exclusion of the "extra days" in 1989 gives the following result: 1988, $R_n = 554$ MJ m^{-2}, $E_{tot} = 290$ MJ m^{-2} and $\bar{T} = 11.0$°C. 1989, $R_n = 525$ MJ m^{-2}, $E_{tot} = 259$ MJ m^{-2} and $\bar{T} = 9.3$°C.

The ratio between Eλ and A is c. 5% higher in 1988 than in 1989. The difference is mainly due to higher summer temperatures in 1988, as mentioned above. It is of interest to note that the ratios of evapotranspiration to the available energy are about the same for the canopy and the undergrowth, respectively. In fact, during 1989, the ratio was even a little higher for the undergrowth than for the birch canopy. It may be noted that A_{can} includes the net radiation absorbed by the woody parts of the birch canopy, which is estimated to 21% of the total net radiation. This value is derived from results of spectral shortwave measurements within 300–1100 nm adjusted for the total shortwave spectrum (Ovhed and Holmgren 1995).

The percentage ratios of E_{can}, E_{vasc} and E_{crypt} are on average 63, 28 and 9%. Comparing dry and wet periods, the ratio of E_{tot} to A_{tot}, (i.e. the efficiency in the rate of evapotranspiration) is 0.46 during dry periods and 0.58 for wet periods. Obviously, the higher ratios during the wet periods are related to the ground vegetation. The birch canopy has the highest ratios in the dry periods while the undergrowth has the highest ratios in the wet periods.

The ratio between E_{can} and E_{tot} drops from 0.71 in the dry to 0.51 in the wet periods, while ratios for the undergrowth show trends in the opposite direction. On average, the cryptogams account for 24% of the undergrowth evapotranspiration. This ratio achieves a maximum of 31%, in the intermediate periods, when the water intercepted by the vascular plants has disappeared, but the cryptogams are still very wet.

Sensitivity analysis

The sensitivity of the results presented in Table 1, to changes of model parameters, was analysed for both summers and are presented in Table 2. Also, the input meteorological parameters were checked individually for their relative effect on the evapotranspiration, keeping other parameters constant. All results in Table 2 are

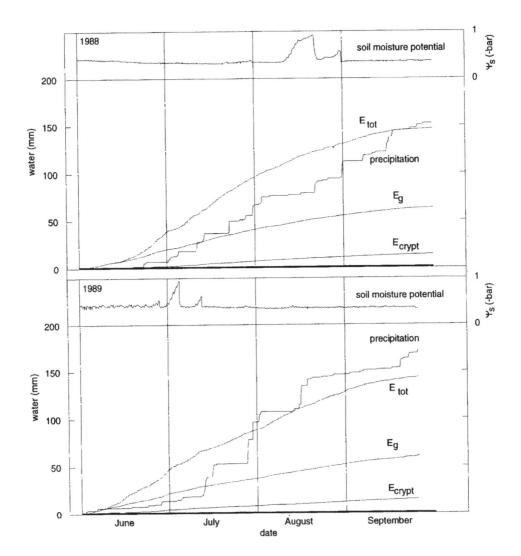

Fig. 3. Accumulated precipitation, accumulated evapotranspiration for the forest (E_{tot}), the undergrowth (E_g) and the cryptogams (E_{crypt}) and the soil moisture potential, for 1988 and 1989.

obtained by summation of hourly values of the indicated periods.

The estimated maximum errors of g_s and g_a, are of the order of $\pm 25-50\%$. The effects on E_{tot} are < c. $\pm 10\%$ (Table 2). Errors in g_s, have greater effects than errors in g_a and, as expected, the influence by the stomatal conductance of the birches is the greatest.

The relative area coverage of the vascular plants and cryptogams has a considerable effect on E_g and also, E_{tot}. The species' distribution varies much from place to place and is one of the factors that is hard to determine. For the present site we have estimated 50% coverage for each of the cryptogams and the vascular plants. Also the maximum water storage of the cryptogams and the value of C_4 have evident effects on E_{tot}, despite the small contribution from E_{crypt}.

The percentage changes in the individual components, E_{can}, E_{vasc} and E_{crypt}, are of course much higher than that in E_{tot}. However, since our method involves separate measurements of E_{can} by a potometric method and E_g by lysimeters, we estimate the maximum error of E_{can} and E_g to $\pm 10\%$ on average. This is on condition that the cut birches and the lysimeters are representative for the conditions at the site. The experiments involve c. 27 cut birches of various sizes used on a total of 54 days of varying weather conditions and 1–2 lysimeters used on 210 days.

Table 2 further shows that the total evapotranspiration is not very sensitive to the changes in the available energy ($\pm 10\%$ of A_{tot} gives $\pm 3\%$ change of E_{tot}). E_{tot} is, however, almost insensitive to the partition of available energy between birch canopy and undergrowth. A

Table 1. The energy and water balance for the period with full birch foliage, from 25 June to 10 September in 1988 and 16 June to 10 September in 1989.

Parameter	1988 (1872 h) (MJ m^{-2})	1988 (mm)	1989 (2113 h) (MJ m^{-2})	1989 (mm)	Dry periods *	Intermediate periods **	Wet periods ***
R_n	553		622				
G	31		22				
S	0		0				
P	5		6				
A_{tot}	520		593				
A_{can}	324		370				
A_g	195		224				
Precipitation		114		144			
$E_{tot}\lambda$	284	114	297	119			
$E_{can}\lambda$	180	72	185	74			
$E_g\lambda$	103	43	112	45			
$E_{vasc}\lambda$	78	31	84	34			
$E_{crypt}\lambda$	25	10	28	11			
Ratios							
$E_{tot}\lambda/A_{tot}$	0.55		0.50		0.46	0.51	0.58
$E_{can}\lambda/A_{can}$	0.56		0.50		0.53	0.52	0.47
$E_g\lambda/A_g$	0.53		0.50		0.35	0.48	0.75
E_{can}/E_{tot}	0.63		0.62		0.71	0.61	0.51
E_{vasc}/E_{tot}	0.28		0.28		0.28	0.27	0.39
E_{crypt}/E_{tot}	0.09		0.09		0.01	0.13	0.10
E_{crypt}/E_g	0.24		0.25		0.03	0.31	0.20

* Periods with total storage of intercepted water <0.1 mm ($\approx 15\%$ of total measuring period). ** Periods with total storage of intercepted water 0.4–0.6 mm ($\approx 15\%$ of total measuring period). *** Periods with total storage of intercepted water >1.0 mm ($\approx 15\%$ of total measuring period).

decrease of the evapotranspiration by the canopy is compensated for by the evapotranspiration by the undergrowth.

Changing D ($=e_s(T) - e$), in Eqs 2 and 3, by $+10\%$ and -10% affects E_{tot} by $+3\%$ and -4%, respectively. These values thus include the balancing effect due to the negative correlation between g_s and D (Eqs 2 and 3). A change of T involves simultaneous changes of e_s and D. An increase of 1°C (keeping e unchanged) results in a 14% increase of E_{tot}. If the change of T is associated with an unchanged relative humidity, which is more realistic than keeping e unchanged according to the climatic records at Abisko, the increase in E_{tot} is only 5%. The climate records from Abisko, 1913–1994, show a high positive correlation between e and T during the summer period. If the simultaneous changes of T and e are applied in the calculations, a change of 1°C results in a change in E_{tot} of 6%.

Discussion

There is an extensive literature on the manifold intricate problems in studies of forest evapotranspiration (see e.g. Denmead 1984 and Stewart 1984). The methods of measuring evapotranspiration from the mountain birch forest are chosen with due regard to the research aim which included separate determination of the canopy and undergrowth evapotranspiration. We

have utilitised some natural features of the birch forest in the area: 1) The leaf sizes are large enough to fit the single leaf chamber of the porometer which facilitates measurements of stomatal conductance. 2) The small sizes of the birches make the potometric method usable in repeated experiments involving many trees during several weeks in varying weather conditions. 3) The root systems of the undergrowth are concentrated in the thin organic layer, which may justify the use of small and shallow lysimeters, for measurements of the undergrowth evapotranspiration. The lack of moisture transfer from soil layers below 0.25 m should partly be compensated for by the fact that there were no functioning birch roots in the lysimeters.

By relating the direct measurements of evapotranspiration and of stomatal conductance to micrometeorological data, it is found possible to estimate the evapotranspiration on a continuous basis.

To neglect horizontal advection is common in energy balance studies, but under some conditions the horizontal energy fluxes may be of importance (Rouse 1984c). The ANS site is situated on the floor of the Abisko Valley c. 300 m from the beach of Lake Torneträsk. In the beginning of the birch foliation period in June, the lake is generally still covered with ice. With on-shore winds there may therefore be a significant divergence of the horizontal flux of energy not incorporated in Eq. 1. However, the temperature and humidity profiles measured inside the canopy at heights <5 m, should accord-

Table 2. Sensitivity analysis of model parameters and meteorological input variables. Single parameters are changed in relation to estimated maximum error, and the influence on the modelled evapotranspiration, for the whole period with foliated canopy, is calculated.

Parameter	Change	E_{tot} (%)	E_{can} (%)	E_g (%)	E_{vasc} (%)	E_{crypt} (%)
g_a (canopy)	+50%	+3	+5			
	−50%	−2	−3			
g_a (undergrowth)	+50%	+4		+9	+6	+20
	−50%	−2		−5	−3	−10
g_s (canopy)	+25%	+9	+15			
	−25%	−11	−18			
g_s (undergrowth)	+50%	+5		+13	+18	
	−50%	−8		−21	−28	
Maximum water storage by cryptograms	+50%	+2		+5		+22
	−50%	−3		−7		−29
Area cover undergrowth (vasc./crypt.)	0.25/0.75	−11		−31	−48	+23
	0.75/0.25	+10		+27	+47	−35
C_4 (= 0.25)	= 1.0	+7		+18		+74
	= 0.1	−4		−11		−45
A_{tot}	+10%	+3	+3	+4	+4	+2
	−10%	−3	−3	−4	−4	−2
A_{can}/A_{tot}	+10%	−1	+2	−6	−7	−4
	−10%	+1	−2	+6	+7	+4
A_g/A_{tot}	+10%	0	−2	+4	+4	+2
	−10%	0	+2	−4	−4	−2
D	+10%	+3	+4	+2	+2	+2
	−10%	−4	−4	−2	−2	−3
T (e unchanged)	+1°C	+14	+17	+10	+11	+8
	−1°C	−17	−20	−11	−12	−9
T (relative humidity unchanged)	+1°C	+5	+6	+3	+3	+2
	−1°C	−5	−6	−3	−3	−2

ing to a model developed by Gryning and Batchvarova (1990), be well inside the internal boundary layer developing over a distance of a few hundred meters.

The method does not rely on flux-gradient relation and no vertical exchange coefficients (Monteith 1973) are utilised for the transfer of moisture and heat. Effects of horizontal advection due to inhomogenities of the forest structure should be of little importance for the present results referring to the conditions some 50 m from the nearest forest edge. Studies of the water balance of treeless heaths in the Abisko Valley suggest that evapotranspiration rates of the heath surfaces that are of the same order of magnitude as for the birch forest (Josefsson 1990) indicating that advection effects may be relatively small for the local variations of the evapotranspiration.

Due to mainly the insulating effects by the snow cover in winter and the organic surface layer in summer, the soil heat flux is, on the average, comparatively small, 4–5% of the net radiation. This is reflected in the low annual temperature amplitude of ±2–3°C at 1 m depth. This may be compared with amplitudes of ±6–8°C below non-vegetated surfaces in the area (Josefsson 1990).

At our study site, the undergrowth has a major role in the water balance accounting for 37% of total forest evapotranspiration, during the period of full birch foliage. The undergrowth is composed of patches with different mixtures of species, indicating differences in light, nutrient, and water resources at the forest floor, as discussed by Sonesson and Lundberg (1974) and Karlsson (1987a, b, 1989). The role of the undergrowth in the water balance may therefore generally be expected to be quite variable, depending on the types of ecosystems involved. However, a detailed study of the undergrowth in the mountain birch forest is beyond the scope of the present study.

The average summer (June, July and August) temperature in 1988 was 10.8°C and in 1989, 9.8°C, compared with a normal value (1913–1994) of 9.9°C. The average cloudiness deduced from measurements of sunshine hours is 74% in 1988 and 78% in 1989 compared with (1913–1994) value of 74%. The summer precipitation in 1988 was 112 mm and 1989 146 mm, compared with the average (1913–1994) of 121 mm. In May–June 1988, the total precipitation was only 10 mm. Only in two years within 1913–1994, has the May–June period had similar low amounts of precipitation. Still, there was no sign of lowering the soil moisture potential, six weeks after snowmelt in 1988.

The annual precipitation in Abisko is c. 300 mm, which is lower than the figure for any other climate station in Sweden. In spite of the low precipitation, the soil conditions in the forest were quite moist most of the time in the summers of 1988 and 1989. Similar conclusions may be drawn from measurements with a large number of gypsum blocks in the summers 1985–1987, in various micro environments in the vicinity of our site (Karlsson pers. comm.). The lowest values measured, in the latter experiments, after extended dry periods are similar to those of the present study. As may be judged from typical ground water depths of >1–2 m in the birch forest in summer (Josefsson 1990), it seems unlikely that a ground water inflow has any effect on the local soil moisture conditions at the site.

Estimates of the water balances for the summer period using the precipitation and air temperature records of 1913–1994, in combination with the results of the sensitivity analysis (Table 2), indicate that evapotranspiration and precipitation are approximately equal, on average. If September is included, the precipitation is, on average, in excess of 12 mm. The maximum deficit in summers of 1988 and 1989 is c. 50 mm (cf. Fig. 4). Using climate records of 1913–1994, the accumulated deficit at the end of August is estimated to be >50 mm, in 9 of the 81 yr. The latter results indicate that severe droughts are not very common, although the accumulated deficits are not directly comparable with the maximum deficits of 1988 and 1989. The main factor for this is the relatively low evapotranspiration and low summer temperatures (cf. Eriksson 1986). Also the water holding capacity of the soils are high (see Site section).

The greatest deficit, 73 mm, was calculated for the summer of 1955, when the summer precipitation was only 46 mm. This estimate, including all other estimates based on the climate records are based on the assumption of constancy of the leaf area index, measured in 1988 and 1989. However, the summer of 1955 happened to coincide with an outbreak year of the caterpillars of *Epirrita autumnata*. In the summers 1954 and 1955 the mountain birch forest in the Abisko valley was completely defoliated and many of the trees were destroyed (Tenow 1972). To restore the foliage to the conditions before the outbreak may take >70 yr (Tenow and Bylund 1995). In these summers the evapotranspiration should have been negligible and probably not have been compensated for by increased evapotranspiration from the undergrowth. The cyclic (c. 10 yr period) insect attacks, which profoundly alter the leaf area index, complicate estimates of the water balance in the past based on climatic variations only.

References

Baldocchi, D., Hutchison, B., Matt, D. and McMillen, R. 1986. Seasonal variation in statistics of photosynthetically active radiation penetration in an oak-hickory forest. – Agric. For. Meteorol. 36: 343—361.

Blanken, P. D. and Rouse, W. R. 1994. The role of willow-birch forest in the surface energy balance of arctic treeline. – Arct. Alp. Res. 26: 403—411.

Carslaw, H. S. and Jaeger, J. C. 1976. Conduction of heat in solids. – Oxford Univ. Press.

Denmead, O. T. 1984. Plant physiology methods for studying evapotranspiration: Problems of telling the forest from the trees. – Agric. Water Manage. 8: 167–189.

Dolman, A. J. and van den Burg, G. J. 1988. Stomatal behaviour in an oak canopy. – Agric. For. Meteorol. 43: 99–108.

Eriksson, B. 1986. Nederbörds och humiditetsklimatet i Sverige under vegetationsperioden. – RMK 46, SMHI Reports, Meteorology and climatology, Norrköping.

Gates, D. M. 1980. Biophysical ecology. – Springer, New York.

Gryning, S.-E. and Batchvarova, E. 1990. Analytical model for the growth of the coastal internal boundary layer during onshore flow. – Quart. J. R. Meteorol. Soc. 116: 187–203.

Halldin, S. and Lindroth, A. 1992 Errors in net radiometry: Comparison and evaluation of six radiometer designs. – J. Atmosph. Oceanic Technol. 9: 762–783.

Holmgren, B., Ovhed, M. and Karlsson, P. S. 1996. Measuring and modelling stomatal and aerodynamic conductances of mountain birch: Implications for tree line dynamics. – Arct. Alp. Res. 28, in press.

Johnson, D. A. and Caldwell, M. M. 1976. Water potential components, stomatal function, and liquid phase water transport resistances of four arctic and alpine species in relation to moisture stress. – Physiol. Plant. 36: 271–278.

Josefsson, M. 1990. The geoecology of subalpine heaths in the Abisko valley, northern Sweden. – Ph.D. thesis, Dept of Physical Geography, Uppsala Univ., Sweden.

Karlsson, S. 1987a. Micro-site performance of evergreen and deciduous dwarf shrubs in a subarctic heath in relation to nitrogen status. – Holarct. Ecol. 10: 114–119.

– 1987b. Niche differentiation with respect to light utilisation among coexisting dwarf shrubs in a subarctic woodland. – Polar Biol. 8: 35–39.

– 1989. In situ photosynthetic performance of four coexisting dwarf shrubs in relation to light in a subarctic woodland. – Funct. Ecol. 3: 481–487.

– 1991. Intraspecific variation in photosynthetic light response and photosynthetic nitrogen utilization in the mountain birch, *Betula pubescens* ssp. *tortuosa*. – Oikos 60: 49–54.

Körner, C. 1985. Humidity responses in forest trees: Precautions in thermal scanning surveys. – Arch. Meteorol. Geophys. Bioclimat. B36: 83–98.

– , Sceel, J. A. and Bauer, H. 1979. Maximum leaf diffusive conductance in vascular plants. – Photosynthetica 13: 45–82.

Lafleur, P. M. and Adams, P. 1986. The radiation budget of a subarctic woodland canopy. – Arctic 39: 172–176.

– and Rouse, W. R.. 1988. The influence of surface cover and climate on energy partitioning and evaporation in subarctic wetland. – Boundary-Layer Meteorol. 44: 327–347.

Larson, D. W. 1977. A method for in situ measurements of lichen moisture content. – J. Ecol. 65: 135–145.

Lhomme, J. P. 1988. Extension of Penman's formulae to multi-layer models. – Boundary-Layer Meteorol. 42: 281–291.

Lindroth, A. and Halldin, S. 1986. Numerical analysis of pine forest evaporation and surface resistance. – Agric. For. Meteorol. 38: 59–79.

Miller, P. C., Oechel, W. C., Stoner, W. A. and Svein-björnsson, B. 1978. Simulation of CO_2 uptake and water relations of four arctic bryophytes of Point Barrow, Alaska. – Photosynthetica 12: 7–20.

Monteith, J. L. 1965. Evaporation and the atmosphere. – In: The state and movement of water in living organisms. Symp. Soc. Exp., Biol. XIX: 205–234.

– , 1973. Principles of environmental physics. – Edward Arnold, London.

Ovhed, M. and Holmgren, B. 1995. Spectral quality and absorption of solar radiation in a mountain birch forest. – Arct. Alp. Res. 27: 381–389.

Péch, G. 1986. Mobile sampling of solar radiation under conifers. – Agric. For. Meteorol. 37: 1528.

Raupach, M. R. and Finnigan, J. J. 1988. Single-layer models of evaporation from plant canopies are incorrect but useful, whereas multilayer models are correct but useless. – Aust. J. Plant Physiol. 15: 705–716.

Rodskjer, N. 1978. Net and solar radiation over bare soil, short grass, winter wheat and barley. – Swedish J. Agric. Res. 8: 195–201.

Ross, J. 1975. Radiation transfer in plant communities. — In: Monteith, J. L. (ed.), Vegetation and the atmosphere. Vol. 1: Principles. – Academic Press, London, pp. 13—55.

Rouse, W. R. 1984a. Microclimate of arctic tree line. 1. Radiation balance of tundra and forest. – Water Resources Res. 20: 57–66.

– 1984b. Microclimate of arctic tree line. 2. Soil microclimate of tundra and forest. – Water Resources Res. 20: 67–73.

– 1984c. Microclimate of arctic tree line. 3. The effects of regional advection on the surface energy balance of upland tundra. – Water Resources Res. 20: 74–78.

Sonesson, M. and Lundberg, B. 1974. Late quarternary forest development of the Torneträsk area, north Sweden. 1. Structure of modern forest ecosystems. – Oikos 25: 121—133.

– , Gehrke, C. and Tjus, M. 1992. CO_2 environment, microclimate and photosynthetic characteristics of the moss *Hylocomium splendens* in a subarctic habitat. – Oecologia 92: 23–29.

Stewart, J. B. 1984. Measurements and predictions of evaporation from forested and agricultural catchments. – Agric. Water Manage. 8: 1–28.

– 1988. Modelling surface conductance of pine forest. – Agric. For. Meteorol. 43: 19–35.

– and Thom, A. S. 1973. Energy budgets in pine forest. – Quart. J. R. Meteorol. Soc. 99: 154–170.

Tenow, O. 1972. The outbreak of *Oporinia autumnata* Bkh and *Operophtera* spp. Bkh. (Lep., Geometridae) in the Scandinavian mountain chain and northern Finland 1862–1968. – Zool. Bidrag Uppsala suppl. 2.

– and Bylund, H. 1995. Recovery of a mountain birch forest after a severe defoliation by *Epirrita autumnata*. – In: Bylund, H. (ed.), Long-term interactions between the autumnal moth and mountain birch: the roles of resources, competitors, natural enemies, and weather. Ph.D. thesis, Swedish Univ. of Agricultural Sciences.

Waggoner, P. E. and Reifsnyder, W. E. 1968. Simulation of the temperature, humidity and evapotranspiration profiles in a leaf canopy. – J. Appl. Meteorol. 7: 400–409.

Weyers, J. and Meidner, H. 1990. Methods in stomatal research. – Longman Sci. and Techn., Harlow.

van Wijk, W. R. and de Vries, D. A. 1963. Periodic temperature variations. – In: van Wijk, W. R. (ed.), Physics of plant environment. North-Holland, Amsterdam.

Zinke, P. J. 1967. Forest interception studies in the United States. – In: Sopper, W. E. and Lull, H. W. (eds), Forest hydrology, pp. 137–161.

Ecological Bulletins 45: 45–52. Copenhagen 1996

Nutrient cycling in subarctic and arctic ecosystems, with special reference to the Abisko and Torneträsk region

Sven Jonasson and Anders Michelsen

Jonasson, S. and Michelsen, A. 1996. Nutrient cycling in subarctic and arctic ecosystems, with special reference to the Abisko and Torneträsk region. – Ecol. Bull. 45: 45–52.

This review of research on nutrient circulation performed in the Torneträsk area, N Sweden, showed that the living phytomass in various ecosystem types there contained between 5 and 30% of the ecosystem pools of biologically fixed N and P. From 4% to a maximum of 14% of the ecosystem N was in leaves and fine root biomass, i.e. in plant tissues with turnover times < 10 yr.

Atmospheric nutrient deposition and nitrogen fixation were important sources of nutrients to annual primary productivity in mires, but were probably relatively unimportant in other ecosystem types. Vascular plants received more than half of their annual requirement of N and P, i.e. the most growth limiting elements, from stored reserves in overwintering organs. The remaining part could be accounted for by uptake of nutrients released by leaching and mineralization of litter and soil organic matter.

The mineralization rate of N and P was low and the soil microbes immobilized nutrients efficiently. About 6% of the total soil N and 35% of the soil pool of P was in the soil microbial biomass. This implies that changes in population sizes of soil microbes could be important for the supply of plant available, inorganic nutrients. Recent observations of large differences in ^{15}N natural abundance in a variety of plant species have, however, given strong evidences that shrubs and dwarf shrubs with ecto- or ericoid mycorrhiza, in addition to using inorganic nutrients, also receive a substantial amount of their N requirement from organic sources through their mycorrhizal symbionts. Since shrubs and dwarf shrubs make up the largest part of the biomass in many low arctic plant communities, other pathways than those which traditionally have been considered in studies of nutrient circulation between plants and soil could be very important for the circulation of N, the nutrient element that is considered to be most limiting for primary productivity in arctic ecosystems.

S. Jonasson and A. Michelsen, Dept of Plant Ecology, Univ. of Copenhagen, Øster Farimagsgade 2D, DK-1353 Copenhagen K, Denmark.

In the Arctic, the amounts of plant available nutrients in the soil, particularly nitrogen (N) and, to some extent, phosphorus (P) exert a strong control on plant growth and productivity (Chapin et al. 1995). This contrasts with the fact that arctic ecosystems contain proportionally more soil N than most other ecosystems, judged from the amounts of deposited soil organic matter (SOM). Post et al. (1982) estimated that the arctic tundra, which covers c. 5% of the world's terrestrial surface (Chapin and Körner 1995), contains c. 14% of the world's total soil organic carbon (C). Since the C/N ratios of the SOM in most tundra soils are comparable to those in other ecosystem types and vary within rather narrow limits (Dowding 1974, Flint and Gersper 1974), the amount of N deposited per unit area in arctic soils should be appreciably above the average for terrestrial ecosystems.

Regardless of these large stores of nutrients, primary productivity at least within the lower latitudes of the arctic is considered to be strongly limited by nutrient deficiency (McKendrick et al. 1978, Ulrich and Gersper 1978, Shaver and Chapin 1986, Jonasson 1992). For instance, annual production of *Sphagnum* mosses doubled after spraying a moss mat at the subarctic Stordalen mire with a fertilizer solution at a rate of 2 g N m^{-2}, while the same amount of N added to a south Swedish mire with higher atmospheric N deposition had no effect on the productivity of the *Sphagnum*

Table 1. Nitrogen pool sizes (g m^{-2}) and proportions (%) in plant biomass, above-ground litter and soil organic matter of four ecosystem types in the Torneträsk area.

Ecosystem type (site)	Pool sizes (g m^{-2})	Proportion (%)		
		Plant	Soil	Litter
Mire to 30 cm (Stordalen)[1]	355	3*	96*	1
Salix shrub (Kamasjaure)[2]	116	8	88	4
Betula shrub (Kamasjaure)[2]	85	34	50	16
Empetrum heath (Kamasjaure)[2]	24	30	56	14

Sources: [1] Rosswall and Granhall (1980) [2] Jonasson (1982, 1983).
* It has been pointed out that the below ground biomass of vascular plants probably was underestimated by a factor of approximately four (Wallén 1986). A part of the N in plant roots is therefore probably included in the soil compartment here. If so, the percentage of N in plants should be increased to 6% and the percentage in soil should be reduced to 93%.

mosses (Aerts et al. 1992). Similarly, addition of N alone, or in combination with P, at rates of 8–10 g N m^{-2} to low arctic heaths and tundra generally causes a flush of plant growth and a change in abundance among the species (Jonasson 1992). This amount of N corresponds to the content in one year's input of anthropogenic, airborne pollution in the most heavily N-polluted farming areas in W Europe (Sutton et al. 1993). It is also well known that species composition and community structure in the Arctic follow a predictable pattern along nutrient or combined nutrient and water gradients (Chapin et al. 1995 and references therein).

The amount of N and other nutrients in the organic matter of arctic soils shows that the nutrient limitation to the plants is not controlled by the quantities of biologically fixed nutrients per se. The limitation is rather a function of low nutrient availability due to long residence time of the nutrients in various ecosystem compartments and, hence, slow turnover between soils and biota. In addition, low rates of chemical weathering and, in comparison to other ecosystems, limited input of nutrients from external sources, and low rates of fixation of atmospheric N (Chapin and Bledsoe 1992) due to adverse climatic conditions further restrict the nutrient availability for tundra organisms. Together with increasing direct temperature limitation to growth at the higher latitudes and altitudes (Jonasson et al. 1996a) and low precipitation in some high arctic regions (Callaghan and Jonasson 1995) all these constraints contribute to the low net primary production and set a low upper limit for the amount of biomass that arctic communities can sustain.

Abisko and the subarctic (used here for areas in the birch forest ecotone) and low arctic (here: treeless areas above the tree-line; see Jonasson 1982) areas surrounding Lake Torneträsk is one of the Arctic regions where most studies on ecosystem nutrient distribution and processes of nutrient circulation has taken place. In this chapter we make an overview of results from some of these studies. We concentrate on the broad-scale nutrient circulation between

ecosystem compartments and on processes of nutrient transformation between the compartments. Other aspects on plant nutrition and the use of nutrients within plants are given elsewhere in this volume.

Nutrient partitioning between plants and soil

In contrast to the large nutrient stores in arctic soil, the biota contain smaller amounts of most nutrients. Table 1 shows data on nitrogen partitioning between plants, litter and SOM in a number of ecosystems in the Torneträsk region. From a subarctic mire at Stordalen, studied intensely during the Swedish Tundra Biome project within the International Biological Programme (IBP), Rosswall and Granhall (1980) reported that the upper 30 cm of peat contained c. 345 g N m^{-2}, compared to 10 and 2.5 g N m^{-2} in plants and litter, respectively. Hence, given the considerable depth of the entire peat column of the mire, which reached >2.5 m at the deepest parts (Sonesson 1970), the vegetation presumably contained <2% of the total pool of mobile and long-term deposited N of the mire; an estimate that probably is representative for most of the vast areas of bogs in the sub- and low Arctic.

Data from the Tsåktso-Kamasjaure plateau in the Torneträsk region gave a similar picture of c. 8% N (Table 1) and P (not shown) in the plant compartment and 92% in the litter plus SOM of mesic Salix shrub tundra with well developed organic soil horizons. The proportion increased to c. 30% in the vegetation on a more dry Betula nana shrub tundra with shallower humus and on Empetrum heaths with discontinuous humus on windswept, dry ridges (Table 1, calculated from Jonasson 1982, 1983). Since most of the biomass N and P was locked and retained into stems and coarse roots, the proportions in leaves and fine roots with rapid turnover ranged from 4 to 14% (calculated from Jonasson 1982, 1983).

The Tsåktso-Kamasjaure area with a mosaic of bogs, mesic shrub heaths and more limited patches of drier shrub and dwarf shrub vegetation is structurally similar to the vast shrub tundra in the European part of the Russian Arctic (Jonasson 1982). Given the large areas of mires and mesic tundra in this region, we therefore estimate that plant tissues with a turnover time of less than a decade only contain 5%–10% of the total ecosystem N and P pool in the European sub and low arctic as a whole.

Nutrient input from external sources

The nutrient input from external sources is low in most places in the Arctic. Rosswall and Granhall (1980) estimated that the mire at Stordalen received 0.4 g N m^{-2} annually from external sources, divided approximately equally between N$_2$-fixation and dry plus wet atmospheric deposition. Due to low levels of precipitation and pollution, the atmospheric deposition of most nutrients at the Stordalen mire probably is below the average for the subarctic and low arctic parts of Scandinavia as a whole (Malmer and Nihlgård 1980). On the other hand, the N$_2$-fixation of c. 0.2 g m^{-2} yr^{-1} (Granhall and Selander 1973) probably is above average for both the ecosystems in the region and for other sub- and low arctic areas (Chapin and Bledsoe 1992), because N$_2$-fixation is stimulated by the prevailing moist conditions in the moss layer of the mire. As a consequence, the low annual input of N from deposition on one hand, and the high N$_2$-fixation on the other, imply that the total of 0.4 g m^{-2} reported by Rosswall and Granhall (1980) may be close to the input from external sources for most ecosystems in the region.

Plant nutrient supply

Nutrient uptake from atmospheric deposition and N$_2$-fixation

In most arctic ecosystems, atmospheric deposition probably is more important for replenishment of lost nutrients from the systems than for direct uptake by plants. However, vegetation dominated by cryptogams may receive a considerable amount of the annually absorbed nutrients directly from atmospheric deposition by dust and rain (Jónsdóttir et al. 1995). Similarly, the supply of atmospheric N through microbial fixation by free-living and symbiontic microorganisms and blue-green algae is vital for the long term accumulation of N in the ecosystems, but probably plays a minor role for the annual nutrient budget of the arctic as a whole, except in places with high abundance of N$_2$ fixing organisms (Granhall and Lid-Torsvik 1975).

The moss-dominated Stordalen mire obviously represents an ecosystem type in which both atmospheric deposition and N$_2$ fixation contribute with appreciable amounts of N$_2$ incorporated annually in the primary producers. Rosswall and Granhall (1980) estimated that 40% of the annual N uptake could be accounted for in equal parts by atmospheric deposition and N$_2$ fixation by blue-green algae. These amounts contrast strongly to the few percent reported for alpine meadows and heaths at Hardangervidda in southern Norway (Chapin and Bledsoe 1992).

Nutrient supply from storage within the plants by translocation

Most of the nutrients needed annually for plant biomass production come from fractions already incorporated in the plants. These nutrients are translocated between sites of active growth and stores (stems, roots, rhizomes or other organs). The translocation is most pronounced in autumn, when nutrients in senescing parts are resorbed and transported to the storage organs, and in spring when the stored nutrients are translocated back to meristems with resumed growth.

However, net movements of nutrients in or out of plant organs can take place at any time during the growing season. For instance, the annual net translocation of N, P and K from the senescing canopy of the graminoid *Eriophorum vaginatum* was c. 60% of the peak season's canopy pools (Jonasson and Chapin 1985). This is probably a conservative estimate of the annual growth support by mobile nutrients, however, because translocation out of the leaves had begun before the leaves approached their full length. These nutrients were allocated to new leaves that appeared successively during the growing season. Hence, due to the sequential leaf development typical for graminoids, the same unit of nutrient could be used several times during a growing season. With this taken into account, the internal circulation of nutrients probably supplied *E. vaginatum* with 80–90% of the annual need of N, P and K for canopy growth (Berendse and Jonasson 1992). Consequently, only 10–20% of the annual incorporation in the biomass needed to be supplied from other sources.

Resorption from leaves of N and P, i.e. the nutrients which most commonly are limiting plant growth, is particularly pronounced and often in the order of 50–80% of the total leaf pool, with occasionally higher or lower values. Resorption of potassium (K) varies from 0 to 90% and some nutrients, e.g. calcium (Ca) are immobile (Malmer and Nihlgård 1980, Jonasson 1989, Berendse and Jonasson 1992). Although the proportion of nutrient resorption in arctic plants probably is similar to that in plants of other regions, as e.g. boreal forests (Chapin and Kedrowski 1983), the internal circulation is of particular significance in arctic regions with strongly nutrient limited plant growth. Efficient retention of nutrients within the plants reduces nutrient losses and minimizes the need for uptake of nutrients in short supply in the soil. In fact, tillers of *E. vaginatum* that were grown from spring to autumn in distilled water produced the same amount of biomass as tillers grown with access to soil nutrients (Jonasson and Chapin 1985). Furthermore, translocation often starts before the soil has thawed in spring and will, thus, allow nutrient transport to the meristems before they can receive nutrients from uptake (Jonasson and Chapin 1985).

Nutrient supply from leaching and decomposition of litter

Although the internal nutrient circulation probably supplies the major part of the most limiting nutrients to the annual arctic vascular plant production, uptake must replenish the nutrient pool for annually lost nutrients. Furthermore, nutrient uptake is necessary for biomass accumulation in juvenile plants and for the nutrient demanding formation of reproductive organs. However, the nutrient uptake needed to sustain growth in many arctic ecosystems probably is proportionally smaller than in most other communities. This is because the vast majority of arctic plants are perennial and the populations often consist of adult specimens. Such plants in the Arctic can subsist on translocated nutrients plus the relatively small amounts that are required to replace the yearly losses. In contrast, non-arctic plant communities often have a higher proportion of young age classes, which are dependent on a large net nutrient uptake for yearly accumulation of mass.

Most nutrients needed annually from external sources must be taken up from the soil, generally in inorganic form. The main sources of these inorganic nutrients are leachates from the plant canopy and litter and nutrients transformed to inorganic form through mineralization of organically bound nutrients in litter and SOM.

To our knowledge, there is no study from the Arctic that has attempted to quantify the amount of leachates that directly enter the soil from the plant canopy. Some measurements on isolated species indicate, however, that the supply of nutrients in leachates probably is relatively small compared with that from litter and the amounts of nutrients released by mineralization processes (Chapin et al. 1980, Jonasson and Chapin 1985). However, the amount of leached nutrients is probably very species specific. For example, addition of extracts of leaves shaken off *Betula pubescens* ssp. *tortuosa*, with relatively high amount of P, reduced the subsequent uptake of ^{32}P by excised roots of graminoids grown in a heath soil from Abisko, as compared to that of plants

grown in soil to which leaf extracts of *Cassiope tetragona* or *Empetrum hermaphroditum*, with traces of P only, were added (Michelsen et al. 1995). Hence, P leached from dying leaves of e.g. *B. tortuosa* may contribute substantially to the P supply of sub-canopy species.

Leaching from the litter probably makes up a substantial part of the annual nutrient supply to the inorganic, plant available pool, and may be of particular relevance for the nutrient uptake by cryptogams in direct contact with the litter (Malmer and Nihlgård 1980).

The magnitude of leaching from plant litter can be inferred from the rate of mass loss and nutrient release from shed *Betula nana* leaves (Jonasson 1983). Whereas C loss was 10–20%, indicating low rate of decomposition in the first year, the initial nutrient losses which occurred during the first year after litterfall were high, except for N, after which the rate of nutrient release declined sharply during the next two years. After one year, the pool of K had declined to c. 15% of the initial amounts, P and magnesium (Mg) had declined to between 40 and 60% whereas almost 80% of the initial Ca remained. In contrast, N was retained in the leaves and its rate of loss was even slower than the loss of weight. The high loss rate of most nutrients presumably was due to leaching of water soluble fractions soon after leaf abscission (Jonasson 1983).

Assuming that the main part of the elemental losses in litter was leachates, resorption of nutrients from the leaves during the period from maximum leaf nutrient content in mid summer plus the first year's leaching losses left 35 and 18% of the peak season's N and P to microbial decomposition. A still lower fraction of only 6% remained of peak season's K (Fig. 1). These figures illustrate that the mobility of the different nutrients and the processes by which they are made available to the plant differ among elements. Thus, the pattern of nutrient element release is similar to that in non-arctic ecosystems, while the rate of mass loss is relatively lower because of temperature limitations on the decomposition process.

Nutrient supply from mineralization of the soil organic matter (SOM)

The rate of nutrient losses from SOM is an appreciably slower process than the nutrient release from litter. While the turnover time of nutrients in primary producers and litter ranges from years in the soft tissues to decades in woody tissues, the turnover of nutrients in SOM normally requires centuries (Rosswall and Granhall 1980, Jonasson 1983) and the most recalcitrant fractions even millennia (Goksøyr 1975). In spite of the slow mineralization, the SOM is an important supplier of nutrients because its large mass outweighs the slow mineralization. For example, at the Stordalen mire,

Rosswall and Granhall (1980) estimated that c. 75% of the annual N uptake of vascular plants came from mineralization of SOM and 25% came from N released by mineralization of litter.

The net nutrient mineralization rate of SOM is, however, highly variable from site to site. Mean daily N mineralization measured by anaerobic incubation of soil at 30°C, i.e., under optimal temperature conditions, varied almost 5-fold in soils of different origin from Abisko and the Torneträsk area (Jonasson 1986). The highest rate of >50 µg N g^{-1} SOM day^{-1} took place in the thin organic crust of frost-heaved soils with high pH and base saturation. The rate declined to c. 25 µg g^{-1} day^{-1} in well developed SOM with high pH and base saturation, and reached the lowest values of 11 µg g^{-1} day^{-1} in shrub heaths with deep organic deposits, low pH and low base saturation.

Measurements of net mineralization under ambient conditions in soils incubated in situ at the Stordalen mire and at a heath and a fellfield at Abisko have shown low rates of N and P mineralization and even negative net mineralization integrated over entire growing seasons (Rosswall and Granhall 1980, Jonasson et

Table 2. Total, inorganic and microbial nitrogen and phosphorus (mg/g SOM; means $\pm$ SE) in soils of a tree-line heath at Abisko, N Sweden, dominated by *Vaccinium uliginosum*, *Empetrum hermaphroditum*, *Rhododendron lapponicum* and other ericaceous dwarf shrubs.

Nutrient	Total in SOM	Inorganic	Microbial
Nitrogen	22.8 ± 0.4	0.02 ± 0.003	1.40 ± 0.05
Phosphorus	1.18 ± 0.02	0.01 ± 0.000	0.41 ± 0.01

al. 1993). Negative mineralization indicates that nutrients from SOM and the soil inorganic pools were immobilized, rather than mobilized, by the soil microbes, and shows that the soil microbial community can be a strong sink for nutrients (Jonasson et al. 1996b). It is likely, therefore, that the plant uptake of nutrients is strongly constrained by immobilization of inorganic nutrients during periods when microbial activity is particularly high. Giblin et al. (1991) found that most of the release of mineralized nutrients to the plant available pool in Alaskan tundra took place during the winter, probably as an effect of leaching of nutrients from microorganisms that died at the end of the summer.

Microbial nutrient utilization

The microbial biomass in arctic soils is small and the microbial activity rather than the biomass determines the rate of decomposition and nutrient mineralization (Rosswall 1975). However, the soil microbial community contains an appreciable amount of the total soil N and P pools despite the low biomass because the C/N and C/P ratios of the microorganisms are low. Jonasson et al. (1996b) found that microbial N made up 6–7% and microbial P c. 35% of the total soil N and P pools of a dwarf shrub heath at Abisko. As a comparison, soil inorganic N was below 0.1%, and inorganic P was lower than 1% of the total pools sizes (Table 2).

Addition of inorganic NPK fertilizers in spring to a heath and a fellfield at Abisko gave no response in microbial C content and therefore probably no change in microbial biomass by the end of the growing season. However, the microbial N and P pools increased strongly. At the fellfield, the microbial N pool had almost doubled, and the microbial P pool had more than doubled in autumn (Jonasson et al. 1996b; Fig. 2). Furthermore, N addition to the heath gave an immediate increase of the soil respiration of c. 30% compared to unfertilized controls. The stimulation started within one day after the application and was maintained at least for several weeks (Illeris unpubl.). This demonstrates that even if the biomass of the microbes did not change, their activity was enhanced by the nutrient addition and they constituted an important sink for nutrients.

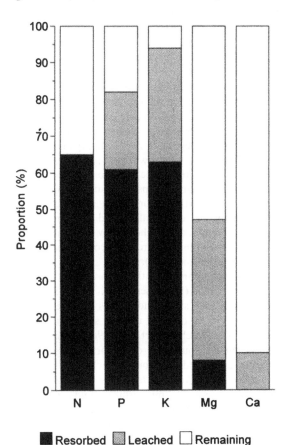

Fig. 1. Proportions of nutrients resorbed and leached from *Betula nana* leaves as percent of peak season's leaf nutrient content. Data are from Jonasson (1983).

The study showed that the microbial community can immobilize large amounts of at least N and P during periods of high activity. On the other hand, a decline of the microbial populations is likely to release large amounts of nutrients, particularly P which is abundant in their biomass (see above). This implies that the transformation of nutrients from organic to plant available, inorganic form is dependent not only on the decomposing activity of the microorganisms per se, but also on changes in their population sizes.

The effect on plant nutrition of nutrients released from declining microbial populations was shown by an experiment in which arctic graminoid species were potted in sterilized and non-sterilized soil from a heath at Abisko (Michelsen et al. 1995). Plant growth rate was higher in sterilized than in non-sterilized soil. Obviously, mobilization of microbially fixed nutrients when the soil was sterilized and the microorganisms were killed caused a strong increase of nutrient availability, which released the plants from nutrient limitation and stimulated their growth.

The growth induced by soil sterilization was reduced when leaf extracts from plants with differing content of carbohydrates were added to the soil. The growth declined continuously as the amount of carbohydrates increased, and correlated with an increasing uptake rate of labelled N and P in excised roots, demonstrating a relative increase of nutrient limitation (Michelsen et al. 1995). Hence, addition of a carbon source to the sterilized soil with an abundance of inorganic nutrients stimulated regrowth of the microbial populations and restored the original balance of nutrient partitioning

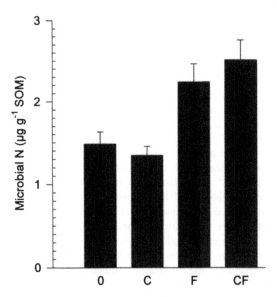

Fig. 2. Microbial nitrogen content (means ± SE) per unit soil organic matter in a fellfield at Abisko after one growing season of factorial carbohydrate (sugar; C) and NPK fertilizer (F) amendment. 0 is control. Data are from Jonasson et al. (1996b).

between microorganisms and plants that existed before the soil sterilization.

The experiment shows that the availability of soil nutrients was too low to support maximum biomass production of plants in soils with a nutrient demanding microbial population. However, the release of microbial nutrients was enough to facilitate plant nutrient uptake and growth. It is possible that nutrient release from microbial populations can explain the pulse of release of inorganic nutrients during winter (Giblin et al. 1991) when the populations presumably decline. It is likely, therefore, that a large part of the annual uptake of nutrients by plants occurs during such pulses in autumn and spring (Chapin et al. 1978), which may explain why plants can accumulate nutrients efficiently in spite of the low, or even lack of, net nutrient release integrated over the entire growing season.

Plant nutrient uptake from soil organic matter through mycorrhiza

A different role in the flow of nutrients from soil to plants is played by mycorrhizal fungi. They transfer nutrients from the large soil volume they exploit to plants through connections with the plant roots. Furthermore, deduced from experiments with temperate isolates, both ectomycorrhizal and ericoid mycorrhizal fungi can function as suppliers to the plant roots of nutrients taken up from the inorganic nutrient pool and also as decomposers of organic matter because these fungi produce proteolytic enzymes that are the key enzymes for the degradation of soil organic nitrogen (Read 1993). Hence, plants with this type of mycorrhiza can circumvent both competition for inorganic nutrients from other plant groups and nutrient immobilization in non-mycorrhizal fungi and soil bacteria and draw directly on organically fixed nutrients (Read 1993). This may explain the highly efficient nutrient uptake observed in the arctic ericoid *Rhododendron lapponicum* (Jonasson 1995).

Most plants on arctic heaths do have ecto- or ericoid mycorrhizal fungi (Michelsen et al. 1996). Studies at Abisko have given strong indications that these mycorrhizal plants utilize different nutrient sources than other species in situ. Michelsen et al. (1996) found that the leaf ^{15}N natural abundance of mycorrhizal species collected at a fellfield at Abisko, including *Empetrum hermaphroditum*, *Cassiope tetragona*, *Vaccinium vitis-idaea*, with ericoid mycorrhiza, and *Polygonum viviparum*, *Betula nana*, *Dryas octopetala*, and *Salix herbacea × polaris* with ectomycorrhiza, was distinctly different from the ^{15}N abundance of graminoids either lacking mycorrhiza or having arbuscular mycorrhiza without known ability to use organic nutrient sources (Fig. 3). This pattern of utilization of different N

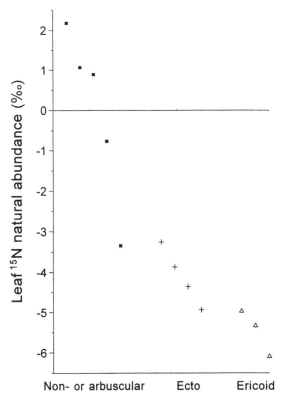

Plant mycorrhizal status

Fig. 3. Leaf ^{15}N abundance of the dominant non- or arbuscular-mycorrhizal, ecto-mycorrhizal, and ericoid mycorrhizal plant species from a fellfield at Abisko. The ^{15}N abundance is expressed in δ^{15}N units. The atmospheric δ^{15}N = 0 by definition. Starting from the left, the non- and arbuscular mycorrhizal species were *Carex bigelowii*, *Lycopodium selago*, *Luzula arcuata*, *Calamagrostis lapponica* and *Festuca vivipara*; the ecto-mycorrhizal species were *Salix herbacea* × *polaris*, *Betula nana*, *Polygonum viviparum* and *Dryas octopetala*; and the ericoid mycorrhizal species were *Cassiope tetragona*, *Vaccinium vitis-idaea* and *Empetrum hermaphroditum*. Data are from Michelsen et al. (1996).

sources by functional plant groups was confirmed by similar results obtained at a tree-line heath at Abisko (Michelsen et al. 1996). We conclude, therefore, that increased weight must be put into investigations of the role of mycorrhiza and other pathways for nutrient uptake, e.g. direct plant uptake of low molecular weight amino acids (Chapin et al. 1993, Kielland 1994) in arctic ecosystems, and into investigations of the interactions between decomposers and plants before the details of the nutrient circulation can be more fully understood.

Acknowledgements – Grants to S. J from the Royal Swedish Academy of Science and the Swedish Natural Science Research Council (no. B-BU 4903-300), to A. M. from the Danish Natural Science Research Council (no. 11-0611-1), and to both authors from the Danish Natural Science Research Council (no. 11-0421-1) and the Swedish Environmental Protection Board (no. 127402) have supported part of the research reviewed in this article.

References

Aerts, R., Wallén, B. and Malmer, N. 1992. Growth-limiting nutrients in *Sphagnum*-dominated bogs subject to low and high atmospheric nitrogen supply. – J. Ecol. 8: 131–140.

Berendse, F. and Jonasson, S. 1992. Nutrient use and nutrient cycling in northern ecosystems. – In: Chapin, F. S. III, Jefferies, R. L., Reynolds, J. F., Shaver, G. R. and Svoboda, J. (eds), Arctic ecosystems in a changing climate, an ecophysiological perspective. Academic Press, San Diego, pp. 337–356.

Callaghan, T. V. and Jonasson, S. 1995. Implications for changes in arctic plant biodiversity from environmental manipulation experiments. – In: Chapin, F. S. III and Körner, C. (eds), Arctic and alpine biodiversity: patterns, causes and ecosystem consequences. Ecol. Stud. 113: 151–166, Springer, Berlin.

Chapin, D. M. and Bledsoe, C. S. 1992. Nitrogen fixation in arctic plant communities. – In: Chapin, F. S. III, Jefferies, R. L., Reynolds, J. F., Shaver, G. R. and Svoboda, J. (eds), Arctic ecosystems in a changing climate, an ecophysiological perspective. Academic Press, San Diego, pp. 301–319.

Chapin, F. S. III and Kedrowski, R. A. 1983. Seasonal changes in nitrogen and phosphorus fractions and autumn retranslocation in evergreen and deciduous taiga trees. – Ecology 64: 376–391.

– and Körner, C. 1995. Patterns, causes, changes, and consequences of biodiversity in arctic and alpine environments. – In: Chapin, F. S. III and Körner, C. (eds), Arctic and alpine biodiversity, patterns, causes and ecosystem consequences. Ecol. Stud. 113: 313–320, Springer, Berlin.

–, Barsdate, R. J. and Barel, D. 1978. Phosphorus cycling in Alaskan coastal tundra: a hypothesis for the regulation of nutrient cycling. – Oikos 31: 189–199.

–, Johnson, D. A. and McKendrick, J. D. 1980. Seasonal movements of nutrients in plants of differing growth form in an Alaskan tundra ecosystem: Implications for herbivory. – J. Ecol. 68: 189–209.

–, Moilanen, L. and Kielland, K. 1993. Preferential use of organic nitrogen for growth by a non-mycorrhizal arctic sedge. – Nature 361: 150–153.

–, Hobbie, S. E., Bret-Harte, M. S. and Bonan, G. 1995. Causes and consequences of plant functional diversity in arctic ecosystems. – In: Chapin, F. S. III and Körner, C. (eds), Arctic and alpine biodiversity, patterns, causes and ecosystem consequences. Ecol. Stud. 113: 225–237, Springer, Berlin.

Dowding, P. 1974. Nutrient losses from litter on IBP tundra sites. – In: Holding, A. J., Heal, O. W., MacLean, S. F. Jr and Flanagan, P. W. (eds), Soil organisms and decomposition in tundra. Tundra Biome Steering Comm., Stockholm, pp. 363–373.

Flint, P. S. and Gersper, P. L. 1974. Nitrogen nutrient levels in arctic tundra soils. – In: Holding, A. J., Heal, O. W., MacLean, S. F. Jr and Flanagan, P. W. (eds), Soil organisms and decomposition in tundra. Tundra Biome Steering Comm., Stockholm, pp. 375–387.

Giblin, A. E., Nadelhoffer, K. J., Shaver, G. R., Laundre, J. A. and McKerrow, A. J. 1991. Biogeochemical diversity along a riverside toposequence in arctic Alaska. – Ecol. Monogr. 61: 415–435.

Goksøyr, J. 1975. Decomposition, microbiology and ecosystem analysis. – In: Wielgolaski, F. E. (ed.), Fennoscandian tundra ecosystems, Part 1, Plants and microorganisms. Ecol. Stud. 16: 230–238, Springer, Berlin.

Granhall, U. and Lid-Torsvik, V. 1975. Nitrogen fixing by bacteria and free-living blue-green algae in tundra areas. – In: Wielgolaski, F. E. (ed.), Fennoscandian tundra ecosystems, Part 1, Plants and microorganisms. Ecol. Stud. 16: 305–315, Springer, Berlin.

– and Selander, H. 1973. Nitrogen fixation in a subarctic mire. – Oikos 24: 8–15.

Jonasson, S. 1982. Organic matter and phytomass on three north Swedish tundra sites, and some connections with adjacent tundra areas. – Holarct. Ecol. 5: 367–375.
– 1983. Nutrient content and dynamics in north Swedish shrub tundra areas. – Holarct. Ecol. 6: 295–304.
– 1986. Influence of frost heaving on soil chemistry and on the distribution of plant growth forms. – Geogr. Ann. 68A: 185–195.
– 1989. Implications of leaf longevity, leaf nutrient re-absorption and translocation for the resource economy of five evergreen plant species. – Oikos 56: 121–131.
– 1992. Plant responses to fertilization and species removal in tundra related to community structure and clonality. – Oikos 63: 420–429.
– 1995. Growth and nutrient allocation in relation to leaf retention time of the wintergreen *Rhododendron lapponicum*. – Ecology 76: 475–485.
– and Chapin, F. S. III 1985. Significance of sequential leaf development for nutrient balance of the cotton sedge, *Eriophorum vaginatum* L. – Oecologia 67: 511–518.
– , Havström, M., Jensen, M. and Callaghan, T. V. 1993. In situ mineralization of nitrogen and phosphorus of arctic soils after perturbations simulating climate change. – Oecologia 95: 179–186.
– , Lee, J. A., Callaghan, T. V., Havström, M. and Parsons, A. 1996a. Direct and indirect effects of increasing temperatures on subarctic ecosystems. – Ecol. Bull. 45: 180–191.
– , Michelsen, A., Schmidt, I. K., Nielsen, E. V. and Callaghan, T. V. 1996b. Microbial biomass C, N and P in two arctic soils and responses to addition of NPK fertilizer and sugar: Implications for plant nutrient uptake. – Oecologia, in press.
Jónsdóttir I. S., Callaghan, T. V. and Lee, J. A. 1995. Fate of added nitrogen in a moss-sedge Arctic community and effects of increased nitrogen deposition. – The science of the total environment 160/161: 677–685.
Kielland, K. 1994. Amino acid absorption by arctic plants: implications for plant nutrition and nitrogen cycling. – Ecology 75: 2373–2383.
Malmer, N. and Nihlgård, B. 1980. Supply and transport of mineral nutrients in a subarctic mire. – In: Sonesson, M. (ed.), Ecology of a subarctic mire. Ecol. Bull. (Stockholm) 30: 63–95.

McKendrick, J. D., Ott, V. J. and Mitchell, G. A. 1978. Effects of nitrogen and phosphorus fertilization on carbohydrate and nutrient levels in *Dupontia fisheri* and *Arctagrostis latifolia* of an Alaskan tundra. – In: Tieszen, L. L. (ed.), Vegetation and production ecology of an Alaskan tundra. Ecol. Stud. 29: 509–537, Springer, Berlin.
Michelsen, A., Schmidt, I. K., Jonasson, S., Dighton, J., Jones, H. and Callaghan T. V. 1995. Inhibition of growth, and effects on nutrient uptake of arctic graminoids by leaf extracts – allelopathy or resource competition between plants and microbes? – Oecologia 103: 407–418.
– , Schmidt, I. K., Jonasson, S., Quarmby, C. and Sleep, D. 1996. Leaf ^{15}N abundance of subarctic plants provides field evidence that ericoid, ectomycorrhizal and non- and arbuscular mycorrhizal species access different sources of soil nitrogen. – Oecologia 105: 53–63.
Post, W. M., Emanuel, W. R., Zinke, P. J. and Stangenberger, A. G. 1982. Soil carbon pools and world life zones. – Nature 298: 156–159.
Read, D. J. 1993. Plant-microbe mutualisms and community structure – In: Schulze, E.-D. and Mooney, H. A. (eds), Biodiversity and ecosystem function. Ecol. Stud. 99: 181–203, Springer, Berlin.
Rosswall, T. 1975. Introduction. – In: Wielgolaski, F. E. (ed.), Fennoscandian tundra ecosystems, Part 1, Plants and microorganisms. Ecol. Stud. 16: 227–229, Springer, Berlin.
– and Granhall, U. 1980. Nitrogen cycling in a subarctic ombrotrophic mire. – In: Sonesson, M. (ed.), Ecology of a subarctic mire. Ecol. Bull. (Stockholm) 30: 209–234.
Shaver, G. R. and Chapin, F. S. III 1986. Effect of fertilizer on production and biomass of tussock tundra, Alaska, U.S.A. – Arct. Alp. Res. 18: 261–268.
Sonesson, M. 1970. Studies on mire vegetation in the Torneträsk area, northern Sweden, III. Communities of the poor mires. – Opera Bot. 26.
Sutton, M. A., Pitcairn, C. E. R. and Fowler, D. 1993. The exchange of ammonia between the atmosphere and plant communities. – Adv. Ecol. Res. 24: 301–393.
Ulrich, A. and Gersper, P. L. 1978. Plant nutrient limitations of tundra plant growth. – In: Tieszen, L. L. (ed.), Vegetation and production ecology of an Alaskan tundra. Ecol. Stud. 29: 457–481, Springer, Berlin.
Wallén, B. 1986. Above and below ground dry mass of the three main vascular plants on hummocks on a subarctic peat bog. – Oikos 46: 51–56.

Ecological Bulletins 45: 53–64. Copenhagen 1996

Resource dynamics within arctic clonal plants

Ingibjörg S. Jónsdóttir, Terry V. Callaghan and Alistair D. Headley

Jónsdóttir, I. S., Callaghan, T. V. and Headley, A. D. 1996. Resource dynamics within arctic clonal plants. – Ecol. Bull. 45: 53–64.

The clonal growth habit is widespread amongst nearly all plant life forms of the Arctic and Subarctic which includes the Abisko area where clonal plant species dominate all major vegetation types. The success of clonal plants can be ascribed to vegetative reproduction and improved nutrient utilisation and acquisition in a variety of ways as a consequence of the clonal growth habit. They do not depend on sexual reproduction and seedling establishment for proliferation in these harsh environments, they can search for nutrients in nutrient-limited and highly heterogeneous habitats, can take up water and nutrients from many sites simultaneously and thus buffer against spatial and temporal fluctuations in availability. They efficiently utilise the once acquired resources through recycling both within individual ramets and between ramets. As a consequence, their populations are stable and this has a stabilising effect on the vegetation as a whole. However, as they conserve nutrients within the living biomass, harmful effects of pollutants can be magnified. The implications of the characteristics and dominance of clonal plants for the resilience of the subarctic ecosystems in the context of environmental change is discussed.

I. S. Jónsdóttir (correspondence), Dept of Botany, Univ. of Göteborg, Carl Skottbergs Gata 22B, S-413 19 Göteborg, Sweden. – T. V. Callaghan, Sheffield Centre for Arctic Ecology, Univ. of Sheffield, 26 Taptonville Rd, Sheffield, U.K. S10 5BR. – A. D. Headley, Dept of Environmental Science, Univ. of Bradford, Bradford, U.K. BD7 1DP.

Clonal plants are by definition those which produce new offspring (ramets) that can complete a life-cycle without going through the sexual cycle. This includes various modes of vegetative spread, such as stolons, rhizomes, root shoots and bulbils. Most creeping plants are clonal as adventitious root production and the death, decay and disintegration of the older parts of the plant allow ramets to become functionally independent. It follows from the definition above that almost all plants are at least potentially clonal and we find a gradient of clonality rather than two distinct groups of clonal and non-clonal plants. Thus, the degree of clonality differs among plants and those with the highest degree of clonality are plants that predominantly proliferate by vegetative reproduction and recruit only infrequently from seed or spores in the vicinity of adult plants. Once established, the genets continue to grow, proliferate and fragment. Potentially they have an indefinite life-span and their fitness has to be viewed in the light of their multiple mode of reproduction and their profound longevity, which can be centuries (Eriksson and Jerling 1990, Fagerström 1992).

In this paper we are mainly concerned with those plants at the extreme end of the clonality continuum that have delayed fragmentation, i.e. persistent connections between ramets. We will provide some evidences of the importance of this trait for the resource dynamics of plants in the Arctic and Subarctic. First we give a brief overview of the resource conditions arctic plants are confronted with and discuss the importance of clonal plants in floras and vegetation of the Subarctic exemplified by the Torneträsk region, Swedish Lapland. Secondly, we describe the consequences of clonal growth for the resource dynamics in arctic plants with persistent ramet connections, mainly based on studies from the Abisko area within the Torneträsk region. Finally, we discuss the implications of the clonal growth habit in the Subarctic for plant population dynamics and the resilience of subarctic vegetation.

Clonal plants in arctic and subarctic environments

The Arctic can be defined in terms of climate or vegetation or both. The climatic definition most commonly used by ecologists is the northern hemisphere area with a mean July temperature of <10°C (Köppen 1936), which roughly sets the southern limits of the treeless tundra. The term Subarctic frequently appears in the literature without a clear definition. Here we will use a definition given by Löve (1970): the natural zone between the treeless arctic tundra and the closed boreal forest to the south. In the maritime climate of the North Atlantic region this zone is rather distinct and is mostly dominated by mountain birch *Betula pubescens* ssp. *tortuosa* forest. Within this zone, however, there are alpine tundra areas with both physical and biotic elements closely resembling those of the arctic tundra due to their location in relatively high altitude at relatively high latitude.

The low temperatures and short growing seasons in the Arctic and Subarctic create environments with low nutrient availability (Jonasson and Michelsen 1996). Also, frost heaving and shallow, poorly developed soils create extreme spatial heterogeneity at different scales (Jonasson 1986, Svensson and Callaghan 1988a) and seasonal variations in nutrient availability can be great (Chapin and Bloom 1976, Chapin et al. 1978, Marion and Kummerow 1990). Interannual and longer term climatic fluctuations are common, which have profound effects on plant nutrient demand and uptake (Chapin and Bloom 1976, McCown 1978, Chapin 1980, Kielland and Chapin 1992) and the chemical composition of plant tissue (Jonasson et al. 1986).

The plants of the Arctic and Subarctic are adapted to these conditions in a variety of ways through their physiology and morphology. To be able to grow and to complete their life-cycle the plants have to be either efficient in acquiring nutrients at low levels of availability and/or to be efficient in utilising the once acquired resources, in terms of new growth per unit resource, to reduce demand. Physiological adaptations for efficient nutrient use are usually considered to be of greater importance for plants on nutrient poor soils than adaptations for efficient uptake (Chapin 1980). Efficient use can be manifested in low tissue nutrient concentrations (Wielgolaski et al. 1975), repeated use of the once acquired resources (Callaghan 1980, Chapin and Shaver 1989, Karlsson and Weih 1996) or reduction of nutrient losses (Berendse and Jonasson 1992). However, efficient resource acquisition can be achieved through other adaptations than physiological. For example, tundra plants adopt a variety of root systems to overcome infertile soils (Callaghan et al. 1991). The extreme situation is represented by many mosses and lichens which, without roots, rely on atmospheric nutrient input. Several other specialisations typical for plants on nutrient

poor soils (e.g. Pate 1994) are also found among arctic plants, such as different forms of mycorrhiza, symbiosis with nitrogen-fixing organisms, hemiparasitism and carnivory (e.g. Callaghan et al. 1991).

In addition to the physiological and morphological characters mentioned above clonal growth with persistent ramet connections is an important and widespread phenomenon among arctic vascular plants. The stature of woody plants is normally limited to the depth of winter snow in the Arctic which means that competition in the vertical plane is not as strong as in temperate or boreal regions. Possibly in response to this limitation of height it might be envisaged that competition for light is occurring in a horizontal direction by the lateral spread of individual plants. The horizontal spread can also greatly improve both utilisation and acquisition of nutrients (Headley et al. 1988a, b, Jónsdóttir and Callaghan 1988, 1990) as will be described in the next section.

In an attempt to assess the importance of clonality in the flora of the Torneträsk region of Swedish Lapland the number of species of vascular plant, excluding hybrids, with different life histories and growth forms is shown in Table 1. For comparison the vascular flora of the county of Durham, England is also shown. In the Torneträsk region there is a somewhat higher frequency of species with presumably a high degree of clonality (perennials: creeping, stoloniferous, rhizomatous, bulbils), 45% compared with 30% in Durham, mainly due to the greater percentage of rhizomatous species. However, the difference between the two regions would probably be much greater if the relative abundance of clonal plants were taken into account.

In the environments within the Abisko area, clonal growth is widespread amongst nearly all major plant life forms. Even the dominant mountain birch tree is often polycormic (multi-stemmed), arising from basal

Table 1. The frequency and proportion of species with different lengths of life-cycle and growth form or type of reproduction in the flora of the Torneträsk region of Swedish Lapland (from Aronsson 1994) and county Durham, England (Graham 1988).

	Torneträsk, Sweden		Durham, England	
Approximate area (km²)	2800		2629	
	n	%	n	%
Annuals	43	7.7	163	14.8
Annuals/Biennials	16	2.9	22	2.0
Biennials	12	2.2	42	3.8
Annuals/Perennials	10	1.8	14	1.3
Perennials – apomictic	38	6.8	145	13.2
Perennials – single stems	47	8.5	172	15.6
Perennials – multi-stemmed	138	24.9	210	19.1
Perennials – creeping	43	7.7	84	7.6
Perennials – stoloniferous	35	6.3	66	6.0
Perennials – rhizomatous	163	29.4	170	15.4
Perennials – bulbils	10	1.8	13	1.2
Total	555	100	1101	100

proliferation of shoots (Fries 1913, Nordhagen 1927) although the degree of clonality cannot be regarded as high due to a frequent seedling establishment relative to vegetative proliferation. Other dominating species, such as graminoids and other herbs, dwarf shrubs, pteridophytes and mosses, have a high degree of clonality. Graminoids and mosses are the dominating life forms in wetlands and in the mid-alpine tundra belt, while dwarf shrubs, club-mosses and mosses are dominant in the birch forest floor and in the low-alpine tundra belt. Other herbs may dominate locally on richer, mesic soils in the forest floor and in the low-alpine belt.

Consequences of clonality for plant resource dynamics

Vegetative reproduction and resource dynamics are two different but closely related aspects of clonal growth of importance in the Arctic and Subarctic. Success in sexual reproduction and establishment of new genets is highly variable and unpredictable due to great fluctuations in flowering intensity from year to year (e.g. Sørensen 1941, Laine and Henttonen 1983), in some years at extremely low levels (e.g. Carlsson and Callaghan 1990a, Havström et al. 1995) and due to often low rates of seedling survival, particularly in closed vegetation (e.g. Jónsdóttir 1995). In such environments it is advantageous to be able to by-pass the sexual cycle through clonal growth (Callaghan and Emanuelsson 1985).

Proliferation of ramets is energy and nutrient demanding, but the competitive advantage of the relatively large offspring is greatly increased in closed vegetation, especially when supported by translocation of resources from older ramets during establishment (Callaghan 1984, Headley et al. 1988a, Jónsdóttir and Callaghan 1988). This means that resources invested in reproduction through clonal growth are less likely to be wasted than investment in sexual reproduction, an important aspect in strongly nutrient limited environments. In addition, hormones may be translocated between ramets where connections are persistent, providing a mechanism for internal control of ramet production through apical dominance (Jónsdóttir and Callaghan 1988, Svensson and Callaghan 1988b). The lateral spread through clonal growth also provides a searching mechanism, where new resource patches are continuously entered and exploited (cf. Hutchings and de Kroon 1994).

The clonal growth habit of plants with persistent connections between ramets allows physiological integration of ramets beyond the establishment process. This improves nutrient and water acquisition through uptake by root systems at a number of different patches simultaneously and translocation into the most actively growing ramets (Headley et al. 1988a, b, Jónsdóttir and Callaghan 1988, 1990). It also improves resource use efficiency of the clone in terms of new growth per unit acquired resource, as nutrients and energy acquired by and stored in old ramets can be remobilised and translocated to actively growing sinks in young ramets (Callaghan 1976, 1980). These mechanisms are particularly important in closed vegetation where plant nutrient availability is low, but where physical disturbance typical of frost heave is not so severe that connections between modules are broken.

Most arctic plant species show a combination of different adaptations to soil infertility of which clonality with persistent ramet connections can be considered as one. It may sometimes be difficult to assess the relative importance of the different traits. For example, both leaf longevity and longevity of ramet connections may enhance resource utilisation through resorption and recycling. The former through recycling between leaves and organs within a ramet, a process available to both unitary and clonal plants, the second through recycling between ramets within a clone, a process only available to clonal plants. A number of studies in the Abisko area and elsewhere have demonstrated the significance of clonal growth for resource acquisition and/ or utilisation in the quantitatively most important life forms among tundra plants, i.e. graminoids (Mattheis et al. 1976, Allessio and Tieszen 1978, Callaghan 1977, 1984, Jónsdóttir and Callaghan 1988, 1989, 1990), dwarf shrubs (Johansson 1974, Tolvanen 1994a) and clubmosses (Callaghan 1980, Headley et al. 1985, 1988a, b).

Graminoids

Beside being one of the dominating life forms in the Subarctic, the graminoid form is perhaps the most characteristic life form of the tundra as a whole. Many arctic graminoids are rhizomatous and the new ramets, termed tillers, are initiated from buds on parent stems (rhizomes and other types of stems). The horizontal distance between ramets varies between species and this controls their ability to spread laterally: rhizomes or stolons may be a couple of decimetres in some species, while extremely short in others. The distance may also vary within some species, either through ramet differentiation or through plastic responses to the environment (e.g. Shaver and Billings 1975, Callaghan 1977, Carlsson and Callaghan 1990b, Jónsdóttir 1991).

On the basis of ramet distance, two main growth patterns may be distinguished among graminoid species: The spreading and the caespitose patterns, often termed guerrilla and phalanx, respectively (sensu Lovett Doust 1981). The two growth patterns may be viewed as two alternative strategies to improve resource acquisition and utilisation in strongly resource-limited habitats. The spreading pattern allows a continuous search for new patches to exploit, while the caespitose growth pattern may resemble unitary perennial plants in many aspects as they are confined to only a limited area and thus forced to an efficient use of limited resources. Therefore, one would expect the caespitose plants to have a greater uptake potential and to utilise the once acquired resources more efficiently than the spreading plants and to have a greater dependence of stored resources to compensate for the limited searching ability.

It is not clear whether graminoids of the two growth patterns differ in uptake efficiency within habitat. Mc-Graw and Chapin (1989) compared two *Eriophorum* species of contrasting growth patterns and found that the spreading *E. scheuchzeri* had a higher nutrient uptake efficiency than the caespitose *E. vaginatum*, which contradicts the hypothesis about a greater uptake efficiency in caespitose and unitary plants. However, this difference was ascribed to adaptations to habitats of contrasting soil fertility rather than to contrasting strategies of nutrient acquisition within habitat. In general, it appears that variation in plant nutrient uptake efficiency correlates more to habitat fertility (Chapin 1980, McGraw and Chapin 1989), root morphology (Callaghan et al. 1991) or possession of and the type of mycorrhiza (Jonasson and Michelsen 1996).

Resorption of nutrients from leaves before senescence varies greatly both between graminoid species of different growth patterns and within species, but all species have a high demand for stored resources (see reviews in Berendse and Jonasson 1992, Shaver and Kummerow 1992). Thus, it is not obvious whether the caespitose growth form is more efficient in nutrient utilisation on a single tiller basis or not. However, the most extreme numbers reported for efficient nutrient resorption come from a study of the caespitose *Eriophorum vaginatum* in Alaska (Jonasson and Chapin 1985).

The arrangement of the vascular bundles in graminoids allows physiological flexibility so that large systems can be integrated. Indeed, an extensive physiological integration, i.e. a large integrated physiological unit (IPU sensu Watson and Casper 1984), has been demonstrated for some spreading arctic and antarctic graminoids. ^{14}C-labelling studies in the arctic grass *Dupontia fisheri* (Allessio and Tieszen 1978), the bipolar grass *Phleum alpinum* (Callaghan 1977), the antarctic sedge *Uncinia meridensis* (Callaghan 1984) and the arctic sedge *Carex bigelowii* (Jónsdóttir and Callaghan 1988) revealed that connected tillers are highly integrated physiologically. Interdependence in arctic graminoids declines with tiller age as found in most temperate species (Marshall 1990), but is usually maintained at a higher degree than commonly found in temperate grass species (Allessio and Tieszen 1978, Jónsdóttir and Callaghan 1989).

Interestingly, the above mentioned studies of physiologically integrated, spreading graminoids all show

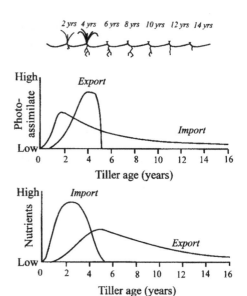

Fig. 1. Changes in the amount of export and import of different resources with age in horizontally spreading tundra graminoids, demonstrating division of labour in time, i.e. developmental division of labour between tiller generations. Based on Jónsdóttir and Callaghan (1988, 1989, 1990).

translocation of ^{14}C into one or more old tiller generations with only live below ground structures. In these species ramet connections persist at least during the life-time of an above ground photosynthetic shoot and often much longer. For example, intact clonal fragments of *Carex bigelowii* usually consist of between 7 and 30 different tiller generations (Jónsdóttir unpubl.) of which only the one or two youngest generations have photosynthesising above ground shoots. ^{14}C applied to the young *C. bigelowii* tillers was translocated into roots and rhizomes over 11 tiller generations suggesting that the below ground organs may be functional for >20 yr after the shoot dies (Jónsdóttir and Callaghan 1988). Up to 14 yr old *C. bigelowii* roots took up both nitrate and ammonium that was translocated mainly acropetally into the young photosynthesising tillers (Jónsdóttir and Callaghan 1990). These studies demonstrated a developmental division of labour (sensu Callaghan 1984 and Jónsdóttir and Callaghan 1990), i.e. a division of labour in time where tillers changed function as they aged (Fig. 1). The young tillers acquire carbon and provide old non-photosynthetic tillers with energy in exchange for nutrients, thereby increasing the number of nutrient acquisition sites and the number of patches that a single spreading clone can exploit. In addition, as the old tillers senesce, resources may be re-mobilised and recycled into the young growing tillers (Callaghan 1976).

The adaptive value of this strategy was shown for *Carex bigelowii* in severing experiments (Jónsdóttir and Callaghan 1988), and is probably a general phenomenon among spreading arctic and antarctic graminoids. When rhizome connections were severed early in the season between the two youngest generations, the survival probability of the youngest generation decreased as did the biomass accumulation in surviving tillers. In additional treatments, survival probability and biomass accumulation of the youngest generation increased with increasing numbers of intact tiller generations and gained the same level of biomass as in completely intact controls when five or more tiller generations were kept intact (Jónsdóttir and Callaghan 1988). Daughter tiller production by the youngest generation was, however, arrested in all the severing treatments (Fig. 2). This clearly demonstrated the dependence of the young photosynthesising tillers on the old below ground tillers.

A high degree of physiological integration in spreading graminoids may buffer the effects of "bad patches" in heterogeneous environments, as resources may be shared among tillers situated in patches of different quality. Localised damage such as grazing can be buffered in a similar way. For example, defoliated *Carex bigelowii* tillers were subsidised by neighbouring undefoliated tillers through increased translocation of photoassimilates (Jónsdóttir and Callaghan 1989). However, the degree of subsidy depended on tiller age and the degree of damage (Fig. 3). When repeatedly

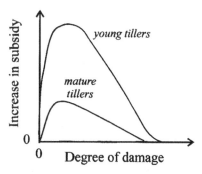

Fig. 3. The effect of localised damage (defoliation) on resource subsidy by neighbouring tillers in *Carex bigelowii* in relation to the degree of damage and the damaged tiller age demonstrating a complex buffering mechanism. Based on Jónsdóttir and Callaghan (1989).

defoliated, subsidy was finally cut off. This probably prevented a weakening of the donor ramets. This shows that integration and translocation of resources is not only governed by simple source-sink relationships which allow local responses, but that there must exist complex mechanisms which prevent detrimental effects to spread within a clone. Thus, environmental heterogeneity will not be totally averaged out through a high degree of physiological integration of clonal fragments (cf. Hutchings and Price 1993).

In spreading graminoids, the search and exploitation may be facilitated by division of labour in space through differentiation of tillers into short and long rhizome tillers. This is seen for example in the arctic sedge *Carex aquatilis* (Shaver and Billings 1975; termed clumping and spreading tillers respectively) and in two south Georgian graminoids, the rush *Rostkoviana magellanica*, and the grass *Phleum alpinum* (Callaghan 1977; termed colonising and pioneer tillers respectively). The pioneer tillers were longer lived and were subsidised with photoassimilates, and probably nutrients and water as well, by their parent tillers throughout a long period while the colonising tillers quickly became self-sufficient for photoassimilates and were shorter lived and could support the growth of the pioneer tillers (Callaghan 1977). Production and prolonged subsidy of pioneer tillers ensures a continuous search for new nutrient pockets and when it happens to root in a "good patch" an increased number of colonising tillers can be produced, hence increasing both energy capture and the probability of successful sexual reproduction (Fig. 4).

In *Carex bigelowii*, tillers with relatively long rhizomes developed from ventral buds whilst tillers developing from dorsal or lateral buds tended to have shorter rhizomes (Carlsson and Callaghan 1990b). There was a greater stimulation of production of dorsal tillers than ventral tillers by nutrient addition, suggesting that *C. bigelowii* responded to the environment through differential breakage of bud dormancy rather

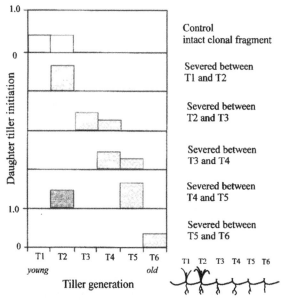

Fig. 2. The effects of severing persistent rhizome connections in the arctic graminoid *Carex bigelowii* on daughter tiller production, demonstrating i) dependence of the youngest tillers on connections with old tillers, ii) intraclonal control of tiller production and iii) the existence of a viable bud bank. Open bars: daughter tiller initiation in intact clonal fragments; shaded bars: daughter tiller initiation basipetal to the site of severing, i.e. the effect of released apical dominance on viable, dormant buds; filled bar: daughter tiller initiation acropetal to the site of severing. Adapted from Jónsdóttir and Callaghan (1988).

than through morphological plasticity. However, the degree of tiller differentiation within this species seems to vary with altitude (Carlsson and Callaghan 1990b) and latitude (Callaghan 1976, Jónsdóttir 1991).

The degree of physiological integration among tillers of caespitose graminoids is poorly known. Studies on American, caespitose bunch grasses have shown that clones consist of an assemblage of small IPU's (Welker and Briske 1992) and the same may be the case in arctic graminoids of this growth pattern. The adaptive significance of this arrangement would mainly be in terms of recycling of resources rather than increasing the number of uptake sites or buffering spatial heterogeneity.

Physiological integration does not only allow translocation of resources between tillers, but hormones as well. Within clonal fragments of *Carex bigelowii* there is a strong apical dominance (Jónsdóttir and Callaghan 1988). Dormancy of up to 12 yr old buds on the rhizomes was broken when released from apical dominance (Fig. 2). This intraclonal control of tiller production may reduce intraclonal competition. It may also improve nutrient utilisation and is probably of importance in both caespitose and rhizomatous spreading patterns: The limited amount of available nutrients within the clone are concentrated into only a few new tillers at a time, thus ensuring successful survival and reproduction. In the case of disturbance there is a long-lived bank of viable buds on the old rhizomes ensuring continued survival of the genet (Jónsdóttir and Callaghan 1988).

It appears that both clonal and non-clonal attributes are important for the resource dynamics in both growth

Table 2. Some characters improving resource acquisition and utilisation in clonal graminoids and their relative importance – a comparison between the spreading and caespitose growth patterns. + indicates that a character is important and – that it is not important. + + indicate that a character important for both growth patterns is of relatively greater importance for one of them. ? indicates that a character is probably important, but evidence is still lacking. See the text for references.

| | Growth pattern | |
	Spreading	Caespitose
Acquisition		
Non-clonal attributes:		
High nutrient uptake potential	+	+
Clonal attributes:		
Search	+	–
Buffer spatial variation in availability	+	–
Division of labour in time	+ +	+ ?
Division of labour in space (tiller differentiation)	+	–
Utilization		
Non-clonal attributes:		
Dependence on storage	+	+ +
Recycling within tiller	+	+ +
Clonal attributes:		
Recycling within clone	+ +	+ ?
Intraclonal control of tiller production	+	+ ?

patterns. However, the clonal attributes appear to be of greater importance in horizontally spreading than in caespitose graminoids (Table 2).

Dwarf shrubs

Dwarf shrubs are the dominating life form of much of the Subarctic, i.e. in the mountain birch forest and up to the low-alpine belt. The majority of the species belong to the Ericaceae and Salicaceae which include both evergreen and deciduous species.

Three main clonal growth patterns can be identified among dwarf shrubs: rhizomatous, stoloniferous and creeping patterns which produce adventitious roots from the stems that become decumbent after winter snows have depressed the shoots. The stoloniferous and rhizomatous growth forms have more extensive horizontal spread compared to the decumbent species, but a combination of two patterns may be found in some species. In general, most rhizomatous species are deciduous (exceptions are: *Vaccinium vitis-idaea* and *Andromeda polifolia*) and have a large proportion of their biomass below ground while most of those species which are stoloniferous or creeping are evergreen (exceptions are: *Arctostaphylos alpina* and some species of *Salix*) with a greater proportion of their biomass above ground. This may be a function of the relative importance of organs for resource storage.

While the importance of clonal growth for vegetative spread in subarctic dwarf shrubs is widely accepted, the importance of the second main aspect of clonality, i.e.

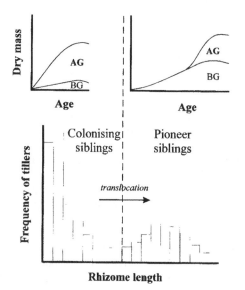

Fig. 4. Schematic illustration of division of labour in space in tundra graminoids through tiller differentation. AG = above ground biomass; BG = Below ground biomass. Based on Shaver and Billings (1975), Callaghan (1976, 1977, unpubl.), Carlsson and Callaghan (1990b).

that of resource dynamics, is rather poorly understood. More research is needed before its importance can be fully assessed. However, resource dynamics of arctic and subarctic dwarf shrubs on a shoot, or a single ramet basis is extensively studied and has mainly been considered in terms of leaf longevity, i.e. evergreen vs deciduous, and more recently in terms of mycorrhizal association rather than in terms of their clonality. The abundance of evergreen shrubs is generally found to be greater in infertile habitats relative to deciduous shrubs (e.g. Miller 1982) and leaf longevity has been related to low soil availability (e.g. Chabot and Hicks 1982, Chapin and Shaver 1989). Several adaptive mechanisms of leaf longevity have been suggested, however, and their relative advantages debated (e.g. Jonasson 1995, Karlsson 1995).

The ramets in dwarf shrubs are not as easily identifiable as the tillers of graminoids. Tolvanen (1994b) defined the ramet in rhizomatous *Vaccinium* species as "the aerial, orthotropic shoot branching from the horizontal below ground stem". The arrangement of the vascular bundles in dicotyledons is generally more fixed compared to graminoids and has been showed to constrain the size of IPU's (e.g. Price et al. 1992). There are only a few studies available that demonstrate resource translocation between ramets in arctic dwarf shrubs or the importance of ramet connections. The evergreen, rhizomatous *Andromeda polifolia* inhabits bogs. It translocated large proportions of assimilated carbon far back into old rhizomes both early (May) and late (August) in the season while great proportions were lost to respiration in June–July (Johansson 1974). This indicates that physiological integration of old and young parts (ramets) of a clone is important at least in the process of storing carbon. In another evergreen species inhabiting bogs, the stoloniferous *Vaccinium microcarpum*, there was a strong acropetal within-branch translocation of ^{14}C in July and no apparent translocation between ramets (Svensson 1995).

Evidence of the importance of rhizome connections for ramet growth and recovery from herbivore damage was found in a two year study of two *Vaccinium* species where herbivory and below ground damage was simulated (Tolvanen 1994a). Rhizome alone severing significantly reduced ramet growth and branching in the deciduous *V. myrtillus* but only in combination with herbivore damage in the evergreen *V. vitis-idaea*. This may reflect the commonly assumed difference in importance of storage organs between deciduous (below ground organs) and evergreen (above ground organs) dwarf shrubs (Tolvanen 1994a), but may also indicate a greater dependence on old ramets for resource acquisition in deciduous shrubs.

The horizontally spreading dwarf shrubs (either below ground or above ground) may also possess a mechanism, although slow, for searching for "rich patches", similar to the horizontally spreading graminoids. For example, Emanuelsson (1984) demonstrated a slow, wave like movement through the landscape of the evergreen, adventitiously rooting *Empetrum hermaphroditum* in the birch forest at Abisko. Such clonal movement may also be an important biotic factor affecting the spatial distribution of soil nutrients in heterogeneous environments through intraclonal redistribution of nutrients that may be released at death (Heal et al. 1989).

Clubmosses

The stoloniferous members of the Lycopodiaceae are good examples of extensive horizontal spread among clonal plants in the Abisko area. The evergreen horizontal branches of *Lycopodium annotinum* spread on average at 0.07 m yr^{-1}, but may spread as fast as 0.2 m yr^{-1} in favourable microsites (Callaghan et al. 1986a). This enables the plant to invade relatively rapidly into gaps in the canopy and studies have shown that growth in *L. annotinum* changes in response to light quality (Svensson et al. 1994). Roots are produced at regular intervals, but many fail to establish due to shallow soils and the presence of large stones. At 5%, *Lycopodium annotinum* has the lowest proportion of biomass allocated to roots of most vascular tundra plants (Callaghan et al. 1986b). The vertical branches represent the largest proportion of the biomass of the plant (65%) and their average life span is 3 yr (Callaghan et al. 1986a), but some may live for as long as 8–13 yr (Callaghan et al. 1986b).

In contrast with the dwarf shrubs, resource dynamics in terms of clonality is rather well known in subarctic clubmosses. The slow rates of nutrient uptake, particularly of phosphate (Headley et al. 1985), are partly due to the thickness of the roots (0.2–1 mm in diameter) and their lack of a mycorrhizal association. The slow rates of nutrient uptake by members of the Lycopodiaceae are compensated for by efficient nutrient utilisation through a high degree of internal nutrient cycling (Callaghan 1980) and intraclonal control of branching through apical dominance (Svensson and Callaghan 1988b). The persistent connection of all parts of the plant as well as their extreme longevity allows the majority of nutrients to be remobilised to furnish the new growth of the apical meristem (Headley et al. 1985). These factors as well as the persistent connection of microphylls and vertical branches to the main horizontal branch allows a large proportion of the nutrients within the vertical branches to be recycled. In the case of phosphorus 75% of the maximum content is recycled (Headley et al. 1985) and this is similar to another common clubmoss, the tufted *Huperzia selago* (Headley 1986) as well as to many arctic tundra plants in general (Berendse and Jonasson 1992). Nitrogen, potassium and other mobile elements are also recycled relatively efficiently, whilst magnesium is not recycled to the same extent and calcium is hardly if at all re-mobilised (Callaghan 1980).

Although the roots of clubmosses have relatively slow rates of nutrient uptake their longevity allows the uptake of water and certain solutes to proceed for as long as 10–13 yr (Headley et al. 1988a, b) (Fig. 5). This allows more distant modules to be subsidised with water and nutrients whilst colonising or crossing inhospitable or unsafe microsites, such as boulders. In the case of transpiration, this is primarily acropetally to the apical meristems of the horizontal branches (Headley et al. 1988b) and the water requirements are largely in the younger parts of the plant, but can be satisfied by root systems up to at least 10 yr old. This is the primary pathway for potassium and calcium movement in the plant, whilst phosphorus is transported primarily in the translocation stream (Headley et al. 1988a) and can either be in an acropetal or basipetal direction from root systems between 1 and 6 yr old.

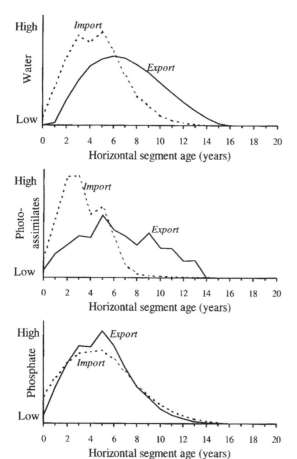

Fig. 5. Division of labour in time between different segments in an evergreen, stoloniferous clubmoss, demonstrated by changes in the amount of export and import of photoassimilates, water and nutrients with age. Based on Callaghan (1980), Headley et al. (1988a, b).

Although the stoloniferous species of clubmoss are superficially most similar to evergreen dwarf shrubs, they are most similar to the rhizomatous graminoids in terms of the degree of physiological integration, the large size of the IPU's and degree of longevity of the modules within those IPU's (Carlsson et al. 1990). Also, they show a certain degree of division of labour in time, i.e. a developmental division of labour, but with greater overlap between the generations than in graminoids, especially in photoassimilate export and import, due to the evergreen and stoloniferous habit (Fig. 5).

Huperzia selago contrasts with the other clubmosses in that it is a tufted species which spreads via bulbils. Individual plants have generally a much shorter lifespan of between 10 and 30 yr compared with hundreds of years for the stoloniferous species. There is virtually no physiological integration between branches in *Huperzia selago* (Headley 1986), but as in caespitose graminoids and evergreen dwarf shrubs, recycling of resources from older sections of the same branch enables plants to continue growing although at much slower rates. The bulbil, however, maintains genets very effectively and can allow a single genetic individual to spread across extremely unsafe sites, such as boulder scree, which is impossible for the caespitose graminoids and dwarf shrubs and even for rhizomatous or stoloniferous species.

Mosses

Mosses grow clonally mainly through branching (Collin and Oechel 1974, Callaghan et al. 1978) and are an important component of the vegetation in the polar regions. In spite of their clonal habit, their resource dynamics have rarely been studied in terms of clonality. In contrast with vascular plants they lack real roots and vascular tissue, except for members of the Polytrichales (e.g. Collins and Oechel 1974). Consequently, they mainly rely on atmospheric nutrient sources and have limited internal cycling of resources. However, considerable quantities may be acquired from the soil and moved upwards, probably by capillary forces both within the moss tissue and along moss stems into the youngest modules (Jónsdóttir et al. 1995). Resorption of nutrients and movement from ageing modules to younger ones – if it occurs – is probably through a similar, but much reduced process.

The difference in the growth form of clonal *Polytrichum commune* and the "colonial" *Hylocomium splendens* has recently been shown to result in individualistic responses of the two mosses to environmental manipulation experiments at Abisko (Potter et al. 1995). *Hylocomium splendens* was shown to suffer from nutrient additions whereas shoot growth and density increased in *P. commune*, possibly due to more extensive nutrient uptake and clonal proliferation.

Implications for population dynamics and ecosystem resilience

The resource dynamics varies both between and within life forms of clonal plants species of the Subarctic, but the clonal attributes are significant. These plants have evolved efficient means for acquiring nutrients at low levels of availability, to utilise the once acquired resources extremely efficiently and to buffer fluctuations in resource availability in space and time. The observed variation in clonal growth patterns in combination with variation in root growth mechanisms (Callaghan et al. 1991) and other specialisation, is probably important in enabling clonal plant species to coexist. Coexisting plants would probably adopt different clonal strategies. For example, the relatively mobile evergreen *Lycopodium annotinum* coexists with the relatively stationary dwarf shrubs of the birch forest. Similarly, horizontally spreading sedges are intermingled with those of the caespitose growth pattern in the graminoid dominated wetlands and in the mid-alpine tundra belt, while in bogs horizontally spreading dwarf shrubs, such as *Andromeda*, coexist with more stationary shrubs.

Genet establishment is infrequent in the closed vegetation of the Subarctic (Callaghan and Emanuelsson 1985). In addition, the establishment process in vascular plants may be extended over several years as they build up a minimum reserve of resources before being capable of normal clonal growth (Jónsdóttir unpubl.) but once established, clones are often extremely long-lived (e.g. Oinonen 1968). Although ramet population size may vary considerably between years at some sites (e.g. Carlsson and Callaghan 1990a, 1994), the clonal growth habit has a profound stabilising effect on the population dynamics in the long term, through both controlled vegetative reproduction and its effects on resource dynamics (Jónsdóttir 1991). As a consequence of the dominance of clonal plants in the Subarctic, the vegetation as a whole is stable under prevailing environmental conditions.

The stabilising effect of clonal plants at population and community levels may even be marked at a ecosystem level: they preserve nutrients within the living biomass with a minimum loss though leaching and seed production. However, as a consequence of clonal growth and internal cycling, nutrients may be spatially redistributed (Heal et al. 1989). Mosses are an important factor in conserving nutrients as they are extremely efficient in intercepting all atmospheric deposition even at anthropogenically increased levels in the Subarctic (Woodin et al. 1985, Aerts et al. 1992, Jónsdóttir et al. 1995).

The conservation and recycling of resources within the living plant biomass may have serious repercussion as a consequence of increased pollution load in the Arctic and Subarctic: pollutants such as radionucleids can be trapped by cushion plants (Svoboda and Hutchinson-Benson 1986) and taken up and accumulated by most plants with roots. Those clonal plants which accumulate and concentrate mobile radionucleids, such as ^{134}Cs, in apical meristems (Tyson et al. 1995) are vulnerable to genetic damage.

Another possible threat to the relative stability of subarctic vegetation is climate change. The Subarctic includes an extensive northern hemisphere vegetation zone, an ecotone between two contrasting vegetation zones, the boreal forest with low albedo and the treeless tundra with high albedo. The geographical position of the zone is not stable and has moved in response to the mean summer position of the Arctic front during the past 6000 yr (Nichols 1967). Beside the notion that the present interglacial period in the northern hemisphere will come to an end within a few thousand years (Imbrie and Imbrie 1980, Ingólfsson and Hjort 1992) it is generally thought that the anthropogenic CO_2-release into the atmosphere will have serious consequences for the climate within the next decades, leading to global warming (Mitchell et al. 1990, Maxwell 1992, Cattle and Crossley 1995). The climate of the subarctic Atlantic region is extremely sensitive to changes in the ocean currents and it is possible that global warming may affect them with cooler climate as a consequence (Broecker 1995). In the event of a warmer climate the Subarctic is the place where the effects of climate warming on vegetation may produce positive feedback to global climate (Callaghan and Jonasson 1995).

It is not too obvious how arctic and subarctic clonal plants, apparently well adapted to prevailing environmental conditions, will respond in the long term to either warmer or cooler climates. Based on our present knowledge on internal resource dynamics of clonal plants we can expect them to buffer the effects of climate changes at least in the short term. Understanding the dynamics of populations and resources in the different life forms of clonal plants is essential for predictions on long term changes at community and ecosystem levels.

Acknowledgements – Many of the studies reviewed here have benefited from the excellent facilities at the Abisko Scientific Research Station. We thank the staff, and particularly the Director M. Sonesson, for support. We thank S. Jonasson and S. Karlsson for valuable comments on the manuscript. I. S. Jónsdóttir wishes to thank Swedish Natural Science Research Council for financial support. T. V. Callaghan wishes to thank the U.K. Natural Environment Research Council for support.

References

Aerts, R., Wallén, B. and Malmer, N. 1992. Growth-limiting nutrients in *Sphagnum*-dominated bogs subject to low and high atmospheric nitrogen supply. – J. Ecol. 80: 131–140.
Allessio, M. L. and Tieszen, L. L. 1978. Translocation and allocation of ^{14}C-photoassimilate by *Dupontia fisheri*. – In: Tieszen, L. L. (ed.), Vegetation and production ecology of

an Alaskan Arctic tundra. Springer, New York, pp. 393–413.

Aronsson, M. 1994. Torneträskområdets kärlväxter. – Märsta.

Berendse, F. and Jonasson, S. 1992. Nutrient use and nutrient cycling in northern ecosystems. – In: Chapin, F. S. III, Jefferies, R. L., Reynolds, J. F., Shaver, G. R. and Svoboda, J. (eds), Arctic ecosystems in a changing climate. An ecophysiological perspective. Academic Press, San Diego, pp. 337–356.

Broecker, W. S. 1995. Chaotic climate. – Sci. Am. 273: 44–50.

Callaghan, T. V. 1976. Growth and population dynamics of Carex bigelowii in an alpine environment. Strategies of growth and population dynamics of tundra plants 3. – Oikos 27: 402–413.

– 1977. Adaptive strategies in the life cycles of south Georgian graminoid species. – In: LLano, G. A. (ed.), Adaptations within Antarctic ecosystems. Proc. 3rd S.C.A.R. symp. Antarctic biol., Smithsonian Inst., Washington, D.C., pp. 891–1002.

– 1980. Age-related patterns of nutrient allocation in Lycopodium annotinum from Swedish Lapland. Strategies of growth and population dynamics of tundra plants 5. – Oikos 35: 373–386.

– 1984. Growth and translocation in a clonal southern hemisphere sedge – Uncinia meridensis. – J. Ecol. 72: 529–546.

– and Emanuelsson, U. 1985. Population structure and processes of tundra plants and vegetation. – In: White, J. (ed.), The population structure of vegetation. Junk Publ., pp. 399–439.

– and Jonasson, S. 1995. Arctic terrestrial ecosystems and environmental change. – Phil. Trans. R. Soc., Ser. A 352: 259–276.

– , Collins, N. J. and Callaghan, C. H. 1978. Photosynthesis, growth and reproduction of Hylocomium splendens and Polytrichum commune in Swedish Lapland. Strategies of growth and population dynamics of tundra plants 4. – Oikos 31: 73–88.

– , Headley, A. D., Svensson, B. M., Lixian, L., Lee, J. A. and Lindley, D. K. 1986a. Modular growth and function in the vascular cryptogam Lycopodium annotinum. – Proc. R. Soc., Ser. B 228: 195–206.

– , Svensson, B. M. and Headley, A. 1986b. The modular growth of Lycopodium annotinum. – Fern Gaz. 13: 65–76.

– , Headley, A. D. and Lee J. A. 1991. Root function related to the morphology, life history and ecology of tundra plants. – In: Atkinson, D. (ed.), Plant root systems: Their effect on ecosystem composition and structure. Blackwell, Oxford, pp. 311–340.

Carlsson, B. Å. and Callaghan, T. V. 1990a. Effects of flowering on the shoot dynamics of Carex bigelowii along an altitudinal gradient in Swedish Lapland. – J. Ecol. 78: 153–166.

– and Callaghan, T. V. 1990b. Programmed tiller differentiation, intraclonal density regulation and nutrient dynamics in Carex bigelowii. – Oikos 58: 219–230.

– and Callaghan, T. V. 1994. Impact of climate change factors on the clonal sedge Carex bigelowii: implications for population growth and vegetative spread. – Ecography 17: 321–330.

– , Jónsdóttir, I. S., Svensson, B. M. and Callaghan, T. V. 1990. Aspects of clonality in the Arctic: a comparison between Lycopodium annotinum and Carex bigelowii. – In: van Groenendael, J. and de Kroon, H. (eds), Clonal growth in plants: regulation and function. SPB Academic Publ., The Hague, pp. 131–151.

Cattle, H. and Crossley, J. 1995. Modelling arctic climate change. – Phil. Trans. R. Soc. Lond., Ser. A 352: 201–213.

Chabot, B. F. and Hicks, D. J. 1982. The ecology of leaf life spans. – Annu. Rev. Ecol. Syst. 13: 229–259.

Chapin, F. S. III, 1980. The mineral nutrition of wild plants. – Annu. Rev. Ecol. Syst. 11: 233–260.

– and Bloom, A. 1976. Phosphate absorption: adaptation of tundra graminoids to a low temperature, low phosphorus environment. – Oikos 26: 111–121.

– and Shaver, G. R. 1989. Difference in growth and nutrient use among arctic plant growth forms. – Funct. Ecol. 3: 73–80.

– , Barsdate, R. J. and Barél, D. 1978. Phosphorus cycling in Alaskan coastal tundra: a hypothesis for the regulation of nutrient cycling. – Oikos 31: 187–199.

Collins, N. J. and Oechel, W. C. 1974. The pattern of growth and translocation of photosynthate in a tundra moss, Polytrichum alpinum. – Can. J. Bot. 52: 355–364.

Emanuelsson, U. 1984. Ecological effects of grazing and trampling on mountain vegetation in northern Sweden. – Ph.D. thesis, Lund Univ., Sweden.

Eriksson, O. and Jerling, L. 1990. Hierarchical selection and risk spreading in clonal plants. – In: van Groenendael, J. and de Kroon, H. (eds), Clonal growth in plants: regulation and function. SPB Academic Publ., The Hague, pp. 79–94.

Fagerström, T. 1992. The meristem-meristem cycle as a basis for defining fitness in clonal plants. – Oikos 63: 449–453.

Fries, T. C. E. 1913. Botanisches Untersuchungen im nördlichsten Schweden. – Vetensk. prakt. Unders. Lappl., Luossavaara–Kirunavaara.

Graham, G. G. 1988. The flora and vegetation of county Durham. – The Durham Flora Comm. Durham county Conserv. Trust, Osborne Kay, Wallsend.

Havström, M., Jonasson, S. and Callaghan, T. V. 1995. Effects of simulated climate change on the sexual reproductive effort of Cassiope tetragona. – In: Callaghan, T. V., Oechel, W. C., Gilmanov, T., Molau, U., Maxwell, B., Tyson, M., Sveinbjörnsson, B. and Holten, J. I. (eds), Global change and arctic terrestrial ecosystems. Ecosystem rep. 10. European Comm., DG XII, Luxembourg, pp. 109–114.

Headley, A. D. 1986. The comparative autecology of some European Lycopodium species sensu lato. – Ph.D. thesis, Univ. of Manchester, U.K.

– , Callaghan, T. V. and Lee, J. A. 1985. The phosphorus economy of the evergreen tundra plant Lycopodium annotinum. – Oikos 45: 235–245.

– , Callaghan, T. V. and Lee, J. A. 1988a. Phosphate and nitrate movement in the clonal plants Lycopodium annotinum L. and Diphasiastrum complanatum (L.) Holub. – New Phytol. 110: 487–495.

– , Callaghan, T. V. and Lee, J. A. 1988b. Water uptake and movements in the evergreen clonal plants Lycopodium annotinum L. and Diphasiastrum complanatum (L.) Holub. – New Phytol. 110: 497–502.

Heal, O. W., Callaghan, T. V. and Chapman, K. 1989. Can population and process ecology be combined to understand nutrient cycling?. – In: Clarholm, M. and Bergström, L. (eds), Ecology of arable land. Kluwer Acad. Publ., pp. 205–216.

Hutchings, M. J. and Price, E. A. C. 1993. Does physiological integration enable clonal herbs to integrate the effects of environmental heterogeneity? – Plant Species Biol. 8: 95–105.

– and de Kroon, H. 1994. Foraging in plants: the role of morphological plasticity in resource acquisition. – Adv. Ecol. Res. 25: 159–238.

Imbrie, J. and Imbrie, J. Z. 1980. Modelling the climatic response to orbital variations. – Science 207: 943–953.

Ingólfsson, Ó. and Hjort, C. 1992. What can the last ice age tell us about the next? – Geol. För. Stockholm Förhandl. 114: 372–374.

Johansson, L.-G. 1974. The distribution and fate of ¹⁴C photoassimilated by plants on a subarctic mire at Stordalen. – Swedish IBP Tundra Biome Project. Techn. Rep. 16: 165–172.

Jonasson, S. 1986. Influence of frost heaving on soil chemistry and on the distribution of plant growth forms. – Geogr. Ann. 68A: 185–195.

– 1995. Nutrient limitation in *Rhododendron lapponicum*, fact or artefact? – Oikos 73: 269–271.
– and Chapin, F. S. III, 1985. Significance of sequential leaf development for nutrient balance of the cotton sedge, *Eriophorum vaginatum* L. – Oecologia. 67: 511–518.
– and Michelsen, A. 1996. Nutrient cycling in subarctic and arctic ecosystems, with special reference to the Abisko and Torneträsk region. – Ecol. Bull. 45: 45–52.
– , Bryant, J. P., Chapin, F. S. III and Andersson, M. 1986. Plant phenols and nutrients in relation to variations in climate and rodent grazing. – Am. Nat. 128: 394–408.
Jónsdóttir, I. S. 1991. Effects of grazing on tiller size and population dynamics in a clonal sedge (*Carex bigelowii*). – Oikos 62: 177–188.
– 1995. Importance of sexual reproduction in arctic clonal plants and their evolutionary potential. – In: Callaghan, T. V., Oechel, W. C., Gilmanov, T., Molau, U., Maxwell, B., Tyson, M., Sveinbjörnsson, B. and Holten, J. I. (eds), Global change and arctic terrestrial ecosystems. Ecosyst. report 10. European Comm., DG XII, Luxembourg, pp. 81–88.
– and Callaghan, T. V. 1988. Interrelationships between different generations of interconnected tillers of *Carex bigelowii*. – Oikos 52: 120–128.
– and Callaghan, T. V. 1989. Localized defoliation stress and the movement of ^{14}C-photoassimilates between tillers of *Carex bigelowii*. – Oikos 54: 211–219.
– and Callaghan, T. V. 1990. Intraclonal translocation of ammonium and nitrate nitrogen in *Carex bigelowii* Torr. ex Schwein. using ^{15}N and nitrate reductase assays. – New Phytol. 114: 419–428.
– , Callaghan, T. V. and Lee, J. A, 1995. Fate of added nitrogen in a moss-sedge Arctic community and effects of increased nitrogen deposition. – Sci. Total Environ. 160/161: 677–685.
Karlsson, P. S. 1995. Nutrient or carbon limitation of shoot growth in *Rhododendron lapponicum* – both or neither! A reply to Jonasson. – Oikos 73: 272–273.
– and Weih, M. 1996. Relationships between nitrogen economy and performance in the mountain birch *Betula pubescens* ssp. *tortuosa* – Ecol. Bull. 45: 71–78.
Kielland, K. and Chapin, F. S. III, 1992. Nutrient absorption and accumulation in arctic plants. – In: Chapin, F. S. III, Jefferies, R. L., Reynolds, J. F., Shaver, G. R. and Svoboda, J. (eds), Arctic ecosystems in a changing climate. An ecophysiological perspective. Academic Press, San Diego, pp. 321–335.
Köppen, W. 1936. Das geographische System der Klimate. – In: Köppen, W. and Geiger, R. (eds), Handbuch der klimatologie, 1. C., Berlin Bornträger.
Laine, K. and Henttonen, H. 1983. The role of plant production in microtine cycles in northern Fennoscandia. – Oikos 40: 407–418.
Löve, D. 1970. Subarctic and subalpine: where and what? – Arct. Alp. Res. 2: 63–73.
Lovett Doust, L. 1981. Population dynamics and local specialisation in a clonal perennial (*Ranunculus repens*). I. The dynamics of ramets in contrasting habitats. – J. Ecol. 69: 743–755.
Marion, G. M. and Kummerow, J. 1990. Ammonium uptake by field-grown *Eriophorum vaginatum* roots under laboratory and simulated field conditions. – Holarct. Ecol. 13: 50–55.
Marshall, C. 1990. Source-sink relations of interconnected ramets. – In: van Groenendael, J. and de Kroon, H. (eds), Clonal growth in plants: regulation and function. SPB Academic Publ., The Hague, pp. 23–41.
Mattheis, P. J., Tieszen, L. L. and Lewis, M. C. 1976. Responses of *Dupontia fischeri* to simulated lemming grazing in an Alaskan arctic tundra. – Ann. Bot. 40: 179–197.
Maxwell, B. 1992. Arctic climate: potential for change under global warming. – In: Chapin, F. S. III, Jefferies, R. L., Reynolds, J. F., Shaver, G. R. and Svoboda, J. (eds), Arctic ecosystems in a changing climate. An ecophysiological perspective. Academic Press, San Diego, pp. 11–34.
McCown, B. H. 1978. The interactions of organic nutrients, soil nitrogen, and soil temperature and plant growth and survival in the arctic environment. – In: Tieszen, L. L. (ed.), Vegetation and production ecology of an Alaskan arctic tundra. Springer, New York, pp. 435–456.
McGraw, J. B. and Chapin, F. S. III, 1989. Competitive ability and adaptation to fertile and infertile soils in two *Eriophorum* species. – Ecology 70: 736–749.
Miller, P. C. 1982. Environmental and vegetational variation across a snow accumulation area in montane tundra in central Alaska. – Holarct. Ecol. 5: 85–98.
Mitchell, J. F. B., Manabe, S., Tokioka, T. and Meleshko, V. 1990. Equilibrium climate change. – In: Houghton, J. T., Jenkins, G. J. and Ephraums, J. J. (eds), Equilibrium climate change. The IPCC scientific assessment. Cambridge Univ. Press, Cambridge, pp. 131–172.
Nichols, H. 1967. Pollen diagrams from sub-arctic central Canada. – Science 155: 1665–1668.
Nordhagen, R. 1927. Die Vegetation und Flora des Sylengebietes – Skr. norske Vidensk-Akad. Mat. naturv. Kl., 1927 (1).
Oinonen, E. 1968. The size of *Lycopodium clavatum* L. and *L. annotinum* L. stands as compared to that of *L. complanatum* L. and *Pteridium aquilinum* (L.) Kuhn stands, the age of the tree stand and the dates of fire on the site. – Acta Forest. Fenn. 87: 5–53.
Pate, J. S. 1994. The mycorrhizal association – just one of many nutrient acquiring specializations in natural ecosystems. – Plant Soil 159: 1–10.
Potter, J. A., Press, M. C., Callaghan, T. V. and Lee, J. A. 1995. Growth responses of *Polytrichum commune* and *Hylocomium splendens* to simulated environmental change in the sub-arctic. – New Phytol. 131: 533–541.
Price, E. A. C., Marshall, C. and Hutchings, M. J. 1992. Studies of growth in the clonal herb *Glechoma hederacea*. I. Patterns of physiological integration. – J. Ecol. 80: 25–38.
Shaver, G. R. and Billings, W. D. 1975. Root production and root turnover in wet tundra ecosystem, Barrow, Alaska. – Ecology 56: 401–409.
– and Kummerow, J. 1992. Phenology, resource allocation and growth of arctic vascular plants. – In: Chapin, F. S. III, Jefferies, R. L., Reynolds, J. F., Shaver, G. R. and Svoboda, J. (eds), Arctic ecosystems in a changing climate. An ecophysiological perspective. Academic Press, San Diego, pp. 193–211.
Sørensen, T. 1941. Temperature relations and phenology of the northeast Greenland flowering plants. – Medd. Grønl. 125: 1–305.
Svensson, B. S. 1995. Carbon allocation patterns in two closely related stoloniferous *Vaccinium* species. – Acta Oecol. 16: 507–517.
– and Callaghan, T. V. 1988a. Small-scale vegetation pattern related to the growth of *Lycopodium annotinum* and variations in its micro-environment. – Vegetatio 76: 167–177.
– and Callaghan, T. V. 1988b. Apical dominance and the simulation of metapopulation dynamics in *Lycopodium annotinum*. – Oikos 51: 331–342.
– , Floderus, B. and Callaghan, T. V. 1994. *Lycopodium annotinum* and light quality: Growth responses under canopies of two *Vaccinium* species. – Folia Geobot. Phytotax. 29: 159–166.
Svoboda, J. and Hutchinson-Benson, E. 1986. Arctic cushion plants as fallout "monitors". – J. Environ. Radioact. 4: 65–76.
Tolvanen, A. 1994a. Differences in recovery between a deciduous and an evergreen ericaceous clonal dwarf shrub after simulated aboveground herbivory and belowground damage. – Can. J. Bot. 72: 853–859.
– 1994b. Recovery ability and plant architecture: a comparison of two ericaceous dwarf shrubs. – Acta Univ. Ouluensis, ser. A Sci. Rerum Nat. 253: 1–38.

Tyson, M. J., Sheffield, L. and Callaghan, T. V. 1995. An overview of [134]Cs and [85]Sr transport, allocation and concentration studies in artificially propagated bracken. – In: Smith, R. T. and Taylor, A. J. (eds), Bracken: an environmental issue. Contr. Int. Conference, Aberystwyth Int. Bracken Group, pp. 38–42.

Watson, A. M. and Casper, B. B. 1984. Morphogenetic constraints on patterns of carbon distribution in plants. – Annu. Rev. Ecol. Syst. 15: 233–258.

Welker, J. M. and Briske, D. D. 1992. Clonal biology of the temperate, caespitose, graminoid *Schizachyrium scoparium*: a synthesis with reference to climate change. – Oikos 63: 357–365.

Wielgolaski, F. E., Kjellvik, S. and Kallio, P. 1975. Mineral content of tundra and forest tundra plants in Fennoscandia. – In: Wielgolaski, F. E. (ed.), Fennoscandian tundra ecosystems. Part 1. Plants and micro-organisms. Springer, Berlin, pp. 316–332.

Woodin, S., Press, M. C., Callaghan, T. V. and Lee, J. A. 1985. Nitrate reductase activity in *Sphagnum fuscum* in relation to wet deposition of nitrate from the atmosphere. – New Phytol. 99: 381–388.

Ecological Bulletins 45: 65–70. Copenhagen 1996

Treeline ecology of mountain birch in the Torneträsk area

Bjartmar Sveinbjörnsson, Heikki Kauhanen, and Olle Nordell

Sveinbjörnsson, B., Kauhanen, H. and Nordell, O. 1996. Treeline ecology of mountain birch in the Torneträsk area. – Ecol. Bull. 45: 65–70.

Birch reproductive success generally diminishes with increasing elevation. Production of inflorescences, seed quality, seed germination and summer seedling survival all decrease. A successful establishment only takes place during a period of several years of favorable weather, presumably as resources can be shifted from growth. Male catkin production is increased both by nitrogen and phosphorus and especially both. Instantaneous rates of photosynthesis and respiration are virtually identical among trees at different elevations but the leafy season is shorter at high elevations and the risk of leaf loss during it is higher, perhaps leading to infrequent carbon limitations for growth. No direct evidence for either a photosynthetic or respiratory limitation at the treeline was found, although high elevation leaf loss resulted in drastically reduced seasonal carbon gain. Nitrogen was found to limit growth of tree height and branch numbers to increasing degree with increasing elevation. This is in spite of increasing foliar nitrogen concentration with increasing elevation. Phosphorus application however, was not found to stimulate growth.

B. Sveinbjörnsson, Dept of Biol. Sciences, Univ. of Alaska Anchorage, 3221 Providence Dr., Anchorage, AK 99508, USA. – H. Kauhanen, Dept of Biology, Univ. of Turku, FIN-20500 Turku, Finland. – O. Nordell, Dept of Ecology, Plant Ecology, Univ. of Lund, Ecology Building, S-223 62 Lund, Sweden.

Ecologists are keenly interested in identifying the environmental factors and physiological processes that determine the distribution of species (Osmond et al. 1987, Graves and Taylor 1988). When the distribution limits are obvious, because they also apply to growth habit and community physiognomy as the case is for treelines, the interest is further enhanced (Koike 1987, Noshiro and Suzuki 1989, Lipscomb and Nilsen 1990).

Because proximity to major universities has facilitated treeline research, alpine treelines have been studied to a greater extent than polar treelines (see however Payette and Gagnon 1979, Black and Bliss 1980, Goldstein 1981, Payette 1983, Cooper 1986, Scott et al. 1987a, b). While the treelines in the Torneträsk area are elevational, these subarctic treelines are at low elevations (c. 650 m a.s.l.). This low elevation simplifies the search for causative environmental factors. This is because there are little or no gradients in many factors, possibly of great importance in mid-latitude alpine situations, such as UV radiation, wind speed, and solar radiation. Thus high latitude elevational treeline research design and interpretation of results is simpler than at mid-latitude elevational treelines and the results are more applicable to polar treelines.

Two major mechanistic categories are generally evoked in discussions on limits to species distribution. The first focuses on establishment, mortality, survival, and tolerance. The latter category addresses the increasing difficulties in material acquisition or material utilization. Although the two categories of mechanisms are obviously interrelated, the first is often seen and evaluated as an either/or mechanism determining presence or absence while the latter is seen as a grading process making existence increasingly difficult and reflected in size and shape of individuals of the tree species. This dichotomy is obviously flawed as there are certainly environmental factors and physiological mechanisms that lead to failure of germination or seedling establishment.

We have carried out studies on mountain birch along elevational gradients in the Torneträsk area of northern Sweden. In these studies we have focused our attention on three elevational zones or belts. The uppermost zone which we refer to as treeline is characterized by widely scattered individual mountain birch trees, clumps or hedges. The forest limit refers to a zone immediately below the treeline zone. This zone represents the upper

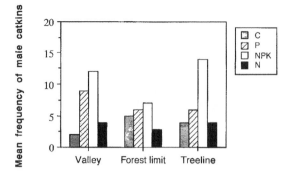

Fig. 1. The frequency of male catkins on birch trees treated with different fertilizers. The trees were growing in situ in different elevational belts on the slopes of Mount Luovarre (see text).

limit of the forest below; the trees are of the same stature and at similar densities as below it. The valley zone refers to the base of the mountain side. It is generally relatively flat and tree height and density is similar to that at the forest border, but sometimes greater, depending on the topography. In each zone, treatments were replicated laterally.

Reproduction

For tree establishment to be successful, the entire reproductive cycle from gamete production to seedling survival must operate. In mountain birch, male catkin production per individual is highly variable but overall similar at different elevations along the slopes of Luovarre (Stenbacken) with the highest mean value at the forest limit (Fig. 1). Male catkin frequency is increased both by phosphorus fertilizer but even more by complete fertilizer and not at all by nitrogen alone. At the same sites (Luovarre), female catkin production varies greatly from year to year, or from 0.87 catkins per tree in 1982 to 15.14 catkins per tree in 1983. There is an overall elevational trend of decreasing female catkin production with increasing elevation. Thus, a five year average is 8.14 catkin per tree for the valley, 8.25 for the forest limit and 5.67 for the tree line. The female catkin production on Luovarre is unaffected by nutrients, while in Iceland, mountain birch female catkin production is stimulated by nitrogen fertilization (Sveinbjörnsson et al. 1993).

Mountain birch seed rain depends both on seed production of each tree and the density of these individuals on the landscape. We have in a previous paper (Sveinbjörnsson et al. 1993) estimated seed rain to be about two orders of magnitude lower at the treeline than in the valley.

Seed production is only the first part of the path to tree establishment, field germination and survival of seedlings is the next step. Age composition studies

(Sonesson and Hoogesteger 1981) indicate that establishment occurrence is sporadic and probably requires a number of favorable years. Presumably, assimilate and nutrient reserves must first reach a minimum pool size, then weather has to allow flower maturation, seed maturation, seed germination and seedling resource acquisition. Hazards such as summer heat and drought as Black and Bliss (1980) have demonstrated for black spruce near the polar treeline in the MacKenzie valley in Canada.

Seeds from the treeline are of lower quality, in terms of the proportion of empty to filled seeds (53% for the valley seeds and 33% for treeline seeds) and also by lighter filled seeds, than lower elevation seeds. As seed germination is positively and linearly dependent on seed weight in filled seeds ($r^2 = 0.30$; $p = 0.03$), decreasing tree establishment could thus both result from increasingly difficult conditions and from decreasing quality of the seed.

While seed germination is uniformly greatest at relatively high temperatures ($>20°C$), the surface temperature regime may not vary much along the gradient (Davis et al. 1990). However, drought may vary significantly and the more exposed treeline site may render seedlings vulnerable to desiccation stress.

Seed germination differs with elevation. The frequency of seed germination is twice as high on average in sixteen valley sites as compared to 12 treeline sites (46% in the valley and 23% at the treeline) and the seedling survival is poor especially at the treeline. During the relative warm and dry summer of 1980 only 2.5% of the seeds germinated within three weeks of snow melt at the treeline on Mount Njulla and none germinated after that time. During the entire summer, 62% of the germinated seeds died, most at the beginning of July, probably from drought. A comparison of mountain birch seed germination in birch forest, snow bed and tree-limit sites near the altitudinal treeline in central Sweden ranged from 6.8% in the forest to 2.1% for the tree-limit (Kullman 1986) again showing elevational decrease in seed germination. In central Sweden, summer mortality at the same sites was 45.1 and 10.9%.

While in central Sweden, summer mortality seems to be lower near the treeline, winter mortality is higher or 27.7% in the birch forest and 75.7% at the tree-limit site (Kullman 1986). On Mount Njulla, conversely, winter treeline survival is highest at the treeline (35%) while it diminishes both above it (10%) and below it (forest limit 20% and valley 12%). Seedling height growth at the end of the summer varied significantly with elevation and ranged from 4.0 mm above treeline (700 m a.s.l.), 4.9 mm at tree line (670 m a.s.l.), 5.6 mm at forest limit (620 m a.s.l.) and 6.8 mm in the valley bottom (at 380 m a.s.l.).

In general then the reproductive process including catkin production, seed rain, seed quality, seed germination, and summer seedling survival all diminish with with increasing elevation, while winter survival does not.

Material balance

Carbon

The carbon metabolism is an important example of the second mechanistic category i.e. material acquisition and utilization, that has received attention. Thus, Tranquillini (1959) demonstrated that *Pinus cembra* trees at the treeline in the Austrian Alps had lower net annual CO_2 gain than trees further down the slope. This was taken to mean that tree growth at the treeline was carbon limited. However, research by Dahl and Mork (1959) indicated instead that respiration, which was correlated with growth, was limited by low temperatures, not by substrate availability. A refinement of the arguments for two mechanistic theories of carbon limitation were later produced by calculating annual CO_2 fixation by black spruce in eastern Canada and demonstrating a certain minimum sum at the treeline (Vowinckel 1975), while in Norway the elevational distribution of Norway spruce was related to certain branch respiration equivalents (Skre 1972). Thus two opposing explanations existed with respect to carbon balance and treeline.

In 1977, a project was initiated by the funding of a proposal from Mats Sonesson, the new director of the Abisko Scientific Research Station, to the Swedish Science Research Council. The project was designed to evaluate the extent to which the photosynthetic process in treeline birches was negatively impacted by the extant conditions there. This was done by comparing the the rates of CO_2 exchange in treeline birch individuals to those at lower elevations.

First the photosynthetic rates of the birches at different elevations were measured. A variety of methods was used to measure branch and leaf net CO_2 gas exchange: simultaneous $^{14}CO_2$ fixation by leaves at different elevations using single leaves of many trees, alternating continuous CO_2 exchange of a few trees in each elevation zone measured from a field laboratory with a site change every 10 days, and alternating measurements on leaves of many trees with portable infrared gas analysis systems with sites measured within hours of each other. Finally, CO_2 exchange of abscised branches was measured under standardized conditions in the laboratory. All these measurements (Sveinbjörnsson unpubl.) showed that trees at different elevations had similar photosynthetic and respiratory rates.

While the rates of net photosynthesis and respiration are similar under similar conditions, different conditions in the different elevational zones would lead to different net CO_2 exchange. So how different are conditions? As expected, given the short distances between the forest border (forest limit) and the uppermost scattered birch individuals or clumps of individuals (treeline), there is little difference in summer air and soil temperature (Sveinbjörnsson 1983). But, leafing out

Table 1. Calculation of seasonal material loss by a late snowstorm at the treeline near Abisko, Swedish Lapland. Assumptions: The example is for a 1 m tall tree with 7 g (dry weight) of leaves (Sveinbjörnsson 1987). The leaves are replaced after 10 d (10 d shorter assimilation season) with 20% lighter leaves (5.6 g). The average daily net CO_2 flux is 50 mg CO_2 g^{-1} dry weight of leaves. The usual season assimilation starts 15 June and ends 10 September and is thus 86 d long, while the shortened season would last 76 d. Retranslocation prior to abscission from the leaves is ignored.

Usual plant	Weight (g)
Gain during the season[1]	19.6 g
Loss of leaves at end of season	7.0 g
Net gain	12.6 g

Damaged plant	Weight (g)
Gain during the season[2]	13.8 g
Loss of leaves at beginning of season	7.0 g
Loss of leaves at end of season	5.6 g
Net gain	1.2 g

[1] (50 mg CO_2 d^{-1} g^{-1})(86 d)(7 g leaves)($\times$ 0.65 g dry wt g $CO_2$$^{-1}$).
[2] (50 mg CO_2 d^{-1} g^{-1})(76 d)(5.6 g leaves)($\times$ 0.65 g dry wt g $CO_2$$^{-1}$).

starts slightly later (c. one week) at treeline than at the forest limit and ends sooner (c. one week). However, a warm period causing leaf-out at low elevations followed by a cooler period inhibiting further leaf out at higher elevations, could increase the difference in the length of the photosynthetic season. Prior to bud break in the spring, terminal stem respiration is high and the time until maximum photosynthetic rate per leaf is achieved depends on leaf expansion rate which is temperature dependent (unpubl.).

Temperature optima for trees at the forest limit and the tree line are similar as are the responses to varying irradiance. The temperature optimum in the late summer-early autumn was 11°C. This optimum is similar to that of treeline snow gum in Australia (Slatyer and Ferrar 1978). The light compensation point was low or c. 10 µmol m^{-2} s^{-1} which is similar to that reported for sugar maple (Hinckley et al. 1978).

Perhaps more important than gaseous CO_2 exchange is tissue turnover. While this turnover may not differ between treeline and forest limit individuals during most years, we have observed greater leaf loss at treeline than at the forest limit, during an early leaf-out in May 1980 which was followed by a snowstorm at the highest elevation. The lost leaves were replaced by a new set of smaller leaves. A rough estimate of the cost of that episode (Table 1) shows that the trees at the treeline may have lost up to 90% of the growing season's carbon assimilation along with other essential nutrients. Hence while summer conditions do not vary greatly between sites, the length of the photosynthetic season may vary and tissue loss increases with elevation relating to the hazard coefficient of Hustich (1979). Thus we see that at the treelines both on-off (leaf loss) and grading processes (reduced carbon gain) affect tree performance.

Although carbon dioxide exchange rates are highest in leaves, all surfaces of the plants exchange carbon dioxide and other gases. We have conducted several studies to quantify the exchange rates of different plant parts (authors' unpubl.). First, the relative surface area of leaves, branches and stems, and coarse and fine roots was measured for birches ranging from 3 cm to 5 m in height (Sveinbjörnsson 1987). These studies demonstrate that a larger part of the Abisko birch is found in roots and leaves than for other subarctic angiosperm trees (Elkington and Jones 1974, Sakai et al. 1979, van Cleve et al. 1983). With increasing tree size there is an increasing difficulty in excavating fine roots and the study therefore most likely underestimated the root component in larger individuals. This study was made on low elevation individuals near the research station and the proportions at higher elevations might be different, probably with an even greater root component.

Next we measured stem, branch, and root respiration rates of birch trees in the field using transparent, water cooled cuvettes. To evaluate the importance of light on stems (the bark contains chlorophyll), stem sections of several trees were measured both under natural light and covered with black non-transparent plastic. No differences were found between carbon dioxide exchange rates of the shaded and lit stems at the same temperature. Stems of differing diameter showed similar rates and rates of coarse roots were similar to those of the stems. Stem and root respiration rate increased exponentially with rising tissue temperature (Fig. 2). There was no interaction between the root respiration rate (manipulated by changing root temperature) and shoot photosynthesis (manipulated by varying photon flux density). Thus, it appears that already in small birch plants, there is no competition for carbon, and that the carbon supply does not limit process rate

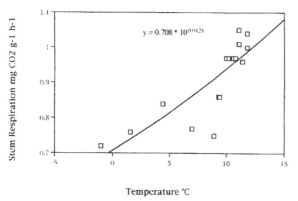

Fig. 2. Temperature relations of respiration of a 4 cm diameter mountain birch stem. The tree was growing at low elevation near the Abisko Scientific Research Station.

Table 2. Tissue nutrient concentration of control trees of mountain birch growing at three elevevations: total non-structural carbohydrate concentration (TNC), N, P and K, all in % of dry weight. Different letters following each compound concentration value are significant ($p < 0.05$) elevational differences. Adopted from Sveinbjörnsson et al. (1992).

Site	Concentration			
	TNC	N	P	K
Tree-line	23.1[a]	2.23[a]	0.25[a]	1.03[a]
Forest limit	21.1[a]	2.21[a]	0.26[a]	1.14[a]
Valley	22.8[a]	1.86[b]	0.20[b]	0.98[b]

arguing against carbon limitation. Root respiration of in situ trees continued for a month after leaf abscission.

While the lack of interaction between above- and below-ground parts in low elevation individuals argues against carbon limitation, it may not apply to high elevation trees. For elevational comparison we analyzed concentration of foliar total non structural carbohydrates in leaves but perhaps unsurprisingly found no differences (Table 2). However, analyses of all plant parts at the onset of winter and beginning of the growing season would be a more unequivocal indicator of the carbon balance. It seems reasonable to assume that carbon is not the major limiting resource during most years but heavy tissue loss or seed set may sometimes require greater than available carbon stores.

Nutrients

While the processes for carbon dioxide exchange have been measured at several treelines, treeline nutrient relations have been much less studied. Unlike the carbon situation however, where supply is fairly constant but the process rate variable, both nutrient availability and acquisition may be important. Substrate availability may be hampered at high elevation by low nitrogen fixation per unit land area because of low frequency of nitrogen fixers and/or low activity rates because of difficult conditions. It is interesting to note that lichen nitrogen fixation has been shown to have a higher temperature optimum than net photosynthesis. Nutrient availability can also be low because of slow turnover due to low decomposer diversity, density or activity, the last one being determined not only by climatic and geological conditions but also by litter quality.

While the literature had considered nutrient limitations, this had been rejected (Tranquillini 1979) because of the high foliar concentrations (see Körner 1989) and often high soil nutrient content (Ehrhardt 1961). However, foliar concentrations are dependent on both uptake rates and dilution by growth. If growth rates decrease more than uptake rates then increased foliar concentrations do not necessarily reflect increased nutrient availability. The only reliable measure of nutrient limitation would then be a comparative response to nutrient addition. Therefore in 1980 and again in 1981, we fertilized trees at treeline, forest limit and valley with nitrogen, phosphorus and a complete fertilizer and monitored the trees in the subsequent years (Sveinbjörnsson et al. 1992).

The results of the fertilizer application experiment (Fig. 3) showed a positive response to nitrogen fertilization, which increased with elevation. No consistent growth increase was found in response to phosphorus fertilizer application. From this we concluded that nitrogen availability was the principal limiting factor and that with elevation it was increasingly so. As in other studies (Körner 1989), we found that tissue nutrient concentrations were higher at higher elevations (Table 2).

Near simultaneous measurements of the net photosynthetic rate of trees at different elevations do not reveal any significant differences and are thus not the likely mechanism of increased growth at high elevation in response to nitrogen. On the other hand leaf size increased in nitrogen fertilized trees and this may have increased growth through increased whole tree photosynthesis, but does not explain the elevational variation in growth response as leaf size increased similarly at all elevations. Thus it seems likely that the elevational variation stems from two sources, stored carbon and shifts in allocation patterns. It is probable that some of the early response may have been affected by stored carbon, but the fact that this elevational fertilizer effect persisted for many years, suggest that allocation may be much more important.

The elevationally increasing nitrogen limitation could be caused by increasing failure of tree nitrogen uptake or by reduced soil nitrogen availability. To evaluate if soil nitrogen availability decreased with elevation and what controlled this decrease, soils at different elevations were sampled and analyzed for extractable nitrogen (Davis et al. 1990). Except for one observation at the treeline, nitrate was not detected in the monthly soil samples, only ammonium. Mean ammonium concentra-

Table 3. Soil ammonium concentrations (means for July 1988 to June 1989) and seasonal ranges (means per zone). Adapted from Davis et al. (1990).

Zone	Mean	Seasonal range
Valley	148 µg g^{-1}	87–218 µg g^{-1}
Forest limit	190 µg g^{-1}	109–328 µg g^{-1}
Treeline	11 µg g^{-1}	3–19 µg g^{-1}

tions from July 1988 to June 1989 were highest at the forest limit, slightly lower in the valley, and significantly lower in the treeline zone, <10% of that in the lower zones (Table 3). While the high forest limit concentrations are surprising, it is worth pointing out that the fertilizer experiments and the soil studies were conducted in different areas; the former at on Mount Luovare at Stenbacken and the latter on Mount Njulla, near Abisko. It appears that while soil nitrogen availability certainly plays a significant role in reducing tree growth at the treeline, differences in tree growth between the forest limit and valley sites may be caused by differences in tree nitrogen acquisition or loss.

What causes the drastically reduced nitrogen availability above the treeline? Mid-season soil temperatures there are higher not lower than those in the valley although they may be lower at the very beginning or end of the season (Table 4). As pointed out above, soil floral and faunal diversity, abundance, and activity rates may all be significantly different (Schinner and Gstraunthaler 1981). The activity rates are partially determined by litter quality and the quality of litter is often said to depend partially on nitrogen concentration. Birch litter from treeline birches decomposed slower when transplanted to the valley sites than did litter from forest limit and valley sites which did not differ. This is in spite of the treeline litter having greater nitrogen concentration than lower elevation litter.

Evergreen conifers dominate at most treelines of the world. Larch, a deciduous conifer is found near the polar treeline in extremely nutrient poor conditions of the Canadian shield and in extremely continental climates of the Siberian polar treeline and the Rocky Mountains elevational treeline. Deciduous angiosperm trees at the treelines include birches and *Choosenia* at

Table 4. Soil temperatures (T) and precipitation from Njulla study sites. Adapted from Davis et al. (1990).

Location	T max (°C)	T min (°C)	Precipitation (cm)
	1988 (2 July to 1 September)		
Valley	8.0	6.8	9.4
Forest limit	16.7	12.0	8.3
Treeline	22.2	7.9	8.0
	1989 (12 July to 1 September)		
Valley	9.0	7.6	8.5
Forest limit	15.1	5.9	11.1
Treeline	16.1	7.5	9.5

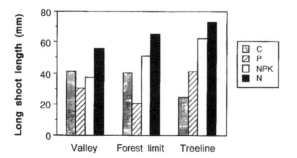

Fig. 3. Total summer height growth of mountain birch treated with different fertilizers. The trees are the same as shown in Fig. 1.

polar and high-latitude elevational treelines in oceanic climates and poplars on south facing slopes at elevational and polar treelines. It is unlikely that the same suite of environmental factors and physiological processes limit all of these tree species but all share shorter growing seasons than trees in the forests. Whether or not this shorter season is in itself the cause of reduced tree growth and the existence of a treeline or whether the short growing season results in differential performance of competing plants which in turn affect soil fertility through their litter quality is not understood. Only experimental manipulation of growing season length can convincingly answer that question.

References

Black, R. A. and Bliss, L. C. 1980. Reproductive ecology of *Picea mariana* (Mill.) BSP., at tree line near Inuvik, Northwest Territories, Canada. – Ecol. Monogr. 50: 331–354.

Cleve, K., van Oliver, L., Schlentner, R., Viereck, L. A. and Dyrness, C. T. 1983. Productivity and nutrient cycling in taiga forest ecosystems. – Can. J. For. Res. 13: 747–766.

Cooper, D. J. 1986. White spruce above and beyond treeline in the Arrigetch Peaks region, Brooks Range, Alaska. – Arctic 39: 247–252.

Dahl, E. and Mork, E. 1959. Om sambandet mellom temperatur, andning, og vekst hos gran (*Picea abies* (L.) Karst.). – Medd. norske Skogforsøksves. 16: 81–93.

Davis, J., Schober, A., Bahn, M. and Sveinbjörnsson, B. 1990. Soil carbon turnover at and below the elevational treeline in northern Fennoscandia. – Arct. Alp. Res. 23: 279–286.

Ehrhardt, F. 1961. Untersuchungen über den Einfluss des Klimas auf die Stickstoffnachlieferung von Waldhumus in verschiedenen Höhenlagen der Tiroler – Alpen. – Forstwiss. Zbl. 80: 193–215.

Elkington, T. T. and Jones, M. G. 1974. Biomass and primary production of birch (*Betula pubescens* s. lat.) in south-west Greenland. – J. Ecol. 62: 821–830.

Goldstein, G. H. 1981. Ecophysiological and demographic studies of white spruce (*Picea glauca* (Moench) Voss) at treeline in the central Brooks Range of Alaska. – Ph.D. thesis, Univ. of Washington.

Graves, J. D. and Taylor, K. 1988. A comparative study of *Geum rivale* L. and *G. urbanum* L. to determine those factors controlling their altitudinal distribution. – New Phytol. 108: 297–304.

Hinckley, T. M., Aslin, R. G., Aubuchon, R. R., Metcalf, C. L. and Roberts, J. E. 1978. Leaf conductance and photosynthesis in 4 species of the oat hickory forest type. – For. Sci. 24: 73–84.

Hustich, I. 1979. Ecological concepts and biogeographical zonation in the North: the need for a generally accepted terminology. – Holarct. Ecol. 2: 208–217.

Koike, T. 1987. The growth characteristics in Japanese mountain birch (*Betula ermanii*) and white birch (*Betula platyphylla* var. *japonica*), and their distribution in the northern part of Japan. – In: Fujimori, T. and Kimura, M. (eds), Human impacts and management of mountain forests. Japan, Forestry and Forest Products Research Institute, pp. 189–220.

Körner, C. 1989. The nutritional status of plants from high altitudes: A worldwide comparison. – Oecologia 81: 379–391.

Kullman, L. 1986. Demography of *Betula pubescens* ssp. *tortuosa* sown in contrasting habitats close to the birch tree-limit in central Sweden. – Vegetatio 65: 13–20.

Lipscomb, M. V. and Nilsen, E. T. 1990. Environmental and physiological factors influencing the natural distribution of evergreen and deciduous ericaceous shrubs on northeast- and southwest-facing slopes of the southern Appalachian Mountains. II. Water relations. – Am. J. Bot. 77: 517–526.

Noshiro, S. and Suzuki, M. 1989. Altitudinal distribution and tree form of *Rhododendron* in the Barun Valley, east Nepal. – J. Phytogeogr. Taxon., 37: 121–127.

Osmond, C. B., Austin, M. P., Berry, J. A., Billings, W. D., Boyer, J. S., Dacey, J. W. H., Nobel, P. S., Smith, S. D. and Winner, W. E. 1987. Stress physiology and the distribution of plants. – Bioscience 37: 38–48.

Payette, S. 1983. The forest tundra and present tree-lines of the northern Québec-Labrador Peninsula. – Nordicana 47: 3–23.

– and Gagnon, R. 1979. Tree-line dynamics in Ungava peninsula, northern Québec. – Holarct. Ecol. 2: 239–248.

Sakai, A., Yoshida, S. and Saito, M. 1979. Biomass and productivity of *Betula papyrifera* near its climatic limit in northwestern Canada. – Low Temp. Sci. Ser. B 37: 33–38.

Schinner, F. and Gstraunthaler, G. 1981. Adaptation of microbial activities to environmental conditions in alpine soils. – Oecologia 50: 113–116.

Scott, P. A., Bentley, C. V., Fayle, D. C. F. and Hansell, R. I. C. 1987a. Crown forms and shoot elongation of white spruce at the treeline, Churchill, Manitoba, Canada. – Arct. Alp. Res. 19: 175–186.

–, Hansell, R. I. C. and Fayle, D. C. F. 1987b. Establishment of white spruce populations and responses to climatic change at the treeline, Churchill, Manitoba, Canada. – Arct. Alp. Res. 19: 45–51.

Skre, O. 1972. High temperature demands for growth and development in Norway spruce (*Picea abies* (L.) Karst.) in Scandinavia. – Medd. Norg. Landbrukshøgskole 51: 1–29.

Slatyer, R. O. and Ferrar, P. J. 1978. Photosynthetic characteristics of treeline populations of the Australian snowgum *Eucalyptus pauciflora*. – Photosynthetica 12: 137–144.

Sonesson, M. and Hoogesteger, J. 1981. Recent tree-line dynamics (*Betula pubescens* Ehr. ssp. *tortuosa* [Ledeb.] Nyman) in northern Sweden. – Nordicana 47: 47–54.

Sveinbjörnsson, B. 1983. Bioclimate and its effect on the carbon dioxide flux of mountain birch (*Betula pubescens* EHRH.) at its altitudinal tree-line in the Torneträsk area, northern Sweden. – Nordicana 47: 111–122.

– 1987. Biomass proportioning as related to plant size in juvenile mountain birch near Abisko, Swedish Lapland. – Rep. Kevo Subarctic Res. Stat. 20: 1–8.

–, Nordell, O. and Kavhanen, H. 1992. Nutrient relations of mountain birch at and below the elevational tree-line in Swedish Lapland. – Funct. Ecol. 6: 213–220.

–, Sonesson, M., Nordell, O. K. and Karlsson, S. P. 1993. Performance of mountain birch in different environments in Sweden and Iceland: Implication for afforestation. Forest development in cold climates. – In: Alden, J. and Mastrantonio, L. (eds), Plenum Press, New York, pp. 79–88.

Tranquillini, W. 1959. Die Stoffproduktion der Zirbe (*Pinus cembra* L.) an der Waldgrenze während eines Jahres. – Planta 54: 107–151.

– 1979. Physiological ecology of the alpine timberline. Tree existence at high altitudes with special references to the European Alps. – Springer.

Vowinckel, T. 1975. The effects of climate on the photosynthesis of *Picea mariana* at the subarctic tree line. – Ph.D. thesis, McGill Univ., Montreal.

Ecological Bulletins 45: 71–78. Copenhagen 1996

Relationships between nitrogen economy and performance in the mountain birch *Betula pubescens* ssp. *tortuosa*

P. Staffan Karlsson and M. Weih

Karlsson, P. S. and Weih, M. 1996. Relationships between nitrogen economy and performance in the mountain birch *Betula pubescens* spp. *tortuosa*. – Ecol. Bull. 45: 71–78.

Relationships between mountain birch, *Betula pubescens* ssp. *tortuosa*, nitrogen economy and performance are discussed based on results mainly from northern Fennoscandia. As is true for most green plants, there is a close relationship between mountain birch seedling growth rate and nitrogen status. In addition to soil nitrogen availability, soil temperature is an important determinant of nitrogen uptake rate and thus of plant nitrogen status and growth. At a soil temperature of 5°C rates of both root nitrogen uptake and growth are close to zero regardless of soil N availability, and above 15°C growth increases only marginally. These results indicate that birch performance could be greatly affected by a changed temperature climate. Results from experiments indicate that the optimum leaf nitrogen concentration with respect to leaf productivity is c. 1.5 mmol N g^{-1}. This value is slightly higher than the mean leaf nitrogen concentration in situ (1.24 mmol N g^{-1}). The major causes of plant nitrogen losses are leaf abscission followed by fine root turnover and herbivory. Losses due to reproduction are marginal during most years, but can occasionally be large.

P. S. Karlsson, Abisko Scientific Research Station, S-981 07 Abisko, Sweden and Dept of Ecological Botany, Univ. of Uppsala, Villavägen 14, S-752 36 Uppsala, Sweden. – M. Weih, Dept of Ecological Botany, Univ. of Uppsala, Villavägen 14, S-752 36 Uppsala, Sweden.

Plant nitrogen economy is commonly the single most important factor limiting the growth of wild plants (Ågren 1985, Field and Mooney 1986). Thus the determinants of plant nitrogen economy and the relationships between nitrogen economy and plant growth are important factors influencing plant performance. Plant nitrogen status affects growth through several processes. One of these involves the carbon dioxide assimilating enzyme, RuBP-carboxylase, which represents the largest nitrogen fraction in a plant and affects the photosynthetic capacity of the leaves (Björkman 1981, Evans 1989). Plant nitrogen status also affects the pattern of dry matter allocation in the plant, mainly that between root and leaf growth (Hirose 1986, 1987). These factors tend to enhance the growth rate of plants as plant N concentration increases.

In the study of plant resource economy, processes by which resources are acquired, such as photosynthesis and the uptake of nutrients by roots, have commonly been studied more thoroughly than the loss compo-

nents and their determinants. However, under nutrient-limiting conditions it may be more important to minimise losses and to efficiently recycle internal resources than to maximise resource gain (Chapin et al. 1993).

At northern latitudes low temperatures restrict plant growth (Chapin 1983, Körner and Larcher 1988). The effects of low temperatures on growth and nutrient uptake can be direct or indirect. In the latter case, the microbial decomposition of litter is retarded, thus slowing down the release of soil nutrients.

In this paper we will attempt to evaluate the relative importance of different environmental factors as well as intrinsic characteristics as determinants of the nitrogen economy and growth performance of the mountain birch. This will be done by combining the results from experimental work on seedlings and in situ studies of mature mountain birch trees. Aspects of birch performance related to tree-line dynamics are discussed by Sveinbjörnsson et al. (1996).

The species

The taxonomic status of the mountain birch has been disputed (cf. e.g. Vaarama and Valanne 1973, Kallio and Mäkinen 1978). Some classify it as a distinct species (*B. tortuosa* [Ledeb.]) but most commonly it is regarded as a subspecies of *B. pubescens* then called *B. p.* ssp. *tortuosa* (Ledeb.) Nyman, or *B. p.* ssp. *czerepanovii* (Orlova) (Hämet-Ahti 1987). It is suggested to have arisen through introgressive hybridisation between *Betula pubescens* and *B. nana* (Elkington 1968). *Betula pubescens* ssp. *tortuosa* is supposed to be the northern element of the same subspecies as the central European *B. p.* ssp. *carpatica* (Gardiner 1984). High variability among mountain birch individuals with respect to both morphological and physiological characteristics has been noted by many authors (Elkington 1968, Vaarama and Valanne 1973, Kallio et al. 1983, Karlsson and Nordell 1988, Karlsson 1991, Suomela and Ayres 1994). The birch forest and vegetation types of the Torneträsk area are described by Sonesson and Lundberg (1974).

Determinants of seedling growth performance

As previously shown for many other plants, the relative growth rate (RGR) of mountain birch seedlings is closely correlated with their nitrogen status (Fig. 1). Since RGR is determined by the leaf productivity (unit leaf rate, ULR) and the leaf area per unit biomass (L_A/W), RGR = ULR $*$ L_A/W (Causton and Venus 1981), the proximate causes for variation in growth rate should be found in these two characteristics. For experimentally grown mountain birch seedlings both

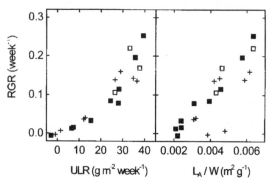

Fig. 2. Relative growth rate (RGR) of experimentally grown mountain birch seedlings as related to unit leaf rate (ULR) and leaf area partitioning (L_A/W). Symbols as in Fig. 1. L_A/W is the mean for the initial and final harvest, i.e. ($L_A/W_{t0} + L_A/W_{t1}$)/2.

these factors contribute to the variation observed in RGR, although ULR is more important (Fig. 2). The variation in ULR and L_A/W are, in turn, related to variations in plant nitrogen status and temperature (ULR, Fig. 3; $L_A/W = 1.46 \times 10^{-4}$ T $- 2.47 \times 10^{-3}$ N $- 5.47 \times 10^{-4}$, where N is leaf nitrogen concentration and T is soil temperature, $r^2 = 0.75$).

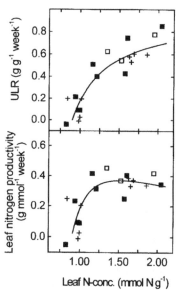

Fig. 3. The unit leaf rate (ULR) as related to plant nitrogen status. Symbols as in Fig. 1. The leaf nitrogen concentration is the mean for the initial and final harvests (cf. Fig. 1). The ULR-leaf nitrogen relationship is fitted to a hyperbolic function that approaches zero as leaf N approaches 0.9 and has a maximum ULR of c. 0.94 (or 0.85/0.9), viz., ULR = 0.85/ (0.9 + 0.38/(N − 0.9)), ($r^2 = 0.75$) where N is leaf nitrogen concentration. Another 5% of the variation in ULR could be attributed to temperature. The nitrogen use efficiency function (ULR/N) is the ULR function divided by the nitrogen concentration.

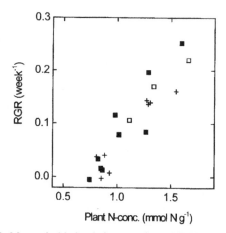

Fig. 1. Mountain birch relative growth rate (RGR) as related to plant nitrogen status. Data from Karlsson and Nordell (1987, the 1984 data, symbol: +), Karlsson and Nordell (1989) (□) and Karlsson and Nordell (1996) (■). The nitrogen status is the mean concentration for the initial (t0) and final (t1) harvest, i.e. (N-conc$_{t0}$ + N-conc$_{t1}$)/2.

ECOLOGICAL BULLETINS 45, 1996

Also, the biomass production per unit leaf nitrogen increases with increasing leaf nitrogen concentration until an optimum point is reached at c. 1.5 mmol N g^{-1} (Fig. 3). In contrast, the leaf CO_2 assimilation rate per unit nitrogen shows an inverse relationship to leaf nitrogen status (Karlsson 1991). The difference between CO_2 assimilation and leaf biomass production per unit N is due to plant respiration (Lambers et al. 1989). Although the CO_2 assimilation per unit nitrogen is high at low leaf nitrogen concentrations, a large fraction of the assimilates seems to be consumed for respiration, leaving nothing or very little for net growth.

Thus, for experimentally grown birch seedlings the most important determinant of growth rate is plant nitrogen status, mainly through its effect on ULR, but also through the effect on dry matter partitioning. The relatively large effect of leaf nitrogen concentration on ULR is, however, partly dependent on the experimental conditions. The higher the nutrient levels used (assuming they result in higher leaf N concentrations) the relatively more important the allocation pattern (L_A/W) becomes as compared with ULR (cf. Hirose 1984, 1986).

Some comments on the applied methods

The patterns discussed above are based on the results from three experiments using mountain birch seedlings c. 2–4 yr old. The growth rate has been computed as the mean RGR for the whole season (c. 10–12 wk). The mean size at final harvest ranged from c. 25 mg to 2 g. The environmental conditions varied both during and between experiments. Light conditions were natural except for two treatments where the natural light was reduced by 50%; two of the experiments used natural temperature conditions, while in one experiment soil temperature was controlled (set to 5, 10 or 15°C). In most cases silica sand was used as substrate, but for three treatments the plants were grown in a organic soil from a birch forest. For plants grown on a sand substrate, the nutrient addition rate varied by a factor of c. 300 (cf. Karlsson and Nordell 1987, 1989 for details of the composition of the nutrient solution), while those grown on natural soil received distilled water only.

The growth analysis technique employed here is usually used for short-term experiments (days or a few weeks) under controlled conditions (cf. Ingestad and Lund 1979, Beadle 1985). For example, Hirose (1984) applied the technique under natural light and temperature conditions. Long time spans between harvests may result in poor relationships between different growth parameters (DeLucia et al. 1992). The results presented here indicate, however, that relatively close relationships can be obtained between growth characteristics and their determinants despite the long time span between harvests and highly variable environmental conditions during the experiments. It should be borne in mind that the purpose of these experiments was to study relationships between the growth, allocation pattern and nitrogen status of the mountain birch in an ecological rather than physiological context.

Determinants of plant nitrogen economy

The growth performance of mountain birch seedlings, as discussed above, can be expected to be closely related to plant nitrogen status. In the following section we will discuss the various components determining the nitrogen status of the mountain birch. First, we will discuss the factors determining nitrogen incomes and thereafter, the major causes of nitrogen acquisition will be considered.

Nitrogen acquisition: root system size and nutrient uptake rate

The ability of a plant to acquire nitrogen is dependent on at least three plant characteristics: 1) the size of the root system, 2) the root system design or architecture and 3) the N uptake efficiency of the roots (including the effect of mycorrhizal infection). For experimentally grown seedlings there is an inverse relationship between the root proportion of total plant biomass (W_R/W) and plant nitrogen status (e.g. Karlsson and Nordell 1996). Although the proportion of growth resources allocated to seedling roots is high under low-nitrogen conditions, the seedling's growth rate is very low; thus, the rate at which the plant can increase its foraging capacity is also low. No information is available about the spatial pattern of mountain birch root foraging for nutrients.

Root nutrient uptake is affected by soil temperature both directly and indirectly. At a soil temperature of 5°C birch nutrient uptake and growth is close to zero (16% of that at 10°C) even if the soil nutrient content is high (Karlsson and Nordell 1996). At a soil temperature of 15°C the N uptake rate is c. 50% higher than at 10°C and almost 10 times higher than at 5°C. The indirect effect is a result of the influence of temperature on the microbial decomposition of soil litter (Van Cleve et al. 1981, 1990).

We are not aware of any studies evaluating the role of mycorrhiza in nutrient uptake and growth in the mountain birch. Results suggesting that mycorrhiza may be important for seedling growth were presented by Magnusson and Magnusson (1992). On severely disturbed Icelandic soils they found that birch seedling size was inversely correlated with the distance to *Salix* shrubs. This relationship was assumed to be due to the higher frequency of mycorrhiza-forming fungi present

Table 1. Causes of nitrogen loss (%) in experimentally grown birch seedlings (Karlsson and Nordell 1987) and c. 30-yr-old trees in situ (Karlsson et al. unpubl.). For seedling data, values are means for four treatment types for each nutrient level (standard deviations are given within parentheses). The data for in situ trees are ranges for three trees over five years estimated by sampling above-ground fractions and computing fine root turnover through the decline rate of a ^{15}N label.

	Experimentally grown seedlings Fertiliser level		C. 30-yr-old trees in situ
	Low	High	
Leaf abscission	34.5	53.1	38
(% of leaf N)	(6.8)	(19.2)	(range 25–55)
Leaf abscission	7.0	15.8	12
(% of whole plane N)	(2.5)	(5.6)	
Leaf herbivory [1]	–	–	2
(% of whole plant N)			
Fine root turnover	?	?	9
(% of whole plant N)			
Reproduction	0	0	0.2
(% of whole plant N)			

[1] Includes leaf chewers only, i.e. a consumption of 6% of leaf biomass (cf. Koponen 1983).

in the neighbourhood of the *Salix* shrubs. Mason et al. (1984) reported that more mycorrhizal fungi were associated with *B. pendula* roots than with *B. pubescens* roots. Studies of *Betula pendula* (Frankland and Harrison 1985, Newton and Pigott 1991) indicate that growth only increases marginally as a result of mycorrhizal infection. Newton (1991) found a correlation between the total number of mycorrhizal tips and *B. pendula* growth, but concluded that there was no consistent relationship between mycorrhizal infection and seedling growth.

Nitrogen loss components

Fine-root dynamics
As for many plant species the below-ground characteristics of the mountain birch are poorly known. In the course of harvesting potted seedlings (inoculated with mycorrhiza) we have noted that the position of most of the fine-root mass in the soil changes between June and August (Weih and Karlsson unpubl.). This suggests that a large part of all fine roots are replaced during the summer.

Leaf abscission
Under experimental conditions birch seedlings lose 35–55% of their leaf nitrogen at leaf abscission (Table 1). The proportion of the plant nitrogen pool lost was more than twice as high in plants given a large supply of fertiliser (16%) as compared with plants given a low supply (7%). By contrast, the evergreen *Pinus sylvestris* loses c. 1% of its nitrogen pool through needle abscis-

sion under similar conditions (Karlsson and Nordell 1987). The lower proportion of the plant nitrogen pool lost by nutrient-stressed plants is partly due to the lower proportion of leaf mass in these plants and partly to more efficient resorption (Table 1).

For mature trees in situ, an average 60–80% of the leaf nitrogen pool is resorbed before abscission (Nordell and Karlsson 1995 and Table 1). There are, however, considerable differences among tree individuals. For instance, trees with low leaf nitrogen concentrations sustain smaller relative losses (Nordell and Karlsson 1995). Climate also influences abscission losses, nitrogen is resorbed more efficiently after cold summers than after warm ones.

Reproduction
The N cost of seed production is, on average, relatively small (< 5% of the plant above-ground N pool). Occasionally the N investment in seed production is of the same magnitude as the amount lost in leaf abscission (Nordell pers. comm.). We know of no reports describing seed mast years in the mountain birch, but for *B. alleghaniensis* and *B. papyrifera*, Gross (1972) reported a massive seed production event and described its consequences for growth.

An average mountain birch stem in the Abisko valley invests c. 46 mg N in male catkins (computations based on c. 200 stems with a mean base diameter of 98 mm, from data in Nordell 1978). For c. 30-yr-old stems (base diameter c. 50 mm) the mean N investment in male catkins is c. 20 mg N, which is about half the magnitude of the investment in female catkins (42 mg N). In comparison, the loss due to leaf abscission is c. 2–3 g N (means for 3 trees over 5 yr; Karlsson et al. unpubl.).

As indicated above, the mountain birch reproductive effort is low during most years; then its consequences for resource economy and plant performance could also be expected to be small. At the shoot level, however, local costs of seed production have been observed for *Betula pendula* (Tuomi et al. 1982). Growth of the female inflorescence of both the mountain birch and *B. pendula* does not seem to depend much on the resources allocated by nearby foliage, i.e. removal of the short-shoot leaves resulted in only c. 15–20% reduction in inflorescence mass (Tuomi et al. 1988a, 1989). In *Betula pendula* the carbohydrates required for inflorescence growth seem to be supplied from reserves in branches. For instance, the carbohydrate content of the branch wood decreased by c. 50% during catkin growth (Sauter and Ambrosius 1986). Some carbohydrates are probably also obtained by photosynthetic assimilation in the catkins themselves. Lehtilä et al. (1994) found no evidence of costs associated with male reproduction in terms of shoot dynamics of the mountain birch in northern Finland. A local cost of male reproduction in terms of a reduced bud production and future male catkin production, have however been found in some, but not all, mountain birch individuals studied at Abisko (Karlsson et al. unpubl.).

Herbivory

At approximately 10-yr intervals the mountain birch forest is subjected to more or less substantial foliage losses caused by *Eperrita autumnata* larvae (Tenow 1972). On the stand level, severe outbreaks have long-lasting effects (Kallio and Lehtonen 1975, Lehtonen 1987, Tenow 1996). The time required for the shoot population to recover following the 1955–1956 outbreak in the Abisko valley has been estimated to be c. 75 yr (Bylund 1995). The time needed for recovery after such a massive defoliation event may be determined not only by a resource limitation but also by the time required to produce new stems that can hold leaf-bearing shoots i.e. new meristems.

After 100% defoliation, it takes 2–3 yr for the tree leaf area and leaf nitrogen concentration to return to pre-defoliation levels (Hoogesteger and Karlsson 1992). Wood formation, measured as annual ring widths, however, is depressed longer. Thus, 3 yr after defoliation no sign of ring width recovery could be observed for trees exposed to 100 or 50 + 100% defoliation (i.e. 50% + 100% defoliation during two subsequent years).

The new foliage emerging a few weeks after defoliation had a higher photosynthetic capacity than the foliage on undefoliated trees which partly compensated for the loss of foliage. The increased photosynthetic rate was due to the elevated leaf nitrogen content of the new foliage. Hoogesteger and Karlsson (1992) found that from one year after defoliation and onwards the nitrogen content remained normal or below normal. In such a situation no compensatory effect on photosynthesis from increased leaf nitrogen content would be expected. However, Prudhomme (1982) reported a case in which photosynthetic capacity remained increased for several years after a defoliation event.

After defoliation by *Eperrita* herbivory, mountain birch leaves were observed to increase their leaf phenol content by almost 40%, from c. 9 to 12.5% of leaf dry matter (Tuomi et al. 1984). Phenol production returned to normal values within c. 4 yr. There seems to be little or no trade-off between this defoliation-induced phenol production and twig growth (Haukioja et al. 1985). The increased phenol production after defoliation has been interpreted as a mechanism for reducing plant carbohydrate levels in response to a nitrogen deficiency rather than an action of defence (Tuomi et al. 1984, 1988b, 1990). However, wood formation is also strongly affected by defoliation and severe herbivory (Eckstein et al. 1991, Hoogesteger and Karlsson 1992). Wood formation probably requires relatively more carbon and less nitrogen than most other plant tissues (cf. Chapin 1989). This observation does not support the assumption of a carbon surplus and nitrogen deficiency during the recovery (cf. above).

It has been estimated that during years with low *Epirrita* densities, leaf chewing insects consume 6% of the total leaf biomass while other insects, such as gall inhabitants and miners, consume another 9% (Haukioja and Koponen 1975, Koponen 1983).

Integrating mountain birch nitrogen economy

For mountain birch nitrogen economy, soil temperature seems to be as important as soil nutrient status for nitrogen acquisition. These results imply that an increase in the summer temperature conditions would result in more vigorous birch growth and possibly the colonisation of new areas, e.g. higher altitudes. A soil temperature induced increase in nutrient release may however be only a transient effect (Bonan and Van Cleve 1992).

Although some components of mountain birch nitrogen economy have yet to be investigated it seems clear from the review above that there are two major causes for nitrogen loss: leaf abscission seems to be the most important followed by fine-root turnover (Table 1). Losses to herbivores are probably relatively small in most years; however, the most severe *Epirrita* population peaks can have long-lasting consequences. Similarly, reproduction does not seem to be a major sink for nitrogen during most years, although substantial investments in reproduction may occur in some individuals or years. To our knowledge no information is available on the extent to which below-ground herbivores feed on the mountain birch. Other potential causes for resource losses are fungal infections. The rust fungus *Melampsoridium betulinum* (cf. Poteri 1992) heavily infects the mountain birch during some years but its effects are, to our knowledge, not investigated. Another fungal infection, the witch broom *Taphrina*, can cause substantial reduction in growth (Spanos and Woodward 1994).

All loss components together seem to cause a mean annual loss of c. 24% of the tree nitrogen pool, i.e. a turnover rate of 0.24. This estimate is based on the turnover rate of a ^{15}N label in three trees over a three-year period (Table 1).

Does the mountain birch optimise nitrogen utilisation?

Due to the relatively large variation in the relationship between ULR and leaf nitrogen concentration, the uncertainty concerning the optimum concentration, indicated in Fig. 3, is relatively large. However, the optimum seems to be close to 1.5 mmol N g^{-1}. For the experimentally grown seedlings, those in 8 treatments (out of 21) clearly had suboptimal nitrogen concentrations. These plants had growth rates of 4% week^{-1} or less. Under such conditions one year of root growth will not produce the amount of roots needed to increase the N incomes in any substantial way. The mean leaf nitrogen concentration for mature trees in situ is 1.24 mmol N g^{-1} (Fig. 4), i.e. slightly less than the optimum concentration for leaf productivity (Fig. 3). If the data in Figs 3 and 4 are combined a mean nitrogen use efficiency in situ of 92% of that at optimum leaf nitrogen concentration is indicated.

Although the birch leaf nitrogen concentration seems close to the optimal for leaf productivity, other factors may also be involved in determining the optimum in situ. Most importantly, *Epirrita* larvae perform better on nitrogen-rich leaves (Ayres et al. 1987). Furthermore, the mountain birch investment in chemical defence decreases with increasing plant N-status (Tuomi et al. 1984, 1988b, Haukioja et al. 1985). Hence, losses to herbivores may be larger for trees with nitrogen-rich leaves than for trees with nitrogen-poor leaves. Thus, one strategy of minimising losses to herbivores could be to lower the nutritional quality of the plant (cf. Haukioja et al. 1991).

Is it important to optimise nitrogen use and growth rate in situ?

This problem may differ between age classes. First, regarding seedling performance, survival has been shown to be related to size (Kullman 1986). During the seedling's first year(s) a high growth rate could thus be expected to enhance its probability of survival. Later, saplings (i.e. plants of a size c. 0.2–1.5 m) are often exposed to winter herbivory by hares and moose (cf. Ruosi 1990, Karlsson unpubl.) Thin birch twigs have a higher nutritional value than thicker ones, although they also contain more phenols (Palo et al. 1992). Thus

for saplings as well a high growth rate decreases the time spent at a vulnerable, pre-reproductive size class. Although growth rate per se may be less important for full grown trees, high carbon incomes may allow trees to more efficiently forage for nutrients since they can establish a more extensive fine-root system. The ability to compete for soil resources could also be expected to increase with fine-root production.

Future research directions

The most obvious lack of knowledge with respect to the resource economy of the mountain birch concerns below-ground processes. The size of the fine-root system, its turnover rate and distribution in the soil, and its interaction with fungi are all factors that need further study in order to better understand mountain birch performance in situ.

Regarding the large intraspecific variation in leaf nitrogen status and other characteristics, several questions could be raised. For example: Do trees with consistently low and high leaf nitrogen concentrations vary with respect to performance, e.g. in terms of nutrient use efficiency, growth rate or susceptibility to herbivores? Do such differences result in selection for particular N-allocation types? Alternatively, do different types vary in their success, e.g., over time, thus maintaining a high variability in the population?

Acknowledgements – This work has been supported by the Swedish Natural Sciences Research Council and the Abisko Research station.

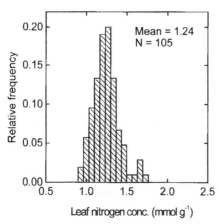

Fig. 4. Frequency distribution of leaf nitrogen concentration for c. 40 mountain birch individuals sampled once per year (in late July or early August) during four different years. Data from Nordell and Karlsson (1995).

References

Ågren, G. I. 1985. Limits to plant production. – J. Theor. Biol. 113: 89–92.
Ayres, M. P., Suomela, J. and MacLean, S. F. Jr, 1987. Growth performance of *Epirrita autumnata* (Lepidoptera: Geometridae) on mountain birch: trees, broods and tree × brood interactions. – Oecologia 74: 450–457.
Beadle, C. L. 1985. Plant growth analysis. – In: Coombs, J., Hall, D. A., Long, S. P. and Scurlock, J. M. O. (eds), Techniques in bioproductivity and photosynthesis. Pergamon Press, Oxford, pp. 20–25.
Björkman, O. 1981. Responses to different quantum flux densities. – In: Encyclopedia of Plant Physiology NS. Vol. 12A, pp. 57–107.
Bonan, G. B and Van Cleve, K. 1992. Soil temperature, nitrogen mineralization, and carbon source sink relationships in boreal forests. – Can. J. For. Res. 22: 629–639.
Bylund, H. 1995. Long-term interactions between the autumnal moth and mountain birch: the roles of resources, competitors, natural enemies, and weather. – Ph.D. thesis, Dept of Entomology, Swedish Univ. Agricult. Sci.
Chapin, F. S. III, 1983. Direct and indirect effects of temperature on arctic plants. – Polar Biol. 2: 47–52.
– 1989. The cost of tundra plant structures: Evaluation of concepts and currencies. – Am. Nat. 133: 1–19.
–, Autumn, K. and Pugnaire, F. 1993. Evolution of suites of traits in response to environmental stress. – Am. Nat. 142: S78–S92.

Causton, D. R. and Venus, J. C. 1981. The biometry of plant growth. – Edward Arnold, London.

Delucia, E. H., Heckathorn, S. A. and Day, T. A. 1992. Effects of soil temperature on growth, biomass allocation and resource acquisition of *Andropogon gerardii* Vitman. – New Phytol. 120: 543–549.

Eckstein, D., Hoogesteger, J. and Holmes, R. L. 1991. Insect-related differences in growth of birch and pine at northern treeline in Swedish Lapland. – Holarct. Ecol. 14: 18–23.

Elkington, T. T. 1968. Introgressive hybridization between *Betula nana* L. and *B. pubescens* Ehrh. in north-west Iceland. – New Phytol. 67: 109–118.

Evans, G. I. 1989. Photosynthesis and nitrogen relationships in leaves of C3 plants. – Oecologia 78: 9–19.

Field, C. and Mooney, H. A. 1986. The photosynthesis-nitrogen relationship in wild plants. – In: Givnish, T. (ed.), On the economy of plant form and function. Cambridge Univ. Press, Cambridge, pp. 25–55.

Frankland, J. C. and Harrison, A. F. 1985. Mycorrhizal infection of *Betula pendula* and *Acer pseudoplatanus*: Relationships with seedling growth and soil factors. – New Phytol. 101: 133–151.

Gardiner, A. S. 1984. Taxonomy of intraspecific variation in *Betula pubescens* Ehrh., with particular reference to the Scottish highlands. – Proc. Royal Soc. Edinburgh 85B: 13–26.

Gross, H. L. 1972. Crown deterioration and reduced growth associated with excessive seed production by birch. – Can. J. Bot. 50: 2431–2437.

Hämet-Ahti, L. 1987. Mountain birch and mountain birch woodland in NW Europe. – Phytocoenologia 15: 449–453.

Haukioja, E. and Koponen, S. 1975. Birch herbivores and herbivory at Kevo. – In: Wielgolaski, F. E. (ed.), Fennoscandian tundra ecosystems part 2: Animal and system analysis. Springer, Berlin, pp. 181–188.

– , Niemelä, P. and Siren, S. 1985. Foliage phenols and nitrogen in relation to growth, insect damage, and ability to recover after defoliation, in the mountain birch *Betula pubescens* ssp. *tortuosa*. – Oecologia 65: 214–222.

– , Ruohomaki, K., Suomela, J. and Vuorisalo, T. 1991. Nutritional quality as a defense against herbivores. – For. Ecol. Manage. 39: 237–245.

Hirose, T. 1984. Nitrogen use efficiency in growth of *Polygonum cuspidatum* Sieb. et Zucc. – Ann. Bot. 54: 695–704.

– 1986. Nitrogen uptake and plant growth II. An empirical model of vegetative growth and partitioning. – Ann. Bot. 58: 487–496.

– 1987. A vegetative plant growth model: adaptive significance of phenotypic plasticity in matter partitioning. – Funct. Ecol. 1: 195–202.

Hoogesteger, J. and Karlsson, P. S. 1992. Effects of defoliation on radial stem growth and photosynthesis in the mountain birch (*Betula pubescens* ssp. *tortuosa*). – Funct. Ecol. 6: 317–323.

Ingestad, T. and Lund, A.-B. 1979. Nitrogen stress in birch seedlings. I. Growth technique and growth. – Physiol. Plant. 45: 137–148.

Kallio, P. and Lehtonen, J. 1975. On the ecocatastrophe of birch forests caused by *Oporinina autumnata* (Bkh.) and the problem of reforestation. – In: Wielgolaski, F. E. (ed.), Fennoscandian tundra ecosystems part 2: Animal and system analysis. Springer, Berlin, pp. 174–180.

– and Mäkinen, Y. 1978. Vascular flora of Inari, Lapland. 4. Betulaceae. – Rep. Kevo Subarctic Res. Stat. 14: 38–63.

– , Niemi S. and Sulkinoja, M. 1983. The Fennoscandian birch and its evolution in the marginal forest zone. – Nordicana 47: 101–110.

Karlsson, P. S. 1991. Intraspecific variation in photosynthetic nitrogen use efficiency in the mountain birch (*Betula pubescens* ssp. *tortuosa*). – Oikos 60: 49–54.

– and Nordell, K. O. 1987. Growth of *Betula pubescens* and *Pinus sylvestris* seedlings in a subarctic environment. – Funct. Ecol. 1: 37–44.

– and Nordell, K. O. 1988. Intraspecific variation in nitrogen status and photosynthetic capacity within mountain birch populations. – Holarct. Ecol. 11: 293–297.

– and Nordell, K. O. 1989. Effects of leaf duration, nutrient supply, and temperature on the seasonal pattern of growth and nitrogen uptake in tree seedlings in a subarctic environment. – Can. J. Bot. 67: 211–217.

– and Nordell, K. O. 1996. Growth of *Betula pubescens* ssp. *tortuosa* and *Pinus sylvestris* seedlings in relation to soil temperature. – Ecoscience 3, in press.

Koponen, S. 1983. Phytophagus insects of birch foliage in northernmost woodlands of Europe and eastern North America. – Nordicana 47: 165–176.

Körner, C. and Larcher, W. 1988. Plant life in cold climates. – In: Long, S. F. and Woodward, F. I. (eds), Plants and temperature. Symp. Exp. Biol. Vol. 42, Cambridge, pp. 25–57.

Kullman, L. 1986. Demography of *Betula pubescens* ssp. *tortuosa* sown in contrasting habitats close to the birch tree-limit in central Sweden. – Vegetatio 65: 13–20.

Lambers, H., Cambridge, M. L., Konings, H. and Pons, T. L. (eds) 1989. Causes and consequences of variation in growth rate and productivity of higher plants. – SPB Acad. Publ., The Hague.

Lehtilä, K., Tuomi, J. and Sulkinoja, M. 1994. Bud demography of the mountain birch *Betula pubescens* ssp. *tortuosa* near tree line. – Ecology 75: 945–955.

Lehtonen, J. 1987. Recovery and development of birch forest damaged by *Epirrita autumnata* in Utsjoki area, north Finland. – Rep. Kevo Subarctic Res. Stat. 20: 35–39.

Magnusson, S. H. and Magnusson, B. 1992. Effect of willow (*Salix*) on the establishment of birch (*Betula pubescens*) from seed. – Natturufraedingurinn 61: 95–108, in Icelandic.

Mason, P. A., Wilson, J. and Last, F. T. 1984. Mycorrhizal fungi of *Betula* spp.: factors affecting their occurence. – Proc. Royal Soc. Edinburgh 85B: 141–151.

Newton, A. C. 1991. Mineral nutrition and mycorrhizal infection of seedling oak and birch III. Epidemiological aspects of ectomycorrhizal infection, and the relationship to seedling growth. – New Phytol. 117: 53–60.

– and Pigott, C. D. 1991. Mineral nutrition and mycorrhizal infection of seedling oak and birch. I. Nutrient uptake and the development of mycorrhizal infection during seedling establishment. – New Phytol. 117: 37–44.

Nordell, O. 1978. Hanhängefrekvens och pollenproduktion hos fjällbjörk, *Betula pubescens* ssp. *tortuosa* i Abisko området, Torne Lappmark. – Honours thesis, Dept of Plant Ecology, Univ. of Lund.

Nordell, K. O. and Karlsson, P. S. 1995. Resorption of nitrogen and dry matter prior to leaf abscission: variation among individuals, sites and years in the mountain birch. – Funct. Ecol. 9: 326–333.

Palo, R. T., Bergström, R. and Danell, K. 1992. Digestibility, distribution of phenols, and fiber at different twig diameters of birch in winter. Implications for browsers. – Oikos 65: 450–454.

Poteri, M. 1992. Screening of clones of *Betula pendula* and *B. pubescens* against two forms of *Melampsoridium betulinum* leaf rust fungus. – Eur. J. For. Pathol. 22: 166–173.

Prudhomme, T. I. 1982. The effect of defoliation history on photosynthetic rates in the mountain birch. – Rep. Kevo Subarctic Res. Stat. 18: 5–9.

Ruosi, M. 1990. Breeding forest trees for resistance to mammalian herbivores – a study based on European white birch. – Acta For. Fenn. 210.

Sauter, J. J. and Ambrosius, T. 1986. Changes in the partitioning of carbohydrates in the wood during bud break in *Betula pendula* Roth. – J. Plant Physiol. 124: 31–43.

Sonesson, M. and Lundberg, B. 1974. Late quaternary forest development of the Torneträsk area, north Sweden. – Oikos 25: 121–133.

Spanos, Y. A. and Woodward, S. 1994. The effects of *Taphrina betulina* infection on growth of *Betula pubescens*. – Eur. J. For. Pathol. 24: 277–286.

Suomela, J. and Ayres, M. P. 1994. Within-tree and among-tree variation in leaf characteristics of mountain birch and its implications for herbivory. – Oikos 70: 212–222.

Sveinbjörnsson, B., Kuahanen, H. and Nordell, O. 1996. Treeline ecology of mountain birch in the Torneträsk area. Ecol. Bull. 45: 65–70.

Tenow, O. 1972. The outbreaks of *Oporinia autumnata* Bkh. and *Operophtera* spp. (Lepidoptera, Geometridae) in the Scandinavian mountain chain and northern Finland 1862–1968. – Zoologiska Bidrag från Uppsala, Suppl. 2.

– 1996. Hazards to a mountain birch forest – Abisko in perspective. – Ecol. Bull 45: 104–114.

Tuomi, J., Niemelä, P. and Mannila, R. 1982. Resource allocation on dwarf shoots of birch (*Betula pendula*): reproduction and leaf growth. – New Phytol. 91: 483–487.

– , Niemelä, P., Haukioja, E., Siren, S. and Neuvonen, S. 1984. Nutrient stress: an explanation for plant anti-herbivore responses to defoliation. – Oecologia 61: 208–210.

– , Nisula, S., Vuorisalo, T., Niemelä, P. and Jormalainen, V. 1988a. Reproductive effort of short shoots in silver birch (*Betula pendula* Roth). – Experientia 44: 540–541.

– , Niemelä, P., Rousi M., Siren, S. and Vuorisalo, T. 1988b. Induced accumulation of foliage phenols in mountain birch: Branch response to defoliation? – Am. Nat. 132: 602–608.

– , Vuorisalo, T., Niemelä, P. and Haukioja, E. 1989. Effects of localized defoliations in female inflorescences in mountain birch, *Betula pubescens* ssp. *tortuosa*. – Can. J. Bot. 67: 334–338.

– , Niemelä, P. and Siren, S. 1990. The Panglossian paradigm and delayed inducible accumulation of foliar phenolics in mountain birch. – Oikos 59: 399–410.

Vaarama, A. and Valanne, T. 1973. On the taxonomy, biology and origin of *Betula tortuosa* Ledeb. – Rep. Kevo Subarctic Res. Stat. 10: 70–84.

Van Cleve, K., Barney, R. and Schlentner, R. 1981. Evidence of temperature control of production and nutrient cycling in two interior Alaska black spruce ecosystems. – Can. J. For. Res. 11: 258–273.

– , Oechel, W. C. and Hom, J. L. 1990. Response of black spruce (*Picea mariana*) ecosystems to soil temperature modification in interior Alaska. – Can. J. For. Res. 20: 1530–1535.

Ecological Bulletins 45: 79–92. Copenhagen 1996

Peat formation and mass balance in subarctic ombrotrophic peatlands around Abisko, northern Scandinavia

Nils Malmer and Bo Wallén

Malmer, N. and Wallén, B. 1996. Peat formation and mass balance in subarctic ombrotrophic peatlands around Abisko, northern Scandinavia. – Ecol. Bull. 45: 79–92.

The apparent, short term litter formation rate in the dominating *Sphagnum* communities which characterise the extensive ombrotrophic parts of mires around Abisko, most of them with permafrost, has been estimated over a 14 yr period using a technique based on ^{14}C-labelling of the vegetation. The losses due to decay in the acrotelm have been estimated from the change in concentration of nitrogen above the upper limit of the catotelm. The litter formation rate in moss hummocks can be as high as 200 g m^{-2} yr^{-1} while the decay losses in the acrotelm are in the range of 40–50 g m^{-2} yr^{-1}. In hollows the decay rate is higher than in hummocks and the residence time of the organic matter is shorter (100 and 170 yr, respectively). The ombrotrophic peat is rather thin (<0.7 m deep), but ^{14}C-datings indicate that the formation of ombrotrophic peat started over 800 yr ago. Until recent time the ombrotrophic peat accumulation rate has been 36–45 g m^{-2} yr^{-1}. However, on the mires larger than c. 10 ha the *Sphagnum*-dominated communities cover only a small part of the total mire surface, often less than one third, and the litter formation rate on the remaining parts of the surface is very low. Therefore, the present over all litter formation rate is only c. 35 g m^{-2} yr^{-1} and does not even compensate for the decay losses in the acrotelm. Although the peat stratigraphy suggests an ongoing peat (and carbon) accumulation the carbon balance in the systems as a whole has changed from sink to source rather recently.

N. Malmer and B. Wallén, Dept of Ecology, Plant Ecology, Lund Univ., Ecology Building, S-223 62 Lund, Sweden.

Studies on the mire vegetation in the Abisko area have been carried out since the decades following the establishment of the Abisko Scientific Station (Fries and Bergström 1910, Fries 1913, Du Rietz 1945, Mårtensson 1956). A monographic treatment focusing on the rich fen vegetation (Du Rietz 1949) and its environment on the northern side of the lake Torne träsk was presented by Persson (1961, 1962, 1965). Ten years later Sonesson (1967, 1969, 1970a, b) presented a similar study focusing on the development of the mires and the poor fen and bog vegetation predominating on the southern side of the lake. A functional approach to the mire ecosystem, particularly the plant productivity and the biogeochemical cycles, was taken when the Stordalen mire east of Abisko became included as a main site in the IBP-Tundra programme. The main results of these studies were compiled in Sonesson 1980.

The variation in the mire vegetation around Abisko is, partly along the poor-rich gradient (due to the variation in soil conditions: Persson 1962, Malmer 1986, Sonesson 1970a, b), partly along the altitudinal gradient, and partly along the climatic gradient from oceanic conditions in the western parts to more continental conditions in the eastern parts. Because of decreasing snow depths from west to east there is a corresponding gradient in the hummock vegetation (Sonesson 1969) and peat deposits with permafrost and palsas do not occur west of Abisko (Andersson 1981, Åkerman and Malmström 1986). Further, in the western parts soligenous fens predominate while around and east of Abisko mires with ombrogeneous and topogenous soil wetness (Sjörs 1950) are widespread. This correspond to the regional differentiation of the mires in northernmost Scandinavia (Eurola and Kaakkinen

1979, Eurola and Vorren 1980, Vorren 1993), viz., various types of bogs along the coast, sloping, soligenous fens on the western side of the mountains and palsa mires in the more continental parts to the east of the mountains. At lower altitudes further to the east and south-east extensive aapa mires (patterned fens) of the northern type predominate (Sjörs 1983).

In the mass balance for a system of peat forming mire vegetation there are three main components, viz., 1) the addition of biomass resulting in the litter formation, 2) the release of mass from the organic matter through the decomposition in the upper, oxic peat layer, the acrotelm (Ingram 1978), and 3) the release of mass through the decay in the anoxic peat below, the catotelm. A positive net mass balance of a whole peat-forming system is assumed if, over a period of time, the net balance in the acrotelm is positive and the transfer of organic matter from the acrotelm to the catotelm exceeds the losses from the catotelm (Clymo 1984).

In the present study we focus on the mass balance in ombrotrophic parts of the mires, particularly their acrotelm, around Abisko. Since these mires represent a climatically conditioned transition between western and eastern types of tundra mires they may be supposed to have been affected by the climatic changes during the last millennia (Sonesson 1968, 1974). Therefore, we also want to see whether the accumulation rate in these systems has changed during recent centuries. We empirically estimate the apparent, short term input of litter to these systems from the net primary production using a method based on labelling of the *Sphagnum* mosses with ^{14}C (cf. Aerts et al. 1992). We also estimate the amount of organic matter remaining for persistent peat accumulation in the catotelm after the decomposition in the acrotelm using the variation of the nitrogen concentration in the surface layers, a method presented earlier (Björck et al. 1991, Malmer and Wallén 1993).

Methods

Study sites

The study area around Abisko (68°21′N, 18°50′E) is in the sub-alpine birch forest region. The annual mean temperature in Abisko is −0.5°C and the length of the vegetation period (daily mean temperature >4°C) is slightly more than three months (Rosswall et al. 1975). The mean annual precipitation around Abisko is rather low, c. 320 mm.

Sampling for the present investigations has been performed on four topogenous mires with extensive ombrogeneous parts, viz., the Western Stordalen mire situated 1.5 km NNW the Stordalen railway station (altitude 360 m; cf. Rosswall et al. 1975, Sonesson et al. 1980), the Eastern Stordalen mire situated 0.8 km NNE of the Stordalen railway station (altitude 360 m; cf.

Sonesson 1970a, p. 97, and 1970b), one mire in the Abisko Valley 2.5 km SW of the Abisko railway station (altitude 450 m), and one close to the Research Station in Abisko (altitude 370 m). This last mire is smaller and is more sheltered than the other ones. On all the mires the sampling has been concentrated on parts with clear ombrotrophic conditions.

The ombrotrophic parts of the mires have a peat depth of 1–3 m with fen peat always underlying the ombrotrophic peat layers. Permafrost is usually found at depths below 0.5–1.0 m. There are three different main types of surface structures: moss hummocks, lichen hummocks and *Sphagnum* hollows (cf. Sonesson 1970a, Rosswall et al. 1975, Sonesson and Kvillner 1980). Patches with bare peat but with a cover of vascular plants also occur. On the elevated moss and lichen hummocks the most abundant vascular plants are dwarf shrubs (*Betula nana* L., *Andromeda polifolia* L., *Empetrum hermaphroditum* Hagerup and *Vaccinium uliginosum* L.) along with *Rubus chamaemorus* L. and *Eriophorum vaginatum* L. On the moss hummocks either *Sphagnum fuscum* (Schimp.) Klinggr. or *Dicranum elongatum* Schleich. dominate the bottom layer. On the lichen hummocks *Cetraria* spp. and *Cladonia* spp. dominate the bottom layer. The total cover of vascular plants there is considerably lower.

The hollows form rather extensive (size: 100–500 m^2) depressions surrounded by the hummock complexes. *Sphagnum balticum* (Russ.) C. Jens. dominates the bottom layer but is often intermixed with *Drepanocladus schultzei* Roth. Scattered shoots of *Eriophorum vaginatum* and *Carex rotundata* Wahlenb. occur, too. In spring and summer during periods of high precipitation, the hollows are filled with water up to the moss surface while during extensive periods of drought the moss surface becomes dry.

On the W Stordalen mire (size 0.25 km^2) four separate ombrotrophic areas have been sampled. The size of these areas ranges between 0.5 and 3 ha. More than 90% of their surface is covered by hummocks leaving only c. 5–7% of the surface for the *Sphagnum balticum* hollows. Point sampling along parallel transects (n = 14, length 18.5 m, distances between points along the transects 0.25 m) showed that mosses and lichens covered 28% (SD ± 6%) and 43% (SD ± 6%), respectively, of the surface with litter or bare peat covering the remaining areas. Vascular plants covered only 60% of the surface. (The percentage cover values presented in Table 5 in Sonesson and Kvillner 1980 are not directly comparable with ours because they refer to the whole mire including its minerotrophic parts.)

The productivity measurements

The short term apparent accumulation of peat litter was studied on the W Stordalen mire and the mire at the Research Station using a method involving labelling of the moss layer with ^{14}C to get an innate marker in the moss plants. Plots with a well developed moss layer and a total cover of vascular plants <40% were selected for the study. In mid-June and mid-August in 1980 and 1981 each plot was covered with a 35 cm high, transparent, perspex chamber with base dimensions of either 20 × 20 or 31.5 × 31.5 cm. The chamber was pressed down c. 6 cm below the surface. Each chamber was supplied with 50–500 µCi ^{14}C by acidification of a $NaH^{14}CO_3$-solution inside the chamber, by injecting 1-M HCl through a rubber membrane in the wall. Each treatment lasted for c. 8 h. By that time there was hardly any ^{14}C left in the air inside the chamber.

The present study includes only samples taken in June 1994 after up to 14 vegetation periods. For the sampling we used a corer, 10 cm in diameter and 50 cm in length (cf. Malmer and Wallén 1993). The corer could be opened and the peat core, up to c. 40 cm in length, transferred intact to a plastic pipe with the same diameter. The core was then transported to the Abisko Research Station, deep frozen within four hours and later sliced (1.0 cm thick slices) by sawing.

The slices were thawed and subsamples (dry weight 10–500 mg) of 1) pieces of woody material >1 mm, 2) fine roots with a diameter <1 mm, 3) pieces of *Eriophorum vaginatum* tussocks and 4) moss material (stems, branches, leaves) were sorted out. These subsamples as well as the remaining part of each slice were then dried at 85°C and weighed to obtain the dry bulk density (dry mass per fresh volume, BD). The ^{14}C-activity in each one of these five fractions was determined by combustion in a Packard tri carb sample oxidizer, trapping the CO_2 in a Carbo Sorb solution with a scintillation fluid added and counting the disintegrations of ^{14}C in a Packard tri carb scintillation counter.

The chemical stratigraphy

From hummocks, cores were taken where either mosses (*Sphagnum fuscum* or *Dicranum elongatum*) or lichens (*Cetraria* spp. or *Cladonia* spp.) formed the bottom layer and the cover of vascular plants was <20%. In the hollows cores were taken in carpets dominated by *Sphagnum balticum* where only a few shoots of vascular plants occurred. Most of the cores (11) were sampled in 1981 and 1984 in the same way as in the labelled plots except that most slices were 2.5 cm thick. Seven of the cores were obtained in 1979 and 1994 from peat monoliths cut out from opened peat sections, sliced (slices 3–6 cm thick in 1979, in 1994 2.5 cm) by cutting with a sharp knife on the site and then deep frozen. During sampling, the peat type (Svensson 1986) and degree of humification (von Post and Granlund 1926) were always noted for the different layers.

After thawing, the living roots and all pieces of wood with a diameter >1 mm were removed. After drying (85°C), the fractions were weighed separately to get the dry bulk density (dry mass per fresh volume, BD) and percentage wood. Then the material (except roots) was ground in a Wiley mill and portions were taken out for the subsequent analyses.

Nitrogen was determined as Kjeldahl-N by a semi-micro method involving digestion with H_2SO_4 ($CuSO_4$ as catalyst), distillation and titration with NaOH. On the samples taken in 1980 and later carbon was determined using a Leco CR-12 autoanalyzer in which the organic material is burnt in oxygen and the CO_2 measured afterwards by an infrared detector. All the analyses of N and C were run in duplicate with good agreement between sample pairs (relative differences nearly always <5%). The values used for organic matter always include its ash content (ash content in the range 0.5–4% in the ombrotrophic peat).

Radiocarbon age determinations

Radiocarbon datings were carried out at the Radiocarbon Dating Lab. in Lund in eight of the profiles from 1979 and 1994. The age calculations are based on a half-life time for ^{14}C of 5568 yr. If not otherwise indicated all ages and lengths of time periods given refer to calendar years before 1980 calibrated according to Stuiver and Reimer (1993).

Calculations

Stratigraphy and depth measurements

In all the cores three structural layers were distinguished, viz. from the surface 1) the moss (or lichen) layer, 2) the litter peat layer (abbreviated to litter layer) and 3) the peat layer (Malmer 1988). Most of the transformation into litter takes place in the upper parts of the litter layer, the litter deposition level (the LDL; cf. Malmer and Wallén 1993). The position of this level was obtained from the combination of stratigraphy and the sample with the lowest concentration of N in the organic matter since the mosses are strongly depleted of N in their basal parts prior to the litter formation (Rydin and Clymo 1989).

All depths are given either below the surface or relative to the LDL. For the calculations in this paper the LDL has been set to the level of the upper surface of the sample representing the uppermost part of the litter layer. (Note, not the midpoint of that sample as in Malmer and Wallén 1993!).

Litter formation rate

The apparent rate of litter formation in the moss hummocks and the *Sphagnum* hollows has been obtained from the ^{14}C-labelled plots. The total amount of organic matter accumulated since labelling has been divided by the number of vegetation periods from the labelling to the sampling.

Decay losses
The decay in the acrotelm has been estimated from the concentrations of N based on the assumption that all nitrogen once supplied to the bog surface becomes incorporated in the organic matter and remains there (Björck et al. 1991, Malmer and Wallén 1993). This is a particularly reasonable assumption for the bogs around Abisko since very conservative nitrogen cycling has been demonstrated on the W. Stordalen mire (Rosswall and Granhall 1980) and the plant growth is N-limited (Aerts et al. 1992). Furthermore, no losses in N by denitrification or in the drainage water have been shown (Rosswall and Granhall 1980), and all N supplied to the *Sphagnum* mosses in the wet and dry deposition has been shown to be readily assimilated (Lee et al. 1986, Lee and Woodin 1988, Woodin and Lee 1987).

The decay of the peat below the LDL leads to loss of mass from the system at a constant proportional rate, viz.:

$$M_t = M_0 \exp(-kt) \tag{1}$$

where M_0 is the amount of organic matter added to the system during a certain period of time, M_t the amount remaining at time t, and k the decay constant (Jenny et al. 1949). As nitrogen is conserved the losses of organic matter in decay will cause an increase in the concentration of nitrogen (as mg g^{-1}) from N_0 in the litter to N_t at time t. In equation (1) M_0 and M_t can then be replaced by the nitrogen free dry weight, i.e., the organic matter except nitrogen. Let this be W_0 and W_t, where

$$W_0 = M_0(1 - N_0) \quad \text{and} \quad W_t = M_t(1 - N_t) \tag{2}$$

The assumption that nitrogen, once it has been incorporated in the organic matter, is conserved further implies that the accumulated amount of nitrogen can be used as a time scale, provided that S_N, the rate of supply of N, has remained constant over time t, viz:

$$t = N_{cum}/S_N \tag{3}$$

where N_{cum} is the amount of nitrogen accumulated down to a defined depth.

Substituting the time t in equation (1) with equation (3), k with k_N as the time scale is changed to N_{cum}, and calculating with W_0 and W_t instead of M_0 and M_t in proportion to nitrogen (eq. 2) gives:

$$W_t/N_t = (W_0 \exp(-k_N * N_{cum}/S_N))/N_0$$
$$= (W_0/N_0)\exp(k_N * N_{cum}/S_N)) \tag{4}$$

Taking the natural logarithm of (4) gives:

$$\ln(W_t/N_t) = \ln(W_0/N_0) - (k_N/S_N)N_{cum} \tag{5}$$

which is the equation for a straight line with cumulative nitrogen (N_{cum}) as the X-scale and the natural logarithm of the quotients W/N ($\ln(W_t/N_t)$) as the Y-scale. From the slope of the line, k_N, it is possible to estimate the decay constant (k yr^{-1}) as

$$k = -k_N * S_N \tag{6}$$

The intercept in eq. (5) will indicate the proportion W_0/N_0 in the litter at the time of its formation and then also, by multiplying by S_N, the litter deposition rate.

All these calculations presuppose that 1) there has been no change in the nitrogen deposition rate during timespan t, 2) the concentration of nitrogen in the litter at LDL (N_0) has been the same over t, viz., there has been no systematic change in productivity or species composition during t, and 3) any decrease in the decay rate (k) with depth as an effect of a differential decay (Svensson and Rosswall 1980, Hogg 1993) is negligible. In the litter layer and the upper part of the peat layer corresponding to the acrotelm the decay rate is much higher than in the layers further down representing the catotelm (Clymo 1984). This means that the slope (k_N) of the regression of $\ln(W/N)$ on N_{cum} in eq. (5) will demonstrate a marked decrease in decay rate at the decay decrease level or DDL in the peat layer, here for simplicity assumed to coincide with the limit between acrotelm and catotelm (Malmer and Wallén 1993).

In the present study calculations according to equation (5) have been carried out separately for each core. The regression line above the DDL has then been obtained from the group of samples below LDL together giving the highest value on R^2 (not least SE for the k_N as in Malmer and Wallén 1993). Because fen peat is reached in the bottom of some profiles the value for $\ln W_t/N_t$ at the DDL has been calculated as an average of the quotients $\ln W_t/N_t$ in the samples in a depth interval of 6 cm downwards from the lowest sample included in the calculation of this regression line. The depth of the DDL is the value for N_{cum} on the regression line corresponding to the $\ln W_t/N_t$ at the DDL calculated in that way. The proportion, a, of the original organic matter which has been lost by decay in the acrotelm down to a specified level (e.g. the DDL) has been calculated in the following way:

$$a * W_0/N_0 = W_t/N_t \quad \text{or, as percentage}$$
$$a = W_t * N_0 * 100/W_0 * N_t \tag{7}$$

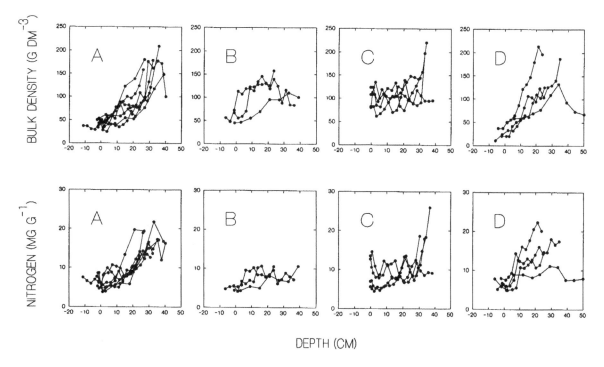

Fig. 1. Dry bulk density (g dm^{-3}, upper row) and concentration of nitrogen (mg g^{-1}, lower row) in relation to distance (cm) from the litter deposition level (LDL; above −, below +) in the cores for chemical stratigraphy from A) the moss hummocks with N > 13 mg g^{-1} at the base of the cores, B) the moss hummocks with N < 13 mg g^{-1} at the base of the cores, C) the lichen hummocks and D) the *Sphagnum balticum* hollows.

Results

Stratigraphy

Moss hummocks

In the moss hummocks the average depth of the moss layer was c. 5 cm (Fig. 1, cf. Tables 1 and 5). The peat type below was always a low humified *Sphagnum* peat with a varying amount of wood pieces (Fig. 2), often with lenses of *Dicranum* peat or *Vaginatum* peat. The gradual transition from the litter layer to the peat layer (Malmer 1988, Malmer and Wallén 1993) occurred 12–20 cm below the surface. In some of the cores the

Sphagnum fuscum-peat was replaced by fen peat (H$_{6-7}$) with remains of *Carex* spp., *Betula*, *Salix*, and "brown mosses" together with the *Sphagnum* spp. at depths > 25 cm below the surface (cf. Sonesson 1970b).

The bulk density increased parallel with the humification (Fig. 1). The concentration of C increased slightly from 480 mg g^{-1} in the surface layers to > 530 mg g^{-1} at the bottom (Tables 1 and 2, cf. Malmer and Wallén 1993). The concentrations of N decreased with depth in the moss layer while it further down always increased with depth (Fig. 1). In most cores the concentration of N in the bottom samples was > 12.5 mg g^{-1}. In three cores, however, it did not exceed this value even though these cores were just as deep as the others.

In all cores from moss hummocks the regression of ln W_t/N_t on N_{cum} (cf. eq. 5) was statistically significant (p < 0.05; R^2 > 0.66, n = 4–8) in the litter layer and the upper parts of the peat layer. The DDL was found at depths ranging from 6 to as much as 26 cm below the LDL (Table 2; cf. Table 3 for the corresponding variation in N_{cum}). However, the average amount of organic matter contained between the LDL and DDL varied less (Table 2). The quotient W_0/N_0 is insignificantly higher if calculated as a mean from the individual cores (Table 3) than if calculated from the cores combined as in Fig. 3.

Table 1. The moss and lichen layers above the litter deposition level (LDL) in the cores for chemical stratigraphy. – Mean values are given together with the 95% confidence limits and number of samples.

	Surface structure		
	Moss hummocks	Lichen hummocks	*Sphagnum* hollows
Depth to the LDL (cm)	4.7 ± 2.2 (n = 10)	0.5 ± 01 (n = 4)	4.8 ± 4.3 (n = 4)
Conc. of C (mg g^{-1})	483 ± 11 (n = 13)	491 ± 20 (n = 4)	476 ± 9 (n = 10)

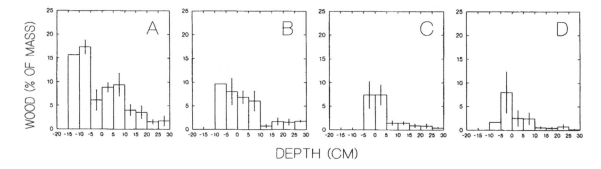

Fig. 2. Content of woody material (as percentage; ± SE indicated) in the organic matter in relation to distance (cm) from the litter deposition level (LDL; above −, below +) in the cores for chemical stratigraphy from A) the moss hummocks with N > 13 mg g^{-1} at the base of the cores, B) the moss hummocks with N < 13 mg g^{-1} at the base of the cores, C) the lichen hummocks and D) the *Sphagnum balticum* hollows.

Lichen hummocks

Due to the reindeer grazing the layer of lichens on the lichen hummocks was thin, not > 0.5 cm (Fig. 1). The peat type immediately below was always the same as that found in the peat layer of the *Sphagnum* hummocks. The LDL is therefore to be set just below the layer of lichens where the small amount of lichen litter is deposited together with some drifting above-ground vascular plant litter. Because of the low abundance of vascular plants the addition of litter below the LDL is negligible. It was also impossible to distinguish separate litter and peat layers although the degree of humification increased with depth, particularly in the bottom of the cores in the same way as in the *Sphagnum* hummocks.

The bulk density was rather high in the lichen hummocks (Fig. 1) with a distinct increase at 30 cm below the surface in all but one of the cores. The concentra-

Table 2. Characteristics of the litter and peat layers in the cores from the moss hummocks and *Sphagnum* hollows for chemical stratigraphy. – Mean values are given together with the 95% confidence limits and the number of samples.

	Surface structure	
	Moss hummocks	*Sphagnum* hollows
Distance between LDL and DDL (cm)	14.3 ± 5.0 (n = 10)	6.3 ± 3.2 (n = 4)
Org. matter between LDL and DDL (kg m^{-2})	10.1 ± 3.2 (n = 10)	4.8 ± 2.5 (n = 4)
Conc. of C (mg g^{-1}) in the layer (2.5 cm) below the LDL	470 ± 17 (n = 6)	460 ± 42 (n = 3)
in the two layers (5 cm) just below the DDL	510 ± 18 (n = 12)	490 ± 18 (n = 6)
Conc. of N (mg g^{-1}) in the layer (2.5 cm) below the LDL	4.8 ± 0.4 (n = 10)	5.8 ± 1.7 (n = 4)
in the two layers (5 cm) just below the DDL	10.6 ± 0.5 (n = 20)	11.8 ± 2.1 (n = 8)

tions of N decreased down to c. 5 cm below the LDL (Fig. 1) but increased then irregularly down to 20–25 cm below the LDL where the mean concentration of N was 9.51 mg g^{-1} (95% conf. limits ± 1.43 mg g^{-1}; n = 12). Below 25 cm the concentration of N increased in three of the cores while in the fourth it remained constant (Fig. 1). In the depth ranges 0–5 cm and 20–25 cm the mean concentrations of C were 493 mg g^{-1} (95% conf. limits ± 10 mg g^{-1}; n = 11) and 487 mg g^{-1} (95% conf. limits ± 8; n = 12) respectively, while in the bottom samples the concentrations were as high as 520–530 mg g^{-1}.

For N_{cum} < 75 g m^{-2} (mean depth below the LDL 9.5 cm; 95% conf. limits ± 5.3 cm; n = 4) in the cores taken together (Fig. 3) the regression of ln W_t/N_t on N_{cum} is barely significant (p = 0.047). In the layers c. 5 cm below the LDL the quotient W_t/N_t is in the range 130–220 while it approached values around 90 in the depth interval 20–25 cm. The average amount of organic matter contained down to the level corresponding to N_{cum} = 75 was 10300 g m^{-2} (95% conf. limits ± 4200 g m^{-2}; n = 4).

Sphagnum hollows

The LDL was c. 5 cm below the surface (Table 1, Fig. 1) and visible as a change in colour from green to light brown. Below the LDL in all cores there were thick litter peat and peat layers consisting of low-humified *Sphagnum* balticum peat usually without woody remains (Fig. 2) down to a distinct transition at 16–27 cm to a more humified layer of peat consisting of *Sphagnum fuscum* or *Dicranum elongatum* peat. In one of the cores the *Sphagnum fuscum*-peat was replaced by fen peat with remains of *Betula* and *Salix* at 35 cm below the surface.

In the moss and litter layers the bulk density was rather low while it in the peat layers approached that of the hummock cores (Fig. 1). The concentration of C was c. 470 mg g^{-1} in the surface layers (Tables 1 and 2) while it was up to 530 mg g^{-1} in the bottom of the

Table 3. Calculations of the coefficients in eq. (5) on the individual cores for chemical stratigraphy from the moss hummocks and the *Spahgnum* hollows. – The values given are means (together with the 95% confidence limits) and (within parentheses) the ranges of values.

	Surface structure	
	Moss hummocks (n = 10)	*Sphagnum* hollows (n = 4)
Decay const. as k_N	-0.0116 ± 0.0026 ($-0.0062-0.0187$)	-0.0212 ± 0.0104 ($-0.011-0.028$)
N_{cum} (g m^{-2}) at the DDL (moss layer excluded)	70 ± 25 ($27-127$)	38 ± 18 ($21-51$)
W_0/N_0 at the LDL	203 ± 26 ($142-246$)	185 ± 49 ($13--224$)
W_t/N_t at the DDL	98 ± 16 ($61-131$)	86 ± 28 ($67-115$)

cores. The increase in the concentration of N below the LDL was very distinct (Fig. 1). In three of the cores the concentrations of N was >13 mg g^{-1} in the bottom while in the fourth core the peak concentration was 11 mg g^{-1}.

In all cores from the hollows the regression of ln W_t/N_t on N_{cum} (cf. eq. 5) was statistically significant (p < 0.05; $R^2 > 0.63$, n = 4–5) for the litter layer and the upper parts of the peat layer with k_N lower than in the moss hummocks (Table 3, Fig. 3). The DDL was found at depths ranging from 4.5 cm to 9.3 cm below the LDL (Table 2) or given as N_{cum} from 21 to 51 g m^{-1} (Table 3). The average amount of organic matter contained between the LDL and DDL was 4820 g m^{-2} (95% conf. limits ± 2500 g m^{-2}; n = 4). At the DDL the concentration of C was not much higher than at the LDL while the concentration of N increased (Table 2). The quotient W_t/N_t at the DDL was half of that for W_0/N_0 at the LDL (Table 3).

The radiocarbon datings and the nitrogen accumulation rates

One of the radiocarbon dated samples (Table 4, Lu-3796) must have been contaminated by radioactive fallout of ^{14}C from tests of nuclear weapons and has therefore to be excluded from the discussions. Furthermore, in the dating of samples with a ^{14}C-age <400 yr there is for methodological reasons usually a choice between alternative ages in the calculations of the calendar years (cf. Stuiver and Reimer 1993). Therefore, because of the SD it is for three samples only possible to state that their age in calendar years is <300 yr before 1980. However, as they are covered by 17000–25000 g m^{-2} of organic matter, only the oldest one of the alternative datings is realistic. An activity, although small, of ^{210}Pb has been found in the samples Lu-1816 and Lu-1817 (Malmer and Holm 1984, Fig. 1, profile A). This could be used as an argument for a dating younger than 270 and 300 yr. However, because of its

mobility ^{210}Pb will easily be moved downwards, thus giving unreliable datings always much younger than the true ones.

There is only a weak relationship between depth and the age of the peat (Table 4). However, the three samples with a concentration of $N > 13$ mg g^{-1} (all >40 cm below the surface) cover the period 2900–300 BP while the other samples are younger than 800 BP.

The estimates of the present nitrogen deposition rate around Abisko range from 0.1 (wet deposition only) to 0.42 g m^{-2} yr^{-1} (Malmer and Nihlgård 1980, Rosswall and Granhall 1980, Malmer and Wallén 1993). For the accumulation of peat with a concentration of $N < 13$ mg g^{-1} the radiocarbon age determinations over three periods with lengths of 280–830 yr (Table 4, cores II, III, and IV) gives the nitrogen accumulation rates 0.42, 0.38 and 0.39 g m^{-2} yr^{-1}, respectively, or as a mean value 0.4 g m^{-2} yr^{-1}. This value will be used in the calculations following from eq. (5) for peat layers with a concentration of $N < 13$ mg g^{-1}.

The litter productivity

In all the cores from the ^{14}C-labelled plots there is a distinct peak in the ^{14}C-activity at 1–4 cm below the LDL (Table 5) for both the moss and root fractions. This suggests that all the organic matter accumulated above this layer must have been added after that the inoculation took place. However, the absolute activity in this one cm thick layer varies depending on the amount of ^{14}C added at the labelling (Fig. 4).

The highest average litter accumulation rate, 200 g m^{-2} yr^{-1} (range 180–270 g m^{-2} yr^{-1}) was found in the moss hummocks on the Research Station mire. Both in the moss hummocks and the hollows on the W Stordalen mire the rate was less than half of that (range 50–158 g m^{-2} yr^{-1}) and 82–148 g m^{-2} yr^{-1}, respectively; cf. also 109 g m^{-2} yr^{-1} in hollows in Aerts et al. 1992). As an accumulation of organic matter does not take place on the lichen hummocks (Rosswall et al.

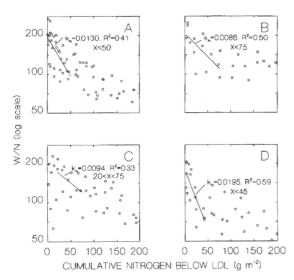

Fig. 3. The variation below the LDL in the quotient W_t/N_t (ln scale) in relation to the depth expressed as cumulative amount of N (N_{cum}) in the cores for chemical stratigraphy from A) the moss hummocks with N > 13 mg g^{-1} at the base of the cores, B) the moss hummocks with N < 13 mg g^{-1} at the base of the cores, C) the lichen hummocks and D) the *Sphagnum balticum* hollows. The regression lines calculated according to eq. (5), k_{Ng} and R^2 are indicated for the samples above the DDL.

1975), the accumulation rate in the hummock complexes on the W Stordalen mire varies from 0 to 160 g m^{-2} yr^{-1}.

With 1 cm thick slices the uncertainty in the determination of the level for the original surface may give an error in the litter production estimates of at most 20 g m^{-2} yr^{-1}. Furthermore, all the organic matter produced since the labelling may not have been included since there may have been both decay losses and a growth of the vascular plants below the level of peak ^{14}C-activity. On the other hand any woody material present above the original surface at the labelling will in the present samples appear as new material since it has been overgrown by the mosses during the experiment.

Discussion

The ombrotrophic peat

Around Abisko the accumulation of ombrotrophic peat began with a fen type of vegetation being replaced by *Sphagnum fuscum* vegetation lacking exclusive fen plants (Sonesson 1970b). Because of higher rates of decomposition, productive mire types such as wooded fens, swamps and minerotrophic *Sphagnum* fens release carbon at a much higher rate (Rosswall et al. 1975, Svensson and Rosswall 1984) and accumulate peat at a slower rate than ombrotrophic bogs (Malmer 1975,

Damman 1979). Therefore the initiation of ombrotrophic peat formation is the result of decreasing decay losses which will increase the peat deposition rate (Svensson 1988).

In subarctic areas the development of ombrotrophic mires may be promoted particularly by climatic conditions giving rise to permanently frozen peat decreasing the decay losses. Such parts will soon form hydrologically isolated structures where *Sphagnum* mosses such as *S. fuscum*, and also dwarf shrubs will expand and eventually exclude the fen plants from the site (Malmer et al. 1994). A few cold winters with a thin snow cover may be enough to initiate such a development, as is the case with the initiation of the formation of palsas (Vorren 1972).

There is a distinct difference between what certainly represents true ombrotrophic peat with a concentration of N < 13 mg g^{-1} and the underlying peat with higher concentrations of N (Fig. 1). The DDL is found in peat layers with concentrations of N < 13 mg g^{-1} and the whole acrotelm therefore always consists of such true ombrotrophic peat. For the peat layers with N > 13 mg g^{-1}, transitional between true ombrotrophic peat and fen peat (Fig. 1), there are no reasons to assume that the rather slow increase with depth in the concentration of N is the result of any ongoing decay processes. Instead these layers form a part of the catotelm. Therefore, the calculations on fewer cores in Malmer and Wallén (1993) suggesting that the DDL in hummock cores was to be placed in these peat layers with N > 13 mg g^{-1} involved misinterpretations.

Disturbances observed by us on the surface of the W Stordalen mire since c. 1975 can be interpreted as the result of melting permafrost. This suggests that the ombrotrophic parts of the mires around Abisko like the palsas along the margins of their distribution in northern Scandinavia (Vorren 1972), at present may be rather unstable. Similar mire structures along the southern limit for permafrost in Canada have collapsed during the last century (Vitt and Halsey 1994).

The shallow layers of true ombrotrophic peat suggest that a widespread occurrence of ombrotrophic systems around Abisko ought to be a rather recent phenomenon. The initiation of the formation of true ombrotrophic *Sphagnum fuscum* peat could in two cores (I and III; cf. Table 4) be radiocarbon dated to AD 1700 and in one (IV) to AD 1550 but there is also one dating (core II) which shows that ombrotrophic peat occurred on the W Stordalen mire 800 yr ago. Below the LDL the true ombrotrophic peat contained 90– >320 g m^{-2} of N in the moss hummocks, 55– >369 g m^{-2} in the hollows, and 289– >332 g m^{-2} in the lichen hummocks. With a deposition of 0.4 g m^{-2} yr^{-1} of N these values cover ages in the range 220– >900 yr. The mean age of the organic matter at the limit between the true ombrotrophic peat and the underlying peat with N > 13 mg g^{-1} was 410 yr (95% conf. limit

Table 4. Radiocarbon datings and recalculated calendar years from cores taken on the Stordalen mires.

Lab. no.	Depth (in cm) below surface	Peat characteristics	^{14}C age BP ($\pm$SD)	Calender year before 1980 (range one SD)
The Western Stordalen mire				
Lu-1816	38–43	Core I. *Sphagnum* peat (H$_4$; N 21.8 mg g^{-1}) just below peat low in N.	50 ($\pm$45)	270[a] 30–300
Lu-1817	43–45	Core I. *Sphagnum* peat (H$_4$; N 16.8 mg g^{-1}) containing mineral soil particles just above fen peat with wood.	170 ($\pm$45)	300[b] 30–320
Lu-1820	50–54	Core II. *Sphagnum fuscum* peat (H$_{3-4}$; N 7.9 mg g^{-1}) 30 cm below hollow peat and just above permafrost.	900 ($\pm$50)	830 780–940
Lu-3796	20–21.5	Core III. *Sphagnum fuscum* peat (H$_4$; N 9.2 mg g^{-1}) at the upper limit to less humified peat.	$\pm$2%* ($\pm$0.7%)	25 20–30
Lu-3798	32.5–34	Core III. *Sphagnum* peat (>H$_5$; N 12.7 mg g^{-1}).	340 ($\pm$60)	410[c] 330–510
The Eastern Stordalen mire				
Lu-3797	22.5–25	Core IV. *Sphagnum fuscum-Dicranum* peat (H$_{3-4}$; N 12.2 mg g^{-1}) just above a layer rich in woody remains at the transition to highly humified peat.	110 ($\pm$60)	280[d] 30–300
Lu-3799	41–42.5	Core IV. Highly humified fen peat (N 16.3 mg g^{-1}) with *Carex* and wood.	2740 ($\pm$110)	2830 2800–2900

*) Increase in activity in relation to standard as the sample has been contaminated by ^{14}C from nuclear weapon tests. The age in calender years has been estimated from what is known about the activity in plant material from about 1955 (Håkansson pers. comm., Skog pers. comm.).
[a]) alternative ages 30 and 100 yr before 1980.
[b]) alternative ages 30, 40, 190, and 230 yr before 1980.
[c]) alternative ages 360 and 460 yr before 1980.
[d]) alternative ages 60, 100, 140 and 250 yr before 1980.

$\pm$130 yr; n = 13). On average the true ombrotrophic peat seems to be older in the lichen hummocks than in the moss hummocks and hollows. Altogether these datings of the transition from fen to bog are much younger than has been found in peat layers on a coastal bog with palsas near the Varangerfjord in Norway (Vorren 1972).

It is only on parts of the W Stordalen mire that the true ombrotrophic peat contains >300 g m^{-2} of N. In these parts the average proportion of organic matter to N in the depth interval (as cumulative N, N$_{cum}$) 70–200 g m^{-2} was 111 g g^{-1} (range 75–161 g g^{-1}; n = 26). It was the same in the deeper peat layers (N$_{cum}$ > 200), viz. 116 g g^{-1} (range 80–164 g g^{-1}; n = 26). With a constant N supply rate this suggests a rather constant overall rate of peat deposition at the DDL of c. 45 g m^{-2} yr^{-1}. Therefore, we cannot see any changes in the ombrotrophic peat accumulation rate during the period AD 1300–1800 although the ombrotrophic systems expanded. For wet and tussock tundra long time average carbon accumulation rates of 27 and 23 g C m^{-2} yr^{-1}, respectively, have been presented (Oechel et al. 1993). If calculated as organic matter that may give the same deposition rates.

The surface structures

The vegetation of the moss hummocks with their dwarf shrubs together with *Sphagnum fuscum* and *Dicranum elongatum* is the main peatforming system on the present ombrotrophic mires. Carbon and nitrogen is continuously incorporated in the living parts of the mosses and vascular plants and then deposited at the LDL as plant litter. The age of the litter at the LDL may be <10 yr in both the moss hummocks and the hollows (Table 5). The subsequent decay losses can then be followed down to the DDL through increasing concentrations of N (Figs 1 and 3).

Summer droughts or a thin snow cover during winter or both together may reduce the vitality of the mosses and thereby the rate of litter formation on exposed moss hummocks. Such an exposure of a moss hummock will also reduce the abundance of vascular plants,

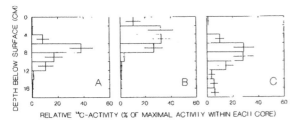

Fig. 4. Distribution of ^{14}C (calculated as % of the highest activity (DPM mg^{-1}) found within each core), 13 or 14 yr after pulse labelling of *Sphagnum fuscum* dominated hummocks at the Research Station mire at Abisko (A, n = 6); *Sphagnum fuscum* dominated hummocks at the Stordalen mire (B, n = 9) and *Sphagnum balticum* dominated hollows at the Stordalen mire (C, n = 6). Means ±SE for successive 2 cm peat layers.

initiate colonisation of the hummocks by lichens, and even open up them for erosion. It is then the surface peat low in N which first disappears but the decomposition in the acrotelm continues to increase the concentrations of N down to the DDL found at the level where N_{cum} was 75 g m^{-2} (Fig. 3) or c. 26 cm below the surface. However, because of the surplus of C in the system any supply of N to the surface of the lichen hummocks may be taken up by the micro-organisms and included in the organic matter at the surface. The result is a concentration of N in the uppermost layers (Fig. 1) which is contrary to what always is seen below the LDL in the cores from the moss hummocks.

The stratigraphy in the *Sphagnum*-hollows shows that they develop within the complexes of hummocks in depressions filled with water because of impeded drainage. The N accumulated in the *Sphagnum balticum* peat overlaying the hummock peat does not exceed 110 g m^{-2} suggesting an age <250 yr for this peat type as well as the hollows themselves. The DDL in the *Sphagnum* hollows is found 10–15 cm above the stratigraphic

limit between the *Sphagnum balticum* peat and the underlying *Sphagnum fuscum* peat. Therefore, the processes of accumulation and decay in the hollows resemble those found on the moss hummocks.

The mass balance

Mass balance of moss hummocks and Sphagnum hollows
There is considerable variability among individual hummocks and hollows in litter productivity (Table 5, Fig. 4). Furthermore, the twice as high productivity found in the moss hummocks on the Research Station mire compared to those on the W Stordalen mire shows that exposure to wind and snowdrift may reduce the plant growth rate. The difference in litter productivity found on the W Stordalen mire between the *Sphagnum balticum* hollows and the moss hummocks (Table 5) is insignificant. Input rates for organic matter calculated from the quotients W_0/N_0 (Table 3) and a N deposition rate of 0.4 g m^{-2} yr^{-1} are somewhat lower than litter productivity, viz. 74 and 82 g m^{-2} yr^{-1}, respectively. As such a calculation (eq. 5) assumes a linear relation between ln W/N and N_{cum}, it will always give a conservative estimate of the litter transfer at the LDL because the slope of the regression line and thus also the intercept, i.e. ln W_0/N_0, is much influenced by its lower parts with a slope less than in the upper parts due to the lower decay rates there (Svensson and Rosswall 1980, Hogg 1993). Furthermore, the sensitivity of the value of the intercept to variation in the concentration of N_0 is low.

A considerable variation is also seen in the decay rates (k_N) and, particularly, in the depths of the DDL both among the moss hummocks and the hollows (cf. Malmer and Wallén 1993). However, in both the moss hummocks and the hollows the concentration of N in the organic matter at the DDL much less variable

Table 5. The depth of the moss layer, the volumetric growth from the labelling in 1980 or 1981 to the sampling in 1994 and the rate of dry mass accumulation during that period in the ^{14}C labelled plots above the ^{14}C peak activity layer. – Mean values and ±95% confidence limits are given.

	Mire and surface structure		
	Research station mire	The W Stordalen mire	
	S. fuscum hummocks (n = 6)	Moss hummocks (n = 9)	*S. balticum* hollows (n = 6)
Depth below surface (cm) to the LDL	4.8 (±0.4)	3.6 (±1.3)	5.8 (±2.4)
The level for the ^{14}C peak activity (cm)	8.7 (±1.8)	4.7 (±1.1)	8.3 (±1.7)
Acc. dry weight (g m^{-2} yr^{-1})	202 (±35)	94 (±25)	112 (±21)

(range 10.5–12.5 mg g^{-1}; cf. Fig. 3 and Table 2). Comparing the quotients W_0/N_0 at the LDL with W_t/N_t at the DDL (Table 3) suggests that in both the moss hummocks and the hollows an equal proportion of originally deposited organic matter (on an average 51% and 52%, respectively,) is lost due to decay. Assuming input rates equal to the litter productivity would, of course, make the losses somewhat higher.

Both the depth of the DDL (Table 3) and the residence time for the organic material above the acrotelm is less in the hollows than in the moss hummocks. With a N supply rate of 0.4 g m^{-2} yr^{-1} the residence time can be estimated to be c. 100 and 170 yr, respectively. The equal percentage loss of organic matter in the acrotelm, therefore depends on the higher decay rate in the hollows (Table 3) compensating for the shorter residence time there compared to the moss hummocks. Greater annual decay losses in the hollows could also be inferred from the fact that the wet depressions annually remain unfrozen for longer periods compared to the hummocks (Rydén and Kostov 1980).

Taken together these calculations indicate a positive mass balance in the acrotelm both in the moss hummocks and the *Sphagnum* hollows. For the moss hummocks it is more positive than earlier calculations (Malmer and Holm 1984, Malmer and Wallén 1993) since the DDL then was considered to be in the transitional peat layers with lower C/N-quotients than in the true ombrotrophic peat. For the moss hummocks the present estimate of the decay rate in the acrotelm (k = 0.0046; cf. eq. 6 and Table 3) as well as the rate of transfer of organic matter to the catotelm, 40 g m^{-2} yr^{-1} at the DDL, is close to what has been estimated from model calculations (Clymo 1978). However, the variation with depth in the concentration of N in the true ombrotrophic peat below the DDL (Fig. 1) suggests a variable rate of transfer (range 30–65 g m^{-2} yr^{-1}) due to a variation in decay conditions, most probably the residence time, similar to that found among the hummocks (Fig. 1, cf. Malmer and Wallén 1993).

Mass balance of lichen hummocks
Because of the low productivity in the lichen hummocks combined with reindeer grazing in spring and autumn, the annual supply of litter can be ignored. It may even be assumed that organic matter now and then is lost from the surface by erosion. This means that there is hardly any litter transferred at the LDL, but down to the DDL the organic matter accumulated in the former moss hummock will continue to decay. Altogether this will make the mass balance negative in the lichen hummocks.

The depth of the DDL in the lichen hummocks should not deviate much from that in the moss hummocks. Although the origin of the peat is the same as in the moss hummocks, the higher humification in the

lichen hummocks may infer a greater proportion of decay resistant material. Therefore, the decay rate ought to be somewhat less than in the moss hummocks (Fig. 3).

The fourth category of structures found in the hummock complexes, the parts covered by litter and bare peat, have not been included in this study but with regard to the peat accumulation processes they ought to resemble the lichen hummocks. As on some lichen hummocks, these parts sometimes have a significant cover of vascular plants, mainly dwarf shrubs. It could then be argued that these parts contribute to the input of organic matter with c. 50 g m^{-2} yr^{-1} (Rosswall et al. 1975). However, except for the 20% consisting of woody parts this litter may decay comparatively fast (Coulson and Butterfield 1978) and will therefore hardly contribute significantly to the organic matter in the deeper parts of the acrotelm (Clymo 1984, Wallén 1992).

Overall mass balance for an ombrotrophic site
The patchiness in the vegetation cover of the studied mires makes the litter deposition over their surface very uneven and thus also the input of organic matter to the peat forming system. Therefore, the proportion of the area covered by productive types of vegetation will be very important for the overall litter accumulation rate. In addition, the quality of the litter formed in a vegetation type may be as important as its productivity for the mass balance. E.g. both *Sphagnum* spp. and woody plant structures form a much more decay resistant litter than leaves and fine roots (Coulson and Butterfield 1978). In the vascular plant biomass on the hummocks the very fine roots (diameter <0.5 cm) form the greatest fraction (Wallén 1986).

Assuming that the organic material contained between the LDL and DDL (cf. Tables 2 and 3 and sect. Lichen hummocks) decomposes at a constant, proportional rate, k, calculated for each of the three surface structures according to eq. (6), suggests rather similar annual mass losses from the acrotelm, viz. 47, 39 and 41 g m^{-2}, for the moss hummocks, lichen hummocks and *Sphagnum* hollows, respectively, or as a weighted average, 43 g m^{-2} yr^{-1}. Furthermore, independently of the variation in surface structure the concentration of N in the organic matter at the DDL is rather uniform and in the range 10.5–12.5 mg g^{-1} (Table 2). With a N supply rate of 0.4 g m^{-2} yr^{-1} the overall amount of organic matter that at the DDL might be transferred to an upwards encroaching catotelm should then be in the range 32–38 g m^{-2} yr^{-1} or somewhat less than that found for the moss hummocks (cf. sect. Mass balance of moss hummocks and *Sphagnum* hollows). Thus, to meet these estimated mass losses in the acrotelm the overall annual litter formation ought to be in the range 71–85 g m^{-2} at least.

The earlier estimate of the overall net annual productivity on the W Stordalen mire, 156 g m^{-2} yr^{-1} (Rosswall et al. 1975, Svensson and Rosswall 1980) refers to the whole mire including its minerotrophic parts and cannot be used for the ombrotrophic segments separately. Moreover, because of the patchiness in the vegetation cover on the ombrotrophic parts the annual litter formation may vary among microsites from 0 to 160 g m^{-2} yr^{-1} (cf. Table 5 and sect. The litter productivity). For the two moss dominated surface structures, together covering roughly 33% of the surface (sect. Study site), an average of at most 100 g m^{-2} yr^{-1} may be inferred. To account for the vascular plant litter on the remaining parts with only vascular plants and covering 30% of the surface, a supply of 10 g m^{-2} yr^{-1} may be inferred (sect. Mass balance of lichen hummocks). This suggests an overall litter production in the ombrotrophic systems of c. 35 g m^{-2} yr^{-1} which would just balance the decay losses from the acrotelm. However, as the long residence time in the acrotelm may disconnect the process of litter accumulation at the surface from the peat formation at the DDL, it is not unrealistic to have such a discrepancy in the relation between input and output from the acrotelm.

The overall litter productivity on the Research Station mire is quite different from that on the Stordalen mires. Not only is the productivity in the moss hummocks higher, but the whole surface is also covered with such hummocks. Therefore, the overall litter formation there will be c. 200 g m^{-2} yr^{-1} (Table 5) which is more than five times that estimated for the W Stordalen mire. We do not have any estimates of the decay rate and the residence time from the Research Station mire but a lesser peat depth suggests that either the decay losses in the acrotelm are higher or the initiation of the peat formation has taken place much later than on the Stordalen mires.

The decay rate in the catotelm is always much less than in the acrotelm (Clymo 1984). From model calculations for a moss hummock system on the W Stordalen mire the value for k in the catotelm has been estimated to be 10^{-5} yr^{-1} (Clymo 1978) or more than two orders of magnitude less than that found for the acrotelm in the present study. Assuming a bulk density of 100 g dm^{-3} (Fig. 1) below the DDL the annual decay losses should then be in the order of 1 g m^{-3} of peat in the catotelm. Even with a peat depth of 3 m such a low value indicates insignificant decay losses compared to the acrotelm.

From measurements of the fluxes of CO_2 and CH_4, Svensson (1974, 1980) estimated the total release of carbon from the ombrotrophic parts of the W Stordalen mire as 44.1 and 33.1 g m^{-2} yr^{-1} in 1973 and 1974, respectively. Of this carbon 30% is assumed to depend on dark respiration in the plants, which means that 23–31 g m^{-2} yr^{-1} ought to refer to the decay in the acrotelm and catotelm combined. In the minero-

trophic parts of the mire, the carbon fluxes may be about twice as high as in the ombrotrophic parts (Svensson 1974). The estimated decay losses in the acrotelm, 43 g m^{-2} yr^{-1} in this study, with an average concentration of C of 480 mg g^{-1} (Table 2), may correspond to a loss of 20 g m^{-2} yr^{-1} of carbon. Then the release of carbon from the catotelm should be in the range 3–11 g m^{-2} yr^{-1}. With a transfer of organic matter (concentration of carbon of 498 mg g^{-1}) to the catotelm in the range 31–38 g m^{-2} yr^{-1} the addition of carbon to the catotelm should be 15–19 g m^{-2} yr^{-1} suggesting a positive net carbon balance for the catotelm in the range 4–16 g m^{-2} yr^{-1} or as a rough average 20 g m^{-2} yr^{-1} of organic matter.

For the whole ombrotrophic system an overall input of litter at the LDL of 36 g m^{-2} yr^{-1} as calculated above, gives a carbon input of 17 g m^{-2} yr^{-1} to the acrotelm (cf. Table 2). With gaseous carbon fluxes in the range 23–31 g m^{-2} yr^{-1} (Svensson 1980), that input does not cover the annual carbon fluxes from the system, not even the lower estimate. At least these parts of the W Stordalen mire may therefore in relation to the atmosphere nowadays function as sources of carbon instead of sinks. A recent change from sink to source for carbon has also been suggested for sedge and tussock tundra in Alaska (Oechel et al. 1993).

To account for both the decay losses and the estimated transfer of organic matter to the catotelm at the DDL the overall deposition rate of litter (conc. of C 470 mg g^{-1}; cf. Table 2) on the ombrotrophic parts of the W Stordalen mire should be in the range 81–106 g m^{-1} yr^{-1}. That is more than twice that which can be estimated from the present study but only half of that found on the Research Station mire. The formation of true ombrotrophic peat has been going on for > 500 yr. The rather small variation in the concentration of N in the peat layers below the DDL suggests that for all that time the annual litter production ought to have been about twice the present one. Therefore, a change to a less productive vegetation cover during the last century may be inferred but the stratigraphy does not give any evidences of that. However, a patchy vegetation like the present one will form a *Sphagnum-Dicranum* type of peat through a regular replacement of lichen hummocks and bare peat with moss hummocks which produce the most decay resistant litter. Most probably, moss hummocks were much more widespread 100 yr ago than today, covering as much as two thirds of the surface instead of one fourth as at present. Abisko is situated in a region with distinct climatic gradients, so climatic shifts in particular may be supposed to have influenced both the initiation of the ombrotrophic peat formation and a recent change in the plant cover and litter formation rate.

Although the calculations in this study are impaired by uncertainties, the results suggest that because of a low biomass production, ombrotrophic mires in the

Abisko area no longer function as sinks for carbon in spite of the fact that the peat stratigraphy suggests an ongoing accumulation of peat. That shows that the present-day carbon balance of a peatforming system cannot be calculated from the peat accumulation rate during past periods as has often been done (e.g., Franzén 1994). Calculations of the present carbon balance for peatlands based on peat accumulation rates during the last few centuries or the last millennium may be irrelevant if there has been changes in the vegetation cover and decay rates in the acrotelm.

Acknowledgements – Tommy Olsson helped us with the ^{14}C-labellings and the construction of the equipment. Mimmi Varga has done all the chemical analyses and the determinations of the ^{14}C-activity. The work has been supported by a grant from the Swedish Natural Science Research Council.

References

Aerts, R., Wallén, B. and Malmer, N. 1992. Growth-limiting nutrients in *Sphagnum*-dominated bogs subject to low and high atmospheric nitrogen supply. – J. Ecol. 80: 131–140.

Åkerman, J. H. and Malmström, B. 1986. Permafrost mounds in the Abisko area, northern Sweden. – Geogr. Ann. 68A: 155–165.

Andersson, L. 1981. Vegetationskarta över de svenska fjällen. Kartblad nr 2 Abisko. – Liber Kartor, Stockholm.

Björck, S., Malmer, N., Hjort, C., Sandgren, P., Ingolfsson, O., Smith, R. I. L. and Liedberg-Jönsson, B. 1991. Stratigraphic and paleoclimatic studies of a 5500-year old moss bank on Elephant Island, Antarctica. – Arct. Alp. Res. 23: 361–374.

Clymo, R. S. 1978. A model of peat bog growth. – In: Heal, O. W. and Perkins, D. F. (eds), Production ecology of British moors and montane grasslands. Ecol. Stud. 27: 187–223.

– 1984. The limits to peat bog growth. – Phil. Trans. R. Soc. Lond. B. 303: 605–654.

Coulson, J. C. and Butterfield, J. 1978. An investigation of the biotic factors determining the rates of plant decomposition on blanket bog. – J. Ecol. 66: 631–650.

Damman, A. W. H. 1979. Geographic patterns in peatland development in eastern North America. – In: Kivinen, E., Heikurainen, L. and Pakarinen, P. (eds), Classification of peat and peatlands. Proc. Int. Peat Soc. Symp. Hyytiälä, Finland, 1979. Int. Peat Soc., Helsinki, pp. 42–57.

Du Rietz, G. E. 1945. Om terminologien för förna och organogen jord samt om circumneutral hedtorv och ängstorv ("Alpenhumus") i de svenska fjällen. – Geol. Fören. Förhandl. 67: 105–113.

– 1949. Huvudenheter och huvudgränser i svensk myrvegetation. – Svensk Bot. Tidskr. 43: 274–309, in Swedish with English summary.

Eurola, S. and Kaakinen, E. 1979. Ecological criteria of peatland zonation and the Finnish mire type system. – In: Kivinen, E., Heikurainen, L. and Pakarinen, P. (eds), Classification of peat and peatlands. Proc. Int. Peat Soc. Symp. Hyytiälä, Finland, 1979. Int. Peat Soc., Helsinki, pp. 42–57.

– and Vorren, K.-D. 1980. Mire zones and sections in North Fennoscandia. – Aquilo Ser. Bot. 17: 39–56.

Franzén, L. G. 1994. Are wetlands the key to the ice-age cycle enigma? – Ambio 23: 300–308.

Fries, T. C. E. 1913. Botanische Untersuchungen im nördlichsten Schweden. Ein Beitrag zur Kenntnis der alpinen und subalpinen Vegetation in Torne Lappmark. – Vetenskapliga och praktiska undersökningar anordnande av LKAB.

– and Bergström, E. 1910. Några iakttagelser öfver palsar och deras förekomst i nordligaste Sverige. – Geol. Fören. Förhandl. 32: 195–205.

Hogg, E. H. 1993. Decay potential of *Sphagnum* peat at different depths in a Swedish raised bog. – Oikos 66: 269–278.

Ingram, H. A. P. 1978. Soil layers in mires: function and terminology. – J. Soil Sci. 29: 224–227.

Jenny, H., Gessel, S. and Bingham, F. T. 1949. Comparative study of decomposition rates of organic matter in temperate and tropical regions. – Soil Sci. 68: 419–432.

Lee, J. A. and Woodin, S. J. 1988. Vegetation structure and the interception of acidic deposition by ombrotrophic mires. – In: Verhoeven, J. T. A., Heil, G. W. and Werger, M. J. A. (eds), Vegetation structure in relation to carbon and nutrient economy. SPB Publ., The Hague, pp. 331–346.

–, Woodin, S. J. and Press, M. C. 1986. Nitrogen assimilation in an ecological context. – In: Lambers, H., Neeteson, J. J. and Stulen, I. (eds), Fundamental, ecological, and agriculturel aspects of nitrogen metabolism in higher plants. Martinus Nijhoff, The Netherlands, pp. 331–346.

Malmer, N. 1975. Development of bog mires. – In: Hassler, A. D. (ed.), Coupling of land and water systems. Ecol. Stud. 10: 85–92.

– 1986. Vegetational gradients in relation to environmental conditions in northwestern European mires. – Can. J. Bot. 64: 375–383.

– 1988. Patterns in the growth and the accumulation of inorganic constituents in the *Sphagnum*-cover on ombrotrophic bogs in Scandinavia. – Oikos 53: 105–120.

– and Nihlgård, B. 1980. Supply and transport of mineral nutrients in a subarctic mire. – In: Sonesson, M. (ed.), Ecology of a subarctic mire. Ecol. Bull. 30: 63–95.

– and Holm, E. 1984. Variation in the C/N-quotient of peat in relation to decomposition rate and age determination with 210-Pb. – Oikos 43: 171–182.

– and Wallén, B. 1993. Accumulation and release of organic matter in ombrotrophic bog hummocks – processes and regional variation. – Ecography 16: 193–211.

–, Svensson, B. and Wallén, B. 1994. Interactions between *Sphagnum* mosses and field layer vascular plants in the development of peat-forming systems. – Folia Geobot. Phytotax. 29: 483–496.

Mårtensson, O. 1956. Bryophytes in the Torneträsk area, Northern Swedish Lappland. II. Musci. – Kungl. Vetenskapsakad. Avhandlingar i naturskyddsärenden 14.

Oechel, W. C., Hastings, S. J., Vourlitis, G., Jenkins, M., Riechers, G. and Grulke, N. 1993. Recent change of Arctic tundra ecosystems from net carbon dioxide sink to a source. – Nature 361: 520–523.

Persson, Å. 1961. Mire and spring vegetation in an area north of Lake Torneträsk, Torne Lappmark, Sweden. I. Description of vegetation. – Opera Bot. 6: 1–187.

– 1962. Mire and spring vegetation in an area north of Lake Torneträsk, Torne Lappmark, Sweden. II. Habitat conditions. – Opera Bot. 6: 1–100.

– 1965. Mountain mires. – Acta Phytogeogr. Suecica 50: 249–256.

Post, L. von and Granlund, E. 1926. Södra Sveriges torvtillgångar. – Sv. Geol. Unders. Avh. C 335.

Rosswall, T. and Granhall, U. 1980. Nitrogen cycling in a subarctic ombrotrophic mire. – In: Sonesson, M. (ed.), Ecology of a subarctic mire. Ecol. Bull. 30: 209–234.

–, Flower-Ellis, J. G. K., Johansson, L. G., Jonsson, S., Rydén, B. E. and Sonesson, M. 1975. Stordalen (Abisko), Sweden. – In: Rosswall, T. and Heal, O. W. (eds), Structure and function of tundra ecosystems. Ecol. Bull. 20: 265–294.

Rydén, B. E. and Kostov, L. 1980. Thawing and freezing in tundra soils. – In: Sonesson, M. (ed.), Ecology of a subarctic mire. Ecol. Bull. 30: 251–281.

Rydin, H. and Clymo, R. S. 1989. Transport of carbon and phosphorus compounds about *Sphagnum*. – Proc. R. Soc. Lond. B. 237: 63–84.

Sjörs, H. 1950. Regional studies in North Swedish mire vegetation. – Bot. Not. 103: 173–222.

– 1983. Mires in Sweden. – In: Gore, A. J. P. (ed.), Ecosystems of the world 4B, Mires: Swamp, Bog, Fen and Moor. Elsevier, pp. 69–94.

Sonesson, M. 1967. Studies on mire vegetation in the Torneträsk area, Northern Sweden. I. Regional aspects. – Bot. Not. 120: 272–296.

– 1968. Pollen zones at Abisko, Torne Lappmark, Sweden. – Bot. Not. 121: 491–500.

– 1969. Studies on mire vegetation in the Torneträsk area, Northern Sweden. II. Winter conditions of the poor mires. – Bot. Not. 122: 481–511.

– 1970a. Studies on mire vegetation in the Torneträsk area, Northern Sweden. III. Communities of the poor mires. – Opera Bot. 26.

– 1970b. Studies on mire vegetation in the Torneträsk area, Northern Sweden. IV. Some habitat conditions of the poor mires. – Bot. Not. 123: 67–111.

– 1974. Late Quaternary forest development of the Torneträsk area, North Sweden. 2. Pollen analytical evidence. – Oikos 25: 288–307.

– (ed.) 1980. Ecology of a subarctic mire. – Ecol. Bull. 30.

– and Kvillner, E. 1980. Plant communities of the Stordalen mire – a comparison between numerical and non-numerical classification methods. – In: Sonesson, M. (ed.), Ecology of a subarctic mire. Ecol. Bull. 30: 113–125.

–, Jonsson, S., Rosswall, T. and Rydén, B. E. 1980. The Swedish IBP/PT Tundra Biome Project. Objectives – planning – site. – In: Sonesson, M. (ed.), Ecology of a subarctic mire. Ecol. Bull. 30: 7–25.

Stuiver, M. and Reimer, P. J. 1993. Radiocarbon calibration program Rev. 3.0.3. – Radiocarbon 35: 215–230.

Svensson, B. H. 1974. Production of carbon dioxide and methane from a subarctic mire. – In: Flower-Ellis, J. G. K. (ed.), Progress report 1973. IBP Swedish Tundra Biome Tech. Rep. 16: 123–143.

– 1980. Carbon dioxide and methane fluxes from the ombrotrophic parts of a subarctic mire. – In: Sonesson, M. (ed.), Ecology of a subarctic mire. Ecol. Bull. 30: 235–250.

– and Rosswall, T. 1980. Energy flow through the subarctic mire at Stordalen. – In: Sonesson, M. (ed.), Ecology of a subarctic mire. Ecol. Bull. 30: 283–301.

– and Rosswall, T. 1984. In situ methane production from acid peat in plant communities with different moisture regimes in a subarctic mire. – Oikos 43: 341–350.

Svensson, G. 1986. Recognition of peat-forming plant communities from their peat deposits in two South Swedish bog complexes. – Vegetatio 66: 95–108.

– 1988. Bog development and environmental conditions as shown by the stratigraphy of Store Mosse mire in southern Sweden. – Boreas 17: 89–111.

Vitt, D. H. and Halsey, L. A. 1994. The bog landforms of continental western Canada in relation to climate and permafrost patterns. – Arct. Alp. Res. 26: 1–13.

Vorren, K.-D. 1972. Stratigraphical investigations of a palsa bog in northern Norway. – Astarte 5: 39–71.

– 1993. The mires of northern Norway. – In: Mörkved, B., Nilssen, A. C., Reymert, P. K. and Graff, O. (eds), Plant life. Way North, Univ. of Tromsø and Tromsø Museum, pp. 52–58.

Wallén, B. 1986. Above- and below-ground dry mass of the three main vascular plants on hummocks on a subarctic peat bog. – Oikos 46: 51–56.

– 1992. Methods for studying below-ground production in mire ecosystems. – Suo 43: 155–162.

Woodin, S. J. and Lee, J. A. 1987. The fate of some components of acidic deposition in ombrotrophic mires. – Environ. Pollut. 45: 61–72.

Ecological Bulletins 45: 93–103. Copenhagen 1996

A cross-continental comparison of phenology, leaf dynamics and dry matter allocation in arctic and temperate zone herbaceous plants from contrasting altitudes

Silvia Prock and Christian Körner

Prock, S. and Körner, C. 1996. A cross-continental comparison of phenology, leaf dynamics and dry matter allocation in arctic and temperate zone herbaceous plants from contrasting altitudes. – Ecol. Bull. 45: 93–103.

Here we investigate, across a wide range of latitudes and altitudes, the commonness of and differences in a number of key features related to growth and development of herbaceous plants of cold climates in comparison to warmer climates. In order to separate genotypic from phenotypic characteristics, we made reciprocal transplants of populations of the same or closely related species. We used plants from high and low altitude in the Tyrolian Alps (47°N, 2600 m and 600 m altitude) and the northern Scandes near Abisko (68°N, 1150 m and 380 m) plus a low altitude site in Spitzbergen (79°N, 50 m). The most important genera in this comparison were *Ranunculus* and *Geum*. In situ, our results show relatively small altitudinal and latitudinal differences in parameters like leaf weight ratio and leaf life span (except Spitzbergen), whereas differences in specific leaf area and fine root weight ratio were pronounced. Reciprocal latitudinal transplants within the alpine or lowland life zone revealed strong but species-specific phenorhythmic disorder related to photoperiod. We conclude that in a warming climate temperature-opportunistic species (e.g. *Geum* spp.) will profit in terms of biomass production and reproduction, while strongly photoperiod-controlled species (e.g. *Ranunculus glacialis*) will benefit little or not. These findings re-emphasize the need for considering phenology in predictive models of vegetation changes.

S. Prock, Inst. of Botany, Univ. of Innsbruck, Sternwartestr. 15, A-6020 Innsbruck, Austria. – C. Körner (correspondence), Inst. of Botany, Univ. of Basel, Schönbeinstr. 6, CH-4056 Basel, Switzerland.

Similar low air temperatures may be found in arctic and alpine regions, but most other components of the climate, in particular those depending on solar angles, differ greatly between these two life zones (Pisek 1960, Billings 1974, Chapin and Shaver 1985). During the short growing season the arctic climate exhibits less extremes of radiative forcing and day/night differences of temperature, and provides a 24 h photoperiod (Table 1). On the other hand regional altitudinal gradients exhibit pronounced thermal differences, whereas radiation sums may be very similar (Körner and Diemer 1987).

In this paper we summarize results of in situ comparisons and a reciprocal cross-continental transplant experiment in which we investigated both, latitudinal and altitudinal effects on plant characteristics of herbaceous perennials. Our comparison of plant traits in-

cluded 1) a high elevation group encompassing the temperate and subarctic, as well as 2) a lowland group of temperate zone, subarctic and high arctic zone species. The combination of 1) and 2) permits an evaluation of 3) altitudinal responses spanning the temperate and subarctic zone (Fig. 1). Populations of the same or congeneric species were investigated for developmental and allometric differences in leaf traits and biomass fractionation as related to the local climate or provenance.

The question of heritability of plant characteristics in cold regions has been addressed in the literature several times, although comparisons were either restricted to latitudinal or altitudinal provenances and/or included only one species. Lohr (1919) and Turesson (1925, 1926) were among the first who proved that ecotypic (i.e. genetic) differentiation does exist between popula-

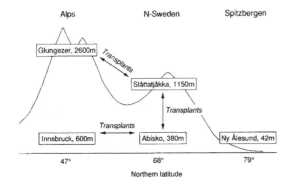

Fig. 1. The latitudinal and altitudinal position of sampling sites.

tions of latitudinally or altitudinally different habitats. They thus disproved Bonnier's (1894) idea that observed differences between morphologies of alpine and lowland taxa are largely phenotypic modifications. In principle, the genetic component of differing morphologies of altitudinal provenances of plants was noted already by Kerner (1898) in his classical transplant garden tests started already in the 1860's. The first statistically solid, fully reciprocal transplant experiment was conducted 80 yr later by Clausen et al. (1948). It took 100 yr from Kerner's pioneering work until a physiological component of such genetic differentiations within species was demonstrated (Mooney and Billings 1961, Mooney and Johnson 1965, followed by the work of Heide 1974, Holzer 1981, Chapin and Chapin 1981, Chapin and Oechel 1983, Sowell and Spomer 1986; see also reviews by Sørensen 1941, Billings 1957, Hiesey and Milner 1965).

The data presented here differ in two respects from earlier work. First, we present data from latitudinal as well as altitudinal comparisons, and transplantation

also includes within-site transplants to test for "excavating-replanting-only effects". Thus, the study also incorporates comparisons between undisturbed plants at all sites. Second, we combined Billings' (1957) idea that the comparison of related taxa is important with Whittaker's (1954) supposition that only multispecies comparisons (community averages) will permit some generalization (Körner 1991). This approach was also employed in an earlier survey on the latitudinal and altitudinal variation in stable carbon isotope discrimination by plants, from which a latitudinally increasing convergence of altitudinal differences emerged (Körner et al. 1991). Using plant material from the same sites as in the forementioned and the current study, it was further documented that leaf nitrogen concentration tends to increase as the environment becomes colder, irrespectively of whether this is due to latitude or altitude (Körner 1989). In all these studies it was impossible to separate genotypic (provenance-specific) from phenoptypic (local environment) effects, which is a central topic of the present study.

Transplant experiments represent one of the possibilities for identifying genotypic features but have intrinsic limitations, because the separation of monofactorial causalities is impossible due to the covariance of many environmental variables. In the present case we had to deal with at least five key variables: temperature, the dosis of quantum supply, the periodicity of quantum supply (photoperiod) and soil properties, including nutrient supply and moisture. By standardization of soils as far as practicable and by watering (when necessary at low altitudes) we tried to minimize the latter two sources of variance.

We tried to separate the temperature from radiation effects by comparing three equally cold, but latitudinally contrasting sites. Since latitudinally increasing day length largely compensates for lower mean quantum flux density (similar dosis), we believe that we can relate

Table 1. Climatic data for the five study sites for the growing season as defined by Fig. 2.

	Alps 47°N	Subarctic 68°N	Arctic 79°N
High altitude:	Glungezer	Slåttatjåkka	
Photoperiod (h)	14.5	>21	
Length of growing season (d)	124	100	
Air temperature (°C)	3.8	4.6	
Soil temperature (°C)	7.1	7.0	
Precipitation (mm)	520	250	
QFD (μmol m^{-2} s^{-1})[a]	748	415	
Low altitude:	Innsbruck	Abisko	Ny Ålesund
Photoperiod (h)	13	21	24
Length of growing season (d)	135[b]	132	72
Air temperature (°C)	12.6	8.1	4.6
Soil temperature (°C)	12.8	11.3	–
Precipitation (mm)	394	163	86
QFD (μmol m^{-2} s^{-1})[a]	769	446	412

[a] only hours with QFD > 30 μmol m^{-2} s^{-1}.
[b] first growing cycle of herbaceous plants only, total season length, including a second flush is c. 270 d.

our results from this comparison to photoperiod. The low altitude comparison between temperate zone and northern latitudes is weaker, since the covariance of photoperiod and temperature is only partly compensated via the two months shift of the major growth period considered here (i.e. the cooler period before the equinox in the temperate zone and after the equinox in the subarctic zone). Altitudinal transplants, which in the ideal case result in temperature changes without a change in photoperiod and quantum supply, may cause a rather severe physiological disorder once thermal gradients become too large. In the current project we had no mid-altitude temperate zone transplant garden that would permit direct comparisons with the thermal climate of northern Sweden. Hence, we refrained from any altitudinal transplants in the Alps (over a 2000 m altitudinal difference, low elevation plants would not have survived the first winter in the alien habitat). At the high arctic site in Spitzbergen only undisturbed plants were studied. Most of the species selected occur in both arctic and temperate life zones.

At each site a suite of key plant traits was measured one to three years after transplanting. These plant properties were selected to account for both developmental as well as growth characteristics. Main emphasis was put on leaf dynamics, flowering and patterns of dry matter allocation in leaves and in the whole plant. With four transplant sites, associated undisturbed sites and one additional high arctic field site, within-site comparisons of 4 or more species and of 5 or more plant characteristics over one to three climatically different years, this investigation reached a degree of complexity that made it virtually impossible to present a complete assessment of results in one short article. We have therefore focussed on observations which reflect the major trends that were observed. We believe this is the minimum complexity required to produce answers of a more general meaning. As illustrated by the results, we would have arrived at rather contrasting conclusions, had we selected only one genus or species, or only two sites. Obviously, it was impossible to include studies of site-specific genotypic variability within the selected species, which has been shown to be substantial (e.g. Callaghan 1974). By analyzing and transplanting propagules of different individuals, the variance of our data is likely to reflect a good deal of such population differences.

The aim of the study was 1) to document in situ functionally important plant differences between latitudinally contrasting life zones (inlcuding high and low altitudes), and 2) to test – by transplanting – to which extent these differences are genotypic. We considered the differences in photoperiod to be most important and made this climatological difference a focus of our discussion.

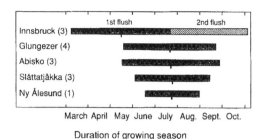

Fig. 2. The length of the growing season (number of years in brackets) at the five study sites as defined by emergence of green leaf tips and completion of leaf senescence in summer-green species. Note that only the main, i.e. the first growing cycle is considered in Innsbruck. Except for Innsbruck, seasons at all other sites are centered around 23 July.

Experimental sites, plant species and methods

Experimental sites and climate

Most of our field work was conducted in and around Innsbruck (Austria, 47°15′N, 11°25′E) and Abisko, northern Sweden (68°21′N, 18°49′E). By including observations and plant collections in Spitzbergen (Ny Ålesund, 78°55′N, 11°56′E) in the last year of the survey, not only altitudinal but also latitudinal limits of higher plant life were included. This project thus covered a latitudinal range of nearly 3000 km and an altitudinal range of 2550 m.

In order to compare the climate and phenology of the various sites in a meaningful way, we first had to define the site-specific growing periods. We used the beginning of leaf appearance (>5 mm leaf size) and whole plant senescence (c. 95% of aboveground tissue dead from visual impression) as markers. At the arctic and high altitude sites this period is similar to the snow-free period. However, at low altitude in the temperate zone many herbaceous plants (and all our experimental plants) complete their first and major growing period in mid-summer, followed by a second late season flush which was disregarded in this comparison. Figure 2 illustrates that "growing periods", as defined above, vary from 136 d at low altitude in Innsbruck to 70 d in Spitzbergen. Except for the low altitude site in Innsbruck the seasons of the remaining four sites are symmetrically centered around 20 July. Hence, except for the low altitude Innsbruck site, climatic comparisons are possible for astronomically identical periods of the year. Although growth occurred in four out of the five plant communities studied during the same calendar period, the photoperiods experienced differ by up to 10 h. Figure 3 illustrates the frequency distribution of quantum flux density for hours with QFD >30 μmol m^{-2} s^{-1} at the study sites. The mean daytime QFD does not differ between low and high altitude in Innsbruck (Table 1: 769 and 748 μmol m^{-2} s^{-1} resp.) which reflects the combined effects of longer day length during the growing season at low altitude (including the longest days in June) and higher maximum radiation at high altitude (Körner and Diemer 1987). Clearly, the two subarctic/arctic sites (seasonal mean QFD for hours >30 μmol m^{-2} s^{-1} of little >400 μmol m^{-2} s^{-1}) receive relatively less radiation in the high QFD range and relatively more hours with weak QFD, reflecting the long light periods of mid-summer nights. However, if one accounts for photoperiod and season length at the Swedish and Austrian sites (Table 1) the resulting seasonal doses of QFD become similar across latitudes.

Transplant gardens were installed at each of the 4 sites in the arctic and temperate zone. For the main transplant experiment we used homogenized local soils without fertilizer addition. In subplots at low altitude we also tested responses to nutrient addition in order to account for possible indirect effects of climate via soil fertility. The two gardens at low altitude (Innsbruck and Abisko) measured c. 4×6 m in area. Soils were mixed from 1/3 local loam, 1/3 sand (<3 mm) and 1/3 of local sub-humus soil (B/C horizon; in Sweden from a slope near Björkliden). At high altitude we inserted wooden frames of 1×2 m (northern Sweden) and 2×2 m (Alps) in the ground and filled them with 20 cm sifted fine substrate from the surrounding area. Frames protruded 5 cm above the ground and were covered with a 1.5 cm wire mesh to prevent feeding by snow mice (Alps) and reindeer (northern Sweden). Alpine

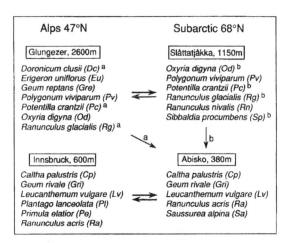

Fig. 4. Observed and (partly) transplanted plant species at the temperate and subarctic sites. Arrows indicate transplantation. Note that fully reciprocal transplants were successful in only 2–3 species per site, but one-sided transplant data exist for all species. a and b indicate which species were transplanted in the a and b direction as well.

soils were virtually identical in parent material at both sites (silicate sand with micaschist in both cases containing large garnet cristals). The two high altitude sites received full sunlight throughout the day. The two low altitude sites were shaded in late afternoon by buildings (Innsbruck) and birch trees (Abisko). Plots were watered when necessary (particularly at low altitude) to exclude drought effects. In order to make valid comparisons we also used natural sites with moist soil for all in situ studies.

Plant species and treatments

We selected typical herbaceous plant species, preferentially those that occur naturally at the various latitudes (Fig. 4). For altitudinal comparisons within the arctic and temperate life zone we used congeneric species (e.g. *Ranunculus glacialis* from high and *R. acris* from low altitude), because intraspecific transplants would be biased towards effects of "optimal vs pessimal" life conditions, not of interest here. Hence, we follow the philosophy of earlier comparisons (cf. Körner and Diemer 1987, Körner and Renhardt 1987) by using related species, each of which has a long evolutionary history in its present habitat (the center of its distribution). This approach runs counter to more stress physiologically oriented studies, which usually aim at comparing plant properties across the full range of a species' distribution.

Plants were collected in the field and transplanted during the last 10 d of June 1987. Clonal propagules from several individuals were used whenever possible and cohorts of the same clone were split between sites.

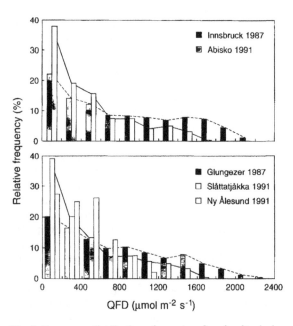

Fig. 3. Frequency distribution of quantum flux density during the growing period (cf. Fig. 2; only periods with QFD >30 μmol m^{-2} s^{-1}) at the five study sites. Dashed lines connect bars for the two temperate zone sites, straight lines connect subarctic data.

Table 2. In situ duration of the active phase of leaf life and its fractions and duration of leaf senescence (days ± SE).

	Leaf lifespan[a]	Leaf expansion	Leaf maturity	Leaf senescence
Innsbruck	78 ± 4 (16)	27 ± 1 (16)	51 ± 4 (16)	11 ± 1 (16)
Glungezer	87 ± 7 (14)[b]	25 ± 4 (5)	57 ± 7 (5)	12 ± 1 (14)
Abisko	95 ± 3 (14)[b]	22 ± 5 (4)	65 ± 5 (4)	11 ± 1 (14)
Slåttatjåkka	67 ± 4 (7)	18 ± 1 (7)	49 ± 4 (7)	10 ± 1 (7)
Ny Ålesund	39 ± 3 (7)	–	–	11 ± 1 (7)

[a] the sum of the duration of leaf expansion plus the mature life phase, excluding the period of leaf senescence (> 5% of leaf area chlorotic or dead until > 95% of leaf tissue is dead).
[b] note that species numbers for which data for lifespan and its fractions were available differ between sites, hence the sums of expansion and maturation do not precisely match the all-species means for lifespan.

Plant collection and planting at each transplant site was completed within 48 h. Ten (in some cases up to 20) individuals per plant species and provenance were planted at each site. Depending on adult plant size plants were spaced so that above-ground competition was minimized. All plots were weeded at regular intervals. The transplanting success was nearly 100% in low altitude plants and in *R. glacialis* at high altitude. In the other high altitude species first winter losses were severe in both transplant gardens. We report only data on species of which at least five individuals per provenance and site survived.

Methods

In order to compare plant development we recorded the time of leaf appearance, leaf length development (rate of expansion), leaf longevity, flowering dynamics, plant maturation and senescence. The dynamics of leaf development were documented by weekly census of labeled leaves following the methods decribed by Diemer et al. (1992). We distinguished three life phases of a leaf: leaf expansion, the mature life phase and leaf senescence. The latter was defined by the onset of leaf chlorosis (minimum 5% leaf area). In this paper we refer to "life span" as the "active life span" and disregard the senescence phase, because, once initiated, it is largely driven by weather randomness. Plant allometry was characterized by specific leaf area (SLA, dm^2 g^{-1}) and the fractionation of total plant dry matter into leaves, stems (including flowers), special storage organs (thick roots, rhizomes, tubers) and fine roots (<2 mm). Following Körner (1994) biomass fractionation was always expressed per total plant biomass, e.g. leaf weight ratio (LWR), above-ground vs total dry weight ratio, because ratios between plant parts such as root:shoot ratios can result in quite misleading numbers. For the determination of biomass fractionation two-thirds (at least five individuals) of all plant individuals transplanted were excavated by the end of the third experimental year (1990).

Results

Phenology of vegetative growth

The onset of vegetative growth at all sites did not differ between local populations and alien provenances, and thus was largely independent of photoperiod. Community means of leaf life span varied between 39 d in Spitzbergen and 95 d in Abisko. A difficulty is that life spans also vary from year to year (depending on snow melt and weather), particularly at high altitude and presumably also in the high arctic, but we have data for one season only from Spitzbergen. For instance, mean leaf life span at Mt Glungezer varied between 68 d (1987; Diemer et al. 1992) and 84 d (1990; this study). In contrast, our 3 yr census on Slåttatjåkka revealed rather similar mean leaf life spans of 69 (1988), 68 (1989) and 67 (1990) days. The variance in Abisko was again substantial: 69 (1988), 87 (1989) and 95 (1990) days. Furthermore, at low altitude in the temperate zone, where several leaf cohorts were produced even within the spring flush, differences of a few days occurred between early and late leaf cohorts. Thus, the overall means presented in Table 2 reflect site- as well as season- and cohort-specific variability. The differences of mean in situ leaf life spans of local populations from Innsbruck, Abisko and Mt Glungezer (78, 95, 87 d) were not significant and represent data for 14 to 16 different species at each site. These results correspond to estimates reported by Diemer et al. (1992) for a comparison within the Alps only. The leaf life spans for Slåttatjåkka (68) and particularly those for Spitzbergen (39) were significantly shorter (7 species per site). Variations in leaf life span were always due to changes in the duration of the mature life phase of leaves. The duration of leaf expansion was relatively invariable, leading to a highly significant correlation between total leaf life span and the length of the mature phase of leaf life (c. 60% or 43–65 d; no data for Spitzbergen) with no strong site- or latitude-specific effects. However, the data in Table 2 indicate a trend towards a shorter duration of the leaf expansion phase in cold sites in the North. It needs to be noted, however, that the final size of leaves differs by >10-fold across our two altitudinal gradients in Sweden and Austria, while cell size is rather invariable (Körner and Pelaez Menendez-Riedl 1989).

Transplants, although initiating vegetative growth at the same time as local populations, partly exhibit provenance-specific differences in the vigour of growth (not in *Geum rivale* and *Potentilla crantzii*). For the comparison of transplant effects our fully reciprocal sample size (species number) is reduced to four or less. We present data for *Ranunculus* and *Geum* (Fig. 5) that illustrate contrasting responses. In the case of *Ranunculus* temperate zone provenances show delayed leaf senescence when planted to the North. Arctic populations show accelerated leaf senescence when planted to the South. No such differences occur in *Geum*, and each further species considered, added a new pattern (cf. Prock 1994). The variation in high altitude provenances from each climatic zone was also substantial. Including all species for which at least part of the transplants were successful (Fig. 4; data not presented here), it appears that the majority of temperate zone provenances experience an extended active period in the North and northern provenances suffer from a shortened active period in the temperate zone. The two Rosaceae species included in our survey adjusted the beginning of leaf senescence to local climate, independent of photoperiod.

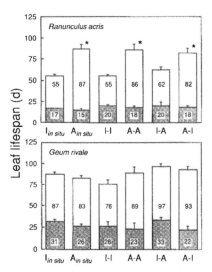

Fig. 5. Examples of species-specific responses of leaf life span to transplantation. I = Innsbruck, A = Abisko; I-I or A-A indicate local "transplanting-only" effects; I-A = plants from Abisko growing in Innsbruck; A-I = plants from Innsbruck growing in Abisko. The shaded part indicates the duration of leaf expansion. Bars indicate standard errors. Note that the site- and provenance-specific responses in *Ranunculus* are contrasted by the rather uniform patterns in *Geum* (means for 5–10 individuals). None of the differences for total life spans between groups of *Geum* individuals were significant, whereas the I-A, I-I/A-A, I-A/A-I, I/A-I and A/I-A differences in *Ranunculus* were all significant (p < 0.05).

An interesting part of the experiment was the joint growth of arctic-alpine (4 species) and temperate-alpine (3 species) provenances at low altitude in Abisko. This "climate warming" treatment for both groups of plants led to a 3-fold increase of leaf number and massive flowering in all species and provenances, except for the subarctic *R. glacialis*. Plants of this species produced only 4 leaves per shoot, exactly as in the natural arctic-alpine site, and the date of senescence was identical with the one in the mountain population reflecting a strict photoperiod control in this species.

While plants from high altitude and high latitude responded to varying seasonal weather conditions by varying their leaf life span, plants at warmer sites (e.g. Innsbruck) responded by varying rates of leaf initiation, and hence leaf number per tiller and season.

Flowering phenology

Flowering behaviour largely parallels leaf dynamics. While the onset of flowering in temperate plants transplanted to the North occurs roughly at the same time as in the natural northern populations, the southern provenances continue flowering for a longer period of time. As a consequence, temperate zone alpine plants with bud preformation in previous years (Larcher 1983), such as in *R. glacialis*, produce a second cohort of flowers when growing in the subarctic-alpine. When transplanted to the subarctic lowland (Abisko; a special experiment, not indicated in Fig. 1), temperate *R. glacialis* flowered three times in one season for the duration 3 yr of the study, although the final cohort of flowers late in the season hardly managed to open and was small. In summary, temperate zone plants growing in the North showed more and prolonged flowering than native populations, with no obvious disadvantages due to the lower temperatures. In contrast, northern provenances have problems in completing their reproductive cycle in the South. In some cases (*Caltha palustris*, *R. acris*, *Saussurea alpina*) even morphological anomalies in leaves and flowers were observed in subarctic populations transplanted to the temperate zone. Thus, as in leaf growth, photoperiodism control of flowering appears to be strong and latitude-specific, and our observations correspond to those of Heide (1985, 1992) in nordic grass ecotypes.

Leaf structure

Leaves are more "expensive" in terms of area produced per dry mass investment at successively colder sites, irrespective of latitude (Fig. 6). For altitudinal gradients, this was documented earlier in the temperate zone (Woodward 1979, Körner et al. 1989). The data presented here show that this altitudinal effect also exists at higher latitudes, although less pronounced. Herba-

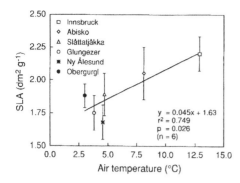

Fig. 6. In situ specific leaf area (SLA ± SE) of the species assemblages studied here as a function of mean air temperature during the growing season. Note that at the alpine and high arctic sites actual leaf temperatures of the rather prostrate plants may deviate from air temperatures more than in the more upright plants of the temperate and subarctic lowland sites.

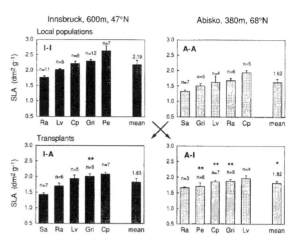

Fig. 7. Transplant effects on specific leaf area (±SE) in temperate zone and subarctic lowland plants. Treatment symbols as in Fig. 5. For species codes see Fig. 4. Arrows indicate the direction of transplantation (* = p < 0.05, ** = p < 0.01, *** = p < 0.001).

ceous plants from cold habitats (independently of latitude) produce only 1.8 dm² of leaf area per gram dry matter compared to c. 2.2 in plants that develop their leaves under relatively warmer climates (Fig. 6). Even though leaf temperatures are warmer than air temperatures during daylight hours, we found a linear correlation between SLA and site temperature (Fig. 6). Due to the low solar angle actual tissue temperatures are possibly substantially lower in Spitzbergen than in the central alps at 47° latitude, which may explain the variation among the cold-end sites of our gradients.

Transplanting low altitude species to the cooler North causes SLA to decrease significantly (by 20%), suggesting a largely phenotypic, possibly temperature-related response. This response is uniform in all five low altitude species tested. However, in subarctic lowland plants SLA did not change significantly when grown in the temperate zone. Their mean SLA (1.8 dm² g⁻¹) in the common garden at Innsbruck is still smaller than is the mean for local provenances, when data for the same species are considered only (2.0 dm² g⁻¹). Furthermore, we found no significant difference between foreign and native populations transplanted to a common garden in Abisko (Fig. 7). When temperate zone alpine plants are transplanted to a northern alpine site, SLA undergoes a small (NS) further reduction (1.78 dm² g⁻¹ on Mt Glungezer vs 1.63 dm² g⁻¹ in the transplants moved to Mt Slåttatjåkka in northern Sweden, 3 species; Fig. 8). In turn, Slåttatjåkka plants produced leaves with slightly higher SLA (NS) when grown on Mt Glungezer. Thus, similar to low altitude transplants, SLA tends to remain lower in the arctic provenances than in the local transplants from Mt Glungezer.

Taken together, we find both a (dominating) phenotypic adjustment of SLA (possibly related to local temperature) and a small, perhaps genotypic residual

causing arctic plants to retain a lower SLA in common gardens both at high and low altitude in the temperate zone. Decreasing SLA with decreasing temperature is likely to be related to greater leaf thickness and/or thicker cell walls. As mentioned above, we have evidence that cell size does not differ between temperate and arctic species (Körner and Pelaez Menendez-Riedl 1989). There is also a possibility that short days (long nights) in the South cause nordic provenances to "retain" lower SLA, since Heide et al. (1985) have shown that photoperiodicity can directly affect SLA in nordic grasses (higher SLA in grasses that receive a long-day treatment at equal total QFD and equal temperature).

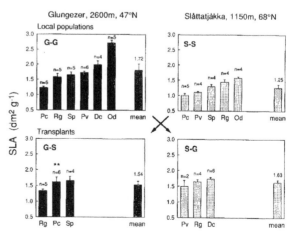

Fig. 8. As Fig. 7, but for alpine sites in the arctic and temperate zone. Symbols: G-G and S-S = local transplants; G-S = plants from Slåttatjåkka but growing at Glungezer; S-G = plants from Glungezer but growing at Slåttatjåkka.

Table 3. In situ plant dry matter allocation (%) at the various study sites at peak season (community means).

Location	n species	Total plant biomass (g)	Leaves	Stems incl. flowers	Flowers only	Storage organs	Fine roots	Total belowgr. fraction
Innsbruck	22[a]	2.83 ± 0.50	21.7	35.3	–	28.5	14.6	43.0
Glungezer	12	1.57 ± 0.55	16.4	17.0	6.9	23.1	43.6	66.7
Abisko	4 (7)[b]	2.50 ± 0.60	19.3	38.9	9.8	20.9	21.0	41.9
Slåttatjåkka	5 (11)[c]	0.90 ± 0.34	18.1	18.4	5.8	25.8	37.7	63.5
Ny Ålesund	8	0.20 ± 0.08	20.8	20.7	10.2	25.6	32.9	58.5

[a] including species from an earlier survey by Körner and Renhardt (1987) in which reproductive parts were included in the stem fraction.
[b] differing n for mean biomass and its fractions are due to the fact that three species exhibited massive vegetative expansion which made it impossible to delimit meaningful "individual" biomass.
[c] total biomass for 5 species only.

Cell division

Although leaf size differs enormously between the three cold-end sites and the lowland sites in the temperate and subarctic zone, cell size does not (see above). Transplants exactly match local leaf sizes. Since leaf expansion periods are similar or slightly shorter in cold compared to warm habitats and cell sizes are similar, one would expect rates of new cell production also to be similar. Although we do not have data on the duration of the cell cycle in developing leaves of each site, in situ mitotic indices for the genus *Ranunculus* are available. Leaves of equal expansion state (5% of final size) were sampled in 4-h intervals over 24 h and were found to have slightly lower fractions of cells in a mitotic stage (6% in high vs 7–8% in low altitude populations) with no latitude-specific differences. In no case did we find a diurnal change of mitotic index or a difference of the relative fraction of cells found in the pro-, meta-, ana- or telophase of mitosis. Hence, we found no evidence that the differences in photoperiod between low and high latitude sites caused any differences in mitosis or its individual steps as far as this can be detected by microscopy (Körner and Pelaez Menendez-Riedl 1989 and unpubl.). Since neither cell size nor the speed of cell division nor the duration of leaf expansion appears to be significantly different to explain leaf size differences, the number of initial cells and/or their participation in cell production must be different (i.e. reduced in cold habitats). In situ, interruption of the cell cycle with colchicine also suggests no substantial difference in the length of the cell cycle in low and high altitude *Ranunculus* (24–36 h; Pelaez Menendez-Riedl and Körner unpubl.).

Biomass and biomass fractionation

Mean total biomass of the herbaceous species varies by more than an order of magnitude among the five sites. While mature individuals accumulate a mean in situ biomass of c. 3 g in Innsbruck, only 0.2 g are attained in Spitzbergen (Table 3). The fractionation to the vari-

ous plant compartments changes much less and not in the same sequence as total plant mass. Remarkably, leaf weight ratio was between 16 and 22% across all sites, and the values for Innsbruck and Spitzbergen are virtually identical (Fig. 9; because the variability between individuals of the same species was larger than that between species, we used the individual plant data in these histograms, except for Innsbruck where we had a large species set from an earlier investigation; Körner and Rendardt 1987). The same holds for storage organs like rhizomes and roots thicker than 2 mm, which comprise between 21 and 29% across all sites. A major difference in overall allocation occurs in stem- and fine root fraction. While 35–39% of total plant biomass were found in stems (flowers included) in Innsbruck and Abisko, this fraction was reduced to 17–21% in the three cold sites. Fine roots represent 15–21% of total plant mass in Innsbruck and Abisko, but reached 33–44% in the cold habitats with Mt Glungezer, always representing the most extreme ratio, followed by Mt Slåttatjåkka and Spitzbergen.

As a result of this general trend of reduced allocation to stems and increased allocation to fine roots (with leaf and storage organ fractions remaining fairly stable), the fraction of below-ground plant parts increased from 42–43% (Innsbruck and Abisko) to 59–67% in the three cold sites. Differences were statistically significant between the lowland sites in Innsbruck and Abisko and the remaining sites, but no significant differences occurred within these two groups of sites. It needs to be noted that the data for Mt Glungezer may represent an extreme situation in the Alps. Data for 25 alpine species collected by Körner and Renhardt (1987; collection sites mostly around 3000 m altitude) were more similar to the rest of the sites considered here. We also found similar leaf weight ratios in 4250 m altitude in the NW-Argentinean Andes (Körner 1994). In other words, plant biomass fractionation appears to be a rather conservative trait, in particular when leaves and storage organs are considered and when comparisons are made within the same (!) life form. It is noteworthy that tussock grasses or dwarf shrubs from cold climates (not

Table 4. Plant dry matter allocation (%) in Ny Ålesund, Spitzbergen.

Species	Total plant biomass (g)	Leaves	Stems incl. flowers	Flowers only	Storage organs	Fine roots	Total belowgr. fraction
Oxyria digyna (5)[a]	0.76	8.5	13.9	7.3	47.0	30.6	77.9
Polygonum viviparum (6)	0.19	14.3	18.9	10.5	52.1	14.8	66.9
Braya purpurascens (4)	0.03	8.8	25.3	16.9	19.9	46.1	66.0
Ranunculus sulphureus (5)	0.08	19.3	22.4	10.2	17.0	41.3	58.3
Papaver dahlianum (4)	0.13	20.5	21.4	8.2	13.5	44.5	58.0
Draba corymbosa (5)	0.12	23.1	20.9	13.7	24.2	31.9	56.1
Saxifraga cernua (4)	0.12	27.8	23.3	5.1	23.9	24.9	48.8
Cerastium spp. (5)	0.17	44.4	19.1	10.0	7.4	29.2	36.6
Mean	0.20	20.8	20.7	10.2	25.6	32.9	58.5
±SE (n = 8)	±0.08	±4.1	±4.1	±1.3	±5.6	±3.8	±4.4

[a] number of individuals harvested.

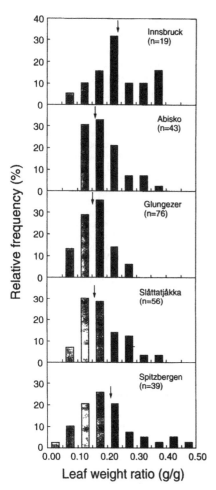

Fig. 9. Frequency distribution of leaf weight ratio (the fraction of leaf dry mass of total plant dry mass). Note that with the exception of Innsbruck, n refers to the number of individuals irrespective of species, because the variability between individuals was greater than between species. Innsbruck data are for 19 species. Arrows indicate means. In no case are group means significantly different.

considered here) often accumulate very large below ground masses, but these life forms are largely restricted to cold climates, not permitting valid comparisons with plants in warmer climates. Within the herbaceous dicotyledonous species studied here, the main change in allocation as the climate gets colder is a shift in biomass from stems to fine roots, confirming the analysis by Körner and Renhardt (1987). This shift could be considered as a compensation for increasingly weaker mycorrhization, particularly at high elevation (cf. Haselwandter and Read 1980). It should be noted that these means include substantial species to species variation, in particular with respect to investment in non-leaf tissue. Thus no particular biomass allocation pattern can be assigned to site-specific climatic conditions. Even at the same extremely cold habitat we may find differences between equally successful species that exceed the differences in community means by a factor of 2 (see also Körner 1991). As an example we present the full data set for Spitzbergen (Table 4) which exhibits a range of leaf weight ratios from 8.5 to 44.4%.

Plants growing in transplant gardens were substantially larger than those growing nearby in the natural environment which possibly is due to the better soil conditions (data not shown; c. 10 g per transplanted individual at low altitude compared to 2 g in nature). The gain in size was largely due to enhanced tillering. When transplanted from Innsbruck to Abisko or vice versa, plants reached only half the size (c. 5 g) of plants that were transplanted locally, but allocation patterns in the different species did not change in a characteristic manner. However, when plants from Innsbruck were transplanted to Abisko, there was a trend of increasing leaf weight ratio, and the opposite trend was found in north-to-south transplants. The long days in the North may signal "continued peak season", and thus allow vegetative growth to continue. *Ranunculus glacialis* for instance showed significantly increased leaf weight ratio

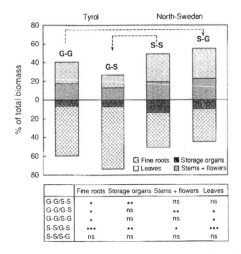

Fig. 10. Responses of biomass allocation to transplantation in *Ranunculus glacialis*. Symbols as in Fig. 8 (* = p < 0.05, ** = p < 0.01, *** = p < 0.001).

(LWR) when transplanted to the North, whereas in the north-to-south transplants (Fig. 10) LWR decreased. Again, photoperiod effects appear to be the most likely explanation.

Conclusions

In this study we compared development and allocation patterns in congeneric or the same herbaceous species from a wide range of latitudes and altitudes. Despite contrasting climates and widely separated evolutionary history we were surprised to find the onset of vegetative growth (except for the lowland site in Innsbruck), leaf life spans (except for Spitzbergen) and overall dry matter allocation to leaves to be very similar in all sites. In contrast, specific leaf area and stem weight ratio decreased, and fine root weight ratio increased as the sites became colder.

Transplantation effects fall into two groups: Changes in temperature cause specific leaf area to change in the same direction as it does when natural populations from thermally different regions are compared, and thus represents a predominantly (but not exclusively) phenotypic characteristic. In contrast, both leaf duration and flowering appear to be ecotypic characteristics in some species (*Ranunculus*) and not in others (the Rosaceae species studied here). The changes in photoperiod due to latitudinal displacement co-affected allocation patterns in photoperiod-sensitive species.

It appears that strong effects of photoperiod in some of the species thwart predictions of their future vigour and/or competitive success based on temperature scenarios alone. Heide (1985 and other papers) documented very high resolution of photoperiodic sensitivity in high latitude populations of herbaceous species

(mainly grasses). Our observations, particularly those in *Ranunculus glacialis*, are in line with Heide's suggestion that photoperiodism becomes more important for growth and survival the more severe the environment becomes.

We did not investigate the genotypic variability within populations of transplanted species (for instance variability in photoperiod sensitivity within original populations of species). It is almost certain that such a variability does exist, and assessing it might reveal a more realistic view of the likelihood of temperature-driven improvements in a species' ability to produce more biomass and perhaps more viable offsprings. If the differential behaviour between the two representatives of Rosaceae and the species from other families holds true, one might expect that the opportunistic behaviour of the Rosaceae herbs could cause their fitness to increase and perhaps also allow their abundance to increase in a warming climate. However, any such predictions also need to account for the changes of seedling survival under new climatic conditions. In many cases, slow vegetative spread may remain the major form of propagation, even in somewhat warmer climates, not permitting the application of simple migration scenarios following isotherm shifts (Steinger et al. 1996).

Acknowledgements – A large number of people had helped us throughout this survey. We are particularly grateful to M. Sonesson who facilitated the installation of the two transplant gardens in the Abisko area, and to him and the ANS staff in Abisko for technical support. E. Vogel, G. Bachmann, B. Schipperges and J. Baddley helped with synchronous leaf census work at the various sites, M. Svensson, K. Zegethofer and M. Engelhard with biomass harvests, R. Schuchter with data handling and statistics, M. Koch (Innsbruck), J. Hanssen-Bauer (Oslo) and N. Å. Anderson (Abisko) supplied meteorological data. M. Diemer provided practical advice and commented to the manuscript. We thank S. Pelaez-Riedl and L. Strasser (Basel) for drawing the figures and for the manuscript work. The final text profited from comments by S. Karlsson and T. Callaghan.

References

Billings, W. D. 1957. Physiological ecology. – Annu. Rev. Plant Physiol. 8: 375–391.
Billings, D. W. 1974. Arctic and alpine vegetation: plant adaptations to cold summer climates. – In: Ives, J. D. and Barry, R. G. (eds), Arctic and alpine environments. Methuen, London, pp. 403–443.
Bonnier, G. 1894. Recherches experimentales. L'adaptation des plantes au climat alpin. – Ann. Sci. Nat. 7e Sér., Bot. 19: 219–257.
Callaghan, T. V. 1974. Intraspecific variation in *Phleum alpinum* L. with specific reference to polar populations. – Arct. Alp. Res. 6: 361–401.
Chapin, F. S. III and Chapin, M. C. 1981. Ecotypic differentiation of growth processes in *Carex aquatilis* along latitudinal and local gradients. – Ecology 62: 1000–1009.
– and Oechel, W. 1983. Photosynthesis, respiration, and phosphate absorption by *Carex aquatilis* ecotypes along latitudinal and local environmental gradients. – Ecology 64: 743–751.

– and Shaver, G. R. 1985. Arctic. – In: Chabot, B. F. and Mooney, H. A. (eds), Physiological ecology of North American plant communities. Chapman and Hall, pp. 16–40.

Clausen, J., Keck, D. D. and Hiesey, W. M. 1948. Experimental studies on the nature of species. III. Environmental responses of climatic races of *Achillea*. – Carnegie Inst. Washington Publ. 581: 1–125.

Diemer, M., Körner, C. and Prock, S. 1992. Leaf life spans in wild perennial herbaceous plants: a survey and attempts at a functional interpretation. – Oecologia 89: 10–16.

Haselwandter, K. and Read, D. J. 1980. Fungal associations of roots of dominant and sub-dominant plants in high-alpine vegetation systems with special reference to mycorrhiza. – Oecologia 45: 57–62.

Heide, O. M. 1974. Growth and dormancy in Norway spruce ecotypes (*Picea abies*). I. Interaction of photoperiod and temperature. – Physiol. Plant. 30: 1–12.

– 1985. Physiological aspects of climatic adaptation in plants with special reference to high-latitude environments. – In: Kaurin, A., Junttila, O. and Nilsen, J. (eds), Plant production in the North. Norwegian Univ. Press, Oslo, pp. 1–22.

– 1992. Flowering strategies of the high-arctic and high-alpine snow bed grass species *Phippsia algida*. – Physiol. Plant. 85: 606–610.

– , Bush, M. G. and Evans, L. T. 1985. Interaction of photoperiod and gibberellin on growth and photosynthesis of high-latitude *Poa pratensis*. – Physiol. Plant. 65: 135–145.

Hiesey, W. M. and Milner, H. W. 1965. Physiology of ecological races and species. – Annu. Rev. Plant Physiol. 16: 203–216.

Holzer, K. 1981. Genetische Zusammenhänge der Fichtenverbreitung in den Alpen. – Allg. Forstzeitung (Forstl. Bundesversuchsanstalt Wien) 207: 421–424.

Kerner, A. von, 1898. Pflanzenleben II. – Bibliographisches Institut, Leipzig.

Körner, C. 1989. The nutritional status of plants from high altitudes. A worldwide comparison. – Oecologia 81: 379–391.

– 1991. Some often overlooked plant characteristics as determinants of plant growth: a reconsideration. – Funct. Ecol. 5: 162–173.

– 1994. Biomass fractionation in plants: a reconsideration of definitions based on plant functions. – In: Roy, J. and Garnier, E. (eds), A whole plant perspective on carbon-nitrogen interactions. SPB Publ., The Hague, pp. 173–185.

– and Diemer, M. 1987. In situ photosynthetic responses to light, temperature and carbon dioxide in herbaceous plants from low and high altitude. – Funct. Ecol. 1: 179–194.

– and Renhardt, U. 1987. Dry matter partitioning and root length/leaf area ratios in herbaceous perennial plants with diverse altitudinal distribution. – Oecologia 74: 411–418.

– and Pelaez Menendez-Riedl, S. 1989. The significance of developmental aspects in plant growth analysis. – In:

Lambers, H., Cambridge, M. L., Konings, H. and Pous, T. L. (eds), Causes and consequences of variation in growth rate and productivity of higher plants. SPB Publ., The Hague, pp. 141–157.

– , Neumayer, M., Pelaez Menendez-Riedl, S. and Smeets-Scheel, A. 1989. Functional morphology of mountain plants. – Flora 182: 353–383.

– , Farquhar, G. D. and Wong, S. C. 1991. Carbon isotope discrimination by plants follows latitudinal and altitudinal trends. – Oecologia 88: 30–40.

Larcher, W. 1983. Ökophysiologische Konstitutionseigenschaften von Gebirgspflanzen. – Ber. dt. Bot. Ges. 96: 73–85.

Lohr, P. L. 1919. Untersuchungen über die Blattanatomie von Alpen- und Ebenenpflanzen. – Ph.D. thesis, Basel.

Mooney, H. A. and Billings, W. D. 1961. Comparative physiological ecology of arctic and alpine populations of *Oxyria digyna*. – Ecol. Monog. 31: 1–29.

– and Johnson, A. W. 1965. Comparative physiological ecology of an arctic and an alpine population of *Thalictrum alpinum* L. – Ecology 46: 721–727.

Pisek, A. 1960. Pflanzen der Arktis und des Hochgebirges. – In: Ruhland, W. (ed.), Handbuch der Pflanzenphysiologie, Band 5. Springer, Berlin, Göttingen, Heidelberg, pp. 377–413.

Prock, S. 1994. Vergleichende ökologische Untersuchungen zur Phänologie, Blattlebensdauer und Biomasseallokation von krautigen Tal- und Gebirgspflanzen aus der Alpenregion, der Subarktis und der Arktis. – Ph.D. thesis, Univ. of Innsbruck, Austria.

Sørensen, T. 1941. Temperature relations and phenology of the northeast Greenland flowering plants. – Medd. Grønland 125.

Sowell, J. B. and Spomer, G. G. 1986. Ecotypic variation in root respiration rate among elevational populations of *Abies lasiocarpa* and *Picea engelmannii*. – Oecologia 68: 375–379.

Steinger, T., Körner, C. and Schmid, B. 1996. Long-term persistence in a changing climate: DNA analysis suggests very old ages of clones of alpine *Carex curvula*. – Oecologia 105: 94–99.

Turesson, G. 1925. The plant species in relation to habitat and climate. Contributions to the knowledge of genecological units. – Hereditas 6: 147–236.

– 1926. Die Bedeutung der Rassenökologie für die Systematik und Geographie der Pflanzen. – In: Fedde, F. (ed.), Repertorium specierum novarum regni vegetabilis 41: 15–37.

Whittaker, R. H. 1954. Plant populations and the basis of plant indication. – Angew. Pflanzensoziologie (Wien), Festschrift Aichinger 1: 183–206.

Woodward, F. I. 1979. The differential temperature responses of the growth of certain plant species from different altitudes. II. Analyses of the control and morphology of leaf extension and specific leaf area of *Phleum bertolonii* D. C. and *P. alpinum* L. – New Phytol. 82: 397–405.

Ecological Bulletins 45: 104–114. Copenhagen 1996

Hazards to a mountain birch forest – Abisko in perspective

Olle Tenow

Tenow, O. 1996. Hazards to a mountain birch forest – Abisko in perspective. – Ecol. Bull. 45: 104–114.

The mountain birch *Betula pubescens* ssp. *tortuosa* forest is a natural, non-managed forest. About every 9–10 yr, stands of this forest are defoliated by caterpillars of the indigenous geometrids *Operophtera brumata* and *Epirrita autumnata*, mainly on the western/maritime and the eastern/continental side of the Fennoscandian mountain chain, respectively. In the Abisko valley-Lake Torneträsk area, the two outbreak types occasionally meet. The great importance of these outbreaks for heath and meadow birch forests in the area is exemplified. Leaf-mining *Eriocrania* spp. and the sawfly *Dineura virididorsata*, known for outbreaks outside the area, occur in the Abisko valley in moderate or low numbers. Weather events ("frosts") have repeatedly decimated the birch bud population locally in the valley and exceptionally long-lasting birch rust *Melampsoridium betulinum* infestations recently caused premature foliage yellowing. Sometimes reindeer browsing of birch foliage has been substantial. Interactions between the forest and main agents are described. A new, or renewed, threat to the birch forest along the Norwegian coast, i.e. attacks of the microlepidopteran *Argyresthia retinella*, is presented. The position of the Abisko valley-Lake Torneträsk area within the outbreak areas of the treated insects in northern Fennoscandia is defined and changes in outbreak distributions are discussed in a time perspective.

O. Tenow, Dept of Entomology, Div. of Forest Entomology, Box 7044, Swedish Univ. of Agricultural Sciences, S-750 07 Uppsala, Sweden.

The mountain birch *Betula pubescens* ssp. *tortuosa* (Ledeb.) forest zone forms the upper border of the boreal forest in the Scandinavian mountain chain (the Scandes). The position of the mountain birch forest exposes itself to severe climatic restraints. It is also the scene for large-scale defoliations by several species of outbreak insects. The Abisko valley-Lake Torneträsk area (68°20′N, 19°00′E) belongs to this forest zone. The area (Fig. 1) is intermediate between the marked maritime region of the Norwegian coast to the west and the climatically continental region to the east. Shifts in weather may be rapid and radical in all seasons. Annual extreme minima of air temperature have been between −20 and −40°C (Holmgren and Tenow 1987). The variability is further enhanced by small scale topography and edaphic conditions.

The Abisko valley-Torneträsk area has access to the Atlantic coast over three pass-points (Fig. 1): through the valley of Bardu-Sördal (420 m a.s.l.), over Björnfjell (500 m a.s.l.) to Rombaken Fjord and from the Abisko valley over Unna Allakas (720 m a.s.l.) to Skjomen Fjord. The only direct passage with a continuous birch forest to the Norwegian coast is through the valley of Bardu-Sördal. A transect over each of these passages runs through the two main facets of the forest, one mostly poor heath birch forest as in the Abisko valley to the east (Sandberg 1965) and one luxuriant meadow birch forest at the Norwegian coast to the west (cf. Hämet-Ahti 1963, Tuhkanen 1984).

In the present paper, some hazards faced by the birch forest in the Abisko-Torneträsk area due to insect outbreaks and weather are focused on. The question of different stresses on the forests along a generalized coast-inland transect is also addressed, as well as risks for the forests in a time perspective.

The 1954–1955 outbreak of the autumnal moth

The birches in the Abisko valley are mostly polycormic (many-stemmed) with stems of rather distinct age-classes (cf. Sonesson and Hoogesteger 1983). Part of the

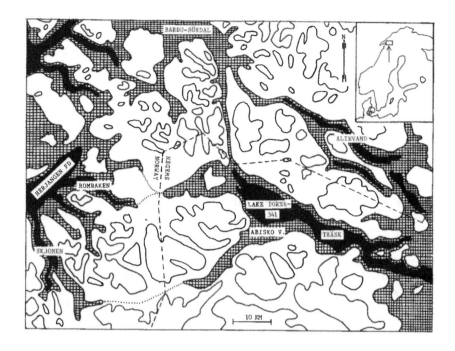

Fig. 1. Birch forest (check hatching) of the Abisko valley (A) – Lake Torneträsk area and the Norwegian coast. Passes to the Norwegian coast: through the valley of Bardu-Sördal, to Rombaken Fjord, and Skjomen Fjord. The 1000 m contour line is indicated above the forest line.

age structure origins from cutting by man (Emanuelsson 1987). Most of it, though, seems to be the result of an interaction with the autumnal moth, *Epirrita* (= *Oporinia*) *autumnata* (Bkh.) (Lep., Geometridae) (Bylund 1995). Outbreaks of this geometrid (Fig. 2A) occur about every 9–10 yr in the Scandes, although the peaks in the local valley reach outbreak levels only at longer intervals (Tenow 1972, Haukioja et al. 1988). In its upswing, the population seems to be driven mainly by feeding-induced amelioration of food quality and release from parasitism, in its downswing by feeding-induced delayed deterioration of food and delayed density dependent parasitism (Haukioja 1991, Ruohomäki 1994, Bylund 1995). Low temperatures in summer (Niemelä 1980) and winter (Bylund 1995) may synchronize outbreaks regionally.

During the 1954–1955 outbreak period, a severe outbreak extended into the Abisko valley. In the peak year 1955, there were 100–200 caterpillars/100 shoots (definition of shoot, see Maillette 1982) of the birches which equalled up to 1000 caterpillars m^{-2} forest floor. The caterpillars defoliated the forest completely and knowing the production of leaves per hectare (Bylund 1995) and the dry weight of a leaf (Haukioja and Iso-Iivari 1976), it is possible to estimate the amount of grazing in the centre of the Abisko valley to have corresponded to c. 780 kg d.w. of mature leaves ha^{-1} (Tenow 1956 and unpubl.). After having grazed all the leaves, the starving caterpillars also defoliated the ground vegetation, mainly *Betula nana*, *Empetrum hermaphroditum*, *Vaccinium myrtillus* and *V. vitis-idaea* (Tenow 1956, 1972).

There was one exception, however. The forest along the lake Abiskojaure-river Abiskojokk water system remained untouched by the outbreak. *Epirrita autumnata* overwinters as eggs on branches and trunks of the birches. Here, on the birches along the watercourse, the eggs froze in cold air accumulations in winter 1954–1955 which protected the birch forest from being defoliated (Tenow 1975, Tenow and Nilssen 1990).

The defoliation caused a 80–90% mortality of the leaf-carrying shoots of the birches and killed many stems (Tenow 1956, Palm 1959). The dwarf shrubs wilted all over the forest and were temporarily substituted by grasses, mainly *Deschampsia flexuosa* (Tenow 1972). Typical birch woodland mushrooms, like *Boletus scaber*, disappeared for several years (Sandberg 1963, 1965). The drastic vegetation changes in turn caused large changes in dependent faunas. Thus, e.g. *Hylecoetus dermestoides*, a beetle that attacks weakened birch trees increased and contributed locally to the killing of birch stems, and beetles living in decaying birch wood flourished (Palm 1959). Some of the insects feeding on birch leaves (Tenow 1963) disappeared for some years (see below). Initially, aquatic ecosystems were also influenced by the caterpillars which fell into brooks and small lakes and ended up as food for fish (cf. Nilsson 1955).

The polycormic mountain birch has a high suckering capacity (Vaarama and Valanne 1973). At the same time as the defoliation in 1955 killed many stems, it triggered a vivid production of basal sprouts which now form a dominating, c. 40-yr-old age-class of stems (Fig. 3A). The rejuvenation through basal sprouts and the recovery of surviving old stems resulted in the foliage growing out to c. 75% of its pre-outbreak leaf biomass by 1987. Still, another 30–40 yr is needed for full recovery, given that the present trend continues (Bylund 1995). In the inner parts of the Abisko valley the recovery has been less successful and the forest is more sparse and open than before.

Winter moth impacts

On the northern side of Lake Torneträsk, the birch forest is of the meadow type with tall monocormic (one-stemmed) birch trees. In 1964–1965, ten years after the *E. autumnata* outbreak, there was a severe defoliation on this side of the lake. The defoliating insects were caterpillars of the winter moth, *Operophtera brumata* L. (Lep., Geometridae) (Fig. 2B), although some damage was attributable to *E. autumnata* (Tenow 1972, Tenow and Bylund 1989).

The outbreak started in 1964 in a horizontal belt in a middle altitude position of the forested slope and spread upward and downward in 1965 (Tenow 1972). The stands at the top and bottom of the slope soon

greened without lasting damage but for most of the birch trees in the middle zone the attack was lethal (Sonesson and Hoogesteger 1983). Previously a luxuriant meadow forest, this zone is now strewn with fallen, decomposing trunks and is with its sparse, surviving trees more like a savanna (Fig. 3b). In autumn 1990, the middle zone produced only 2–3% of the pre-outbreak number of leaves (Tenow unpubl.).

Thus, while much of the forest south of Lake Torneträsk recovered almost completely in 35 yr, the damaged zone north of the lake had improved very little after 25 yr. One reason for the different ability to recover could be a weaker suckering capacity of the monocormic than the polycormic growth form of birch (cf. Kallio 1984). Another reason could be that reindeer browsing prevents establishment of birch seedlings in the damaged zone.

During outbreak periods, outbreaks of *O. brumata* are synchronized with those of *E. autumnata* (Tenow 1972). Usually the outbreaks occur only along the west and north coast of Norway and, therefore, this attack north of the lake was a temporal eastern expansion into an area where mass-occurrence of *E. autumnata* otherwise have dominated (Tenow 1972, Bylund 1995). The

Fig. 2. A) Caterpillar of the autumnal moth *Epirrita autumnata*. B) Caterpillar of the winter moth *Operophtera brumata*.

Fig. 2. C) Blotch mine of the birch leaf miner *Eriocrania* sp. D) Larvae of the sawfly *Dineura virididorsata*. All c. ×2.

origin of the outbreak may have been immigrants from the western *O. brumata* population via the birch stands of Bardu-Sördal (Fig. 1) where caterpillars of this species were common in 1964 (Tenow unpubl.). In 1994, a new outbreak of *O. brumata* started on the northern side of the lake in much the same altitudinal position as in 1964.

Operophtera brumata has been found very rarely in the Abisko valley on the southern side of the lake and only during years of *E. autumnata* peaks. Thus, in 1965 only 0.03 caterpillars/100 shoots were found in the Abisko valley, compared with 10.7 *E. autumnata* caterpillars/100 shoots (Bylund 1995). Why do outbreaks of *O. brumata* not occur on the southern side of Torneträsk? Firstly, outbreaks are not typical even for the northern side and may arise only when certain conditions are met. One prerequisite is a south-facing slope with an early budburst to which egg hatch seems to be best synchronized (Tenow 1972). The *O. brumata* female cannot fly due to stunted wings and dispersal of the species is mainly from ballooning by young caterpillars (Edland 1971). The Abisko valley lacks a direct, forested passage to the coast that could link larval populations (Fig. 1) and has few south- or southwest-facing slopes (with an early budburst) such that a lack of synchrony between budburst and egg hatch may hamper colonization and mass propagation.

Another factor preventing outbreaks in the Abisko valley could be low winter air temperatures. Like *E. autumnata*, *O. brumata* overwinters in the egg stage on the trees. The eggs freeze at a higher temperature (mean supercooling point: −35.0°C) than those of *E. autumnata* (SCP: −36.0°C) and cannot sustain prolonged chilling equally well (MacPhee 1967, Nilssen and Tenow 1990). Most eggs freeze after 16 h at an exposure to −33°C (MacPhee 1967); this and lower temperatures occur on average every 3–4 yr in the valley (records of the Abisko Scientific Reseach Station).

Other outbreak insects

In the Abisko valley two more insect species occur which are known for occasional outbreaks, however, not in the Abisko-Torneträsk area: leaf-miners of the microlepidopterous genus *Eriocrania* and the sawfly *Dineura virididorsata* (Retz.).

Fig. 3. A) Mountain birch individual with basal sprouts triggered by the 1954–1955 *Epirrita autumnata* outbreak. In centre of the sprouts, remnants of old stems killed by the outbreak. The Abisko valley, July 1985. B) Meadow birch forest on the northern side of Lake Torneträsk with middle zone devastated by the 1964–1965 *Operophtera brumata* outbreak: 7 September 1990. Photo: R. Axelsson. C) Birch trees damaged by the microlepidopterous moth *Argyresthia retinella*. Gullesfjord, Troms, Norway, 9 July 1991.

Eriocrania spp. (Lep., Eriocraniidae), at Abisko mainly *E. semipurpurella* (Stephens) and *E. sangii* (Wood) (Tenow 1963, Bylund and Tenow 1994), are small moths that fly at budburst and insert their eggs in the tips of young leaves. The larva lives inside the leaf and forms a yellowish or reddish blotch mine (Fig. 2C) from which it drops to the soil for overwintering when full-grown. In the Abisko valley the number of mines fluctuates with an alternate-year occurrence which

probably is due to a semivoltine life-cycle (Bylund and Tenow 1994).

During outbreaks of *Eriocrania*, the large number of mined leaves colours the birch stands yellowish brown. Mass occurrences are often more restricted than those of *E. autumnata* and seem not to cause severe damage. Outbreaks in northern Fennoscandia have been observed at both high (Koponen 1974, 1981) and low altitudes (Tenow unpubl.).

Fig. 4. A) Basal sprouts of mountain birch triggered by the 1954–1955 *Epirrita autumnata* outbreak and browsed by reindeer. The Abisko valley, July 1963. B) Top branches of birch damaged by frost in winter 1962–1963. The Abisko valley, July 1963.

Larvae of the sawfly *D. virididorsata* (Hym. Symphyta, Tenthredinidae) (Fig. 2D) feed in late summer on mature leaves. They graze the upper layers of the leaves without gnawing holes. The intact lower cuticula wilts and during outbreaks the attacked stands look reddish brown in the distance. The prepupa overwinters on the ground. Outbreaks are known from northeastern Fennoscandia (Koponen 1973, 1981): in 1896 over a large area at Karasjok in inner Finnmark, Norway (Schöyen 1897, cf. Koponen 1981) and in 1973 in a limited area in NE Finland (Koponen 1981). Because defoliations occur in late summer to autumn when the next year's buds have already formed, the birch trees do not suffer severe damage.

A new (or renewed) threat

A fifth insect has recently appeared as an outbreak species in the birch forests along the north Norwegian coast. This is the microlepidopterous moth *Argyresthia retinella* Zell. (Yponomeutidae) (Johansen and Kobro 1996) which previously has not been reported to occur in masses in Scandinavia or elsewhere (Svensson 1994). The small larvae tunnel in the petioles of developing birch leaves and in the buds of next year's foliage. As a result, the leaves wilt and the entire shoot dies. At least since 1990, birch trees have accumulated damage through this attack in stands along the coast from Narvik in the west to Alta and possibly farther to the northeast (cf. Solberg et al. 1994). Initially, an attack triggers a prolific compensatory production of adventi-

tious shoots which causes a highly clumped distribution of the leaves over the tree crown (Fig. 3C). With a long-lasting attack, however, the gradual elimination of leaf-carrying shoots results in death of branches until the whole tree finally succumbs. Because of the many bare branches, a severely attacked forest looks blackish in the distance.

In spite of a lack of older reports, *A. retinella* attacks are probably not a new phenomenon in the region. This is also indicated by old photographs (not shown) which show birch trees of the same severely damaged appearance as seen during the present outbreak (cf. Fig. 3C).

Reindeer browsing

Reindeer *Rangifer tarandus* L. regularly browse mountain birch leaves (Skuncke 1958, Haukioja and Heino 1974). When feeding, they strip off the leaves without biting off the twigs (Skjenneberg and Slagsvold 1968). Basal sprouts are preferably browsed when available, mainly at the tips (Fig. 4A) (cf. Haukioja and Heino 1974). In spring, the reindeer herds of the Talma and Gabna Saame villages move through the mountain birch region north and south of Lake Torneträsk, respectively, from their winter grounds in the coniferous region to the east (Emanuelsson 1987). Mostly they pass the birch forest rapidly on their migration to areas above the forest line. Occasionally, however, spring to early summer browsing on birch may become substantial. This happened in the Abisko valley in 1963. As a

Table 1. Approximate extent of biotic and abiotic damages to the leaf biomass of the mountain birch forest of the Abisko valley in 1955–1967 and 1984–1995. n.d. = no data. Figures within parentheses = probable but not checked.

Agents	Year	Destroyed leaf biomass (kg d.w. ha^{-1})	Area (ha)	Comments
E. autumnata	1955	780[a]	6000–7000	outbreak, all area
E. autumnata	1965	30[a]	(6000–7000)	non-outbreak peak
E. autumnata	1987	12[a]	(6000–7000)	non-outbreak peak
Frost belt	1962	n.d.	10–15[b]	630 m a.s.l.
Frost belt	1963	24[b]	n.d.	locally, 410 m a.s.l.
Frost belt	1985	n.d.	c. 40[c]	390–410 m a.s.l.
Frost belt	1991	20–>60[c]	c. 80[c]	410–430 m a.s.l.
Reindeer	1963	10[b]	n.d.	locally
Birch rust	1989–1991	n.d.	6000–7000	peak 1990, all area

[a] Estimated from Bylund (1995); [b] Tenow unpubl.; [c] Tenow and Holmgren unpubl.

result, 86% of the basal sprouts were browsed and 11% of shoots in the sprout stratum were locally deprived of their leaves in the centre of the valley (Tenow unpubl.). This corresponds to the removal of c. 10 kg d.w. ha^{-1} of mature leaves (cf. Haukioja and Iso-Iivari 1976, Bylund 1995) which is four times the amount browsed from sprouts by reindeers in NE Finland in 1973 in a forest which was severely damaged by E. autumnata caterpillars during an outbreak in 1965 (Haukioja and Heino 1974). However, the most important impact of reindeer browsing should be the suppression of growth of birch seedlings which may greatly delay the regeneration of birch stands devastated by E. autumnata or O. brumata (cf. Kallio and Lehtonen 1973, Haukioja and Heino 1974).

"Frost" damage

In early summer 1991, the trees in a horizontal zone at c. 410–430 m a.s.l. in the Abisko valley did not leaf out. This zone was narrow on the steep slopes of Mt Njulla and spread out on flatter ground in the centre of the valley. It stayed grey and distinct all summer and was still discernible in summer 1995. Most of the buds had died in their winter stage (cf. Fig. 4B), others survived but developed leaves that were pale green, deformed, wrinkled or with holes and ruptures. The injury was at the tips of branches and sprouts. It was hypothesized that the damage arose at a cold backlash in the middle of April after a very warm first half of the month when buds presumably were dehardened in a topoclimatically sensitive belt of the forest (Tenow et al. 1992). The foliage of Scots pine trees growing within the damaged zone turned reddish, while pine trees outside the zone remained green (Tenow et al. 1992). The zone was similar to the "red belts", which have been described from coniferous forests on valley slopes in north Europe and Canada and given various explanations (Langlet 1929, Venn 1962, Wellington 1976, Jalkanen and Nähri 1993).

Frost belts of the same type were observed in the Abisko valley in 1962 along c. 800 m at the forest line on Mt Njulla, in 1963 just north of the lake Vuolep Njakajaure in the centre of the valley (estimates, see below), and in 1985 in a zone of similar extension as in 1991 (Table 1). The 1963 frost damage seems to have been less extensive and severe than in 1985 and 1991.

A leaf pathogen

In 1989–1991, the birch stands of the Abisko valley and most of northern Fennoscandia were highly infested with birch rust, Melampsoridium betulinum (Fr.) Kleb., causing yellowing of all foliage already in late summer or early autumn. Heavy rust infections may affect photosynthesis negatively and cause a premature abscission of leaves (Lappalainen et al. 1995). Melampsaridium betulinum overwinters as spores in fallen leaves or as hyphae in dormant buds (Dooley 1984, Bennell and Millar 1984). It thrives in moist weather and the overwintering hyphae in the buds benefit from warm winters (Jalkanen and Nikula 1993).

Interactions

The most important interaction in the Abisko valley, between agents treated here, seems to be that between E. autumnata and the birch forest. The 1954–1955 outbreak triggered the growth of the now dominating, c. 40-yr-old age class of stems (see above). Older stems that survived the outbreak form two additional age classes, one rather small c. 70–90-yr-old and one still smaller c. 150–180-yr-old (Tenow et al. unpubl.).

These three age-classes of stems may all have been triggered by E. autumnata outbreaks with an average interval of c. 60 yr. It appears from independent data that E. autumnata attacks only mature and old forests severely (Tenow 1972, Bylund 1995). In the Abisko valley, E. autumnata outbreaks therefore seem to have a rejuvenating role from time to time. By this process and the protection of birch stands by freezing of eggs in cold air accumulations in winter, a large scale mosaic of differently aged stands is maintained in the Torneträsk area (Tenow and Bylund 1989).

There are also short-lasting and erratic interactions. One example is the indirect effect of *M. betulinum* infections (see above) on *E. autumnata* caterpillars. Caterpillars that were fed leaves from birch trees which were heavily infected by the rust the previous year experienced a longer larval period and gave lighter pupae than controls (Lappalainen et al. 1995). Lighter female pupae suggests that the resulting females will lay fewer eggs which should retard population growth the following year.

In the same respect, 1963 was somewhat special year for the mountain birch and its grazers. In the cold winter 1962/1963, the air temperature fell to the super-cooling point of *E. autumnata* eggs (see above). Because of that, 99% of the overwintering eggs died which resulted in a temporal reversal of the growth of the *E. autumnata* population (Bylund and Tenow 1994, Tenow unpubl.). The year continued with damage caused to the birch foliage by frost and reindeer (see above). Both types of damage and some of their effects on the *E. autumnata* and *Eriocrania* populations were quantified in the Abisko valley (Tenow unpubl.). The frost damage was concentrated to a horizontal belt, part of which occurred in the forest at the lake Vuolep Njakajaure. Here, 40, 24 and 15% of the birch buds in the upper and lower half of crowns and in the sprout stratum, respectively, were injured or killed. The early and warm spring caused reindeer herds to arrive earlier and stay longer in the Abisko valley than usual and browse more intensive. At Vuolep Njakajaure, the proportion of browsed sprouts on one plot inside the frost belt was as large as on one plot outside the belt. However, only 7% of the leaves were removed inside the belt compared with 14% outside it. This difference was statistically significant and should be due to that the frost had destroyed the most preferred leaves and, possibly, had induced deterrent, chemical resposes in remaining leaves (cf. Haukioja et al. 1988).

In summer 1963, the density of *Eriocrania* mines was lower inside the frost belt than outside it and more so in the upper half of birch crowns than in the lower half of crowns. There was a significant negative correlation between combined frost and reindeer damages and number of *Eriocrania* mines.

For *E. autumnata* the situation was reversed: the density of caterpillars was reduced only in the sprout stratum and mainly outside the frost-damaged area, i.e. where reindeer browsing only or mainly had occurred. Caterpillars were found only on sprouts browsed of <10–15% of their shoots.

Conclusively, *Eriocrania* eggs or young larvae were fewer or succumbed in frost-damaged buds and leaves,

and mines and caterpillars were unintentionally devoured or dislodged by browsing reindeer.

Epirrita autumnata and *Eriocrania* spp. themselves seem to be antagonistic. Because the *E. autumnata* caterpillars and the *Eriocrania* larvae both feed on birch foliage in early summer they are potential competitors and, actually, in the years immediatelly following the 1955 outbreak of *E. autumnata*, *Eriocrania* mines were not observed in the Abisko valley and the population did not appear at normal levels until 1959–1960. At intermediate population levels, the two insects avoid occupying the same branches of the host tree (Bylund and Tenow 1994).

We know from long-term studies that the population density of *Eriocrania* spp. in the Abisko valley was up to ten times higher in the newly rejuvenated, c. 10-yr-old birch forest of the 1960's than in the same c. 30-yr-old forest of the 1980's (Bylund and Tenow 1994). Thus, *Eriocrania* seems to prefer young birch trees and stands (cf. Heath 1976) while *E. autumnata* thrives in mature and old stands. These circumstances should tend to separate outbreaks of the two species in space and time.

The approximate extent of different damages inflicted on the birch forest in the Abisko valley is listed for comparison in Table 1 and some of the interactions are summed up in Fig. 5.

Large-scale and long-term perspectives

In some of the insects, the distribution of outbreaks is clearly zoned in a coast-inland direction (Fig. 6). Attacks of *A. retinella* are restricted to the near-coast birch forest, the most severe damage often being concentrated on the lowest part of the slopes (Fig. 7). Outbreaks of *O. brumata* occur in the same zone but extend further inland. Thus, in 1993–1995, in the overlap zone, birch stands severely damaged by *A. retinella*, were completely defoliated (Tenow unpubl.). The outbreak area of *E. autumnata*, in turn, partly overlaps with that of *O. brumata* but has its largest extent far into the local continental climate region to the east. One exception is for the Finnmarksvidda highland plain (Fig. 6) where extreme cold winters usually kill *E. autumnata* eggs and so prevents outbreaks (Tenow and Nilssen 1990). Where outbreaks of both geometrids occur in the same locality, the attacks of *E. autumnata* centre in the upper part of the slope, and those of *O. brumata* in the middle or lower parts. In 1995, e.g. at Gratangen close to the coast, 55 km NW of Abisko, there were mass-occurrences of *E. autumnata* above and *O. brumata* below in the slope (cf. Tenow 1965). In addition, in the lowermost part of the slope there was an incipient attack of *A. retinella* (Fig. 7).

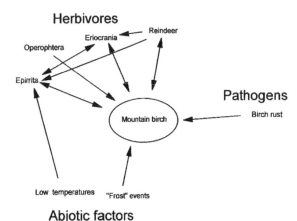

Herbivores

Abiotic factors

Pathogens

Fig. 5. Schematic representation of interactions and abiotic perturbations in the mountain birch-folivor system in the Abisko-Torneträsk area. Arrows from reindeer to insects indicate unintentional "predation" in the sprout stratum.

Eriocrania outbreaks seem to develop in relatively restricted areas, isolated in time in both maritime and continental climate regions (Fig. 6), at both high and low altitudes (see above) and do not show any horizontal zoning.

Dineura virididorsata outbreaks, finally, have been reported from eastern areas only, remote from the Abisko valley (Fig. 6).

The zoning of outbreaks should to a significant extent depend on the maritime-continental climate gradient. Hence, long-term climate changes will have consequences for insect sub-populations along this gradient. *Epirrita autumnata* and *O. brumata* occur in low numbers outside their outbreak areas. *Epirrita autumnata* is present on the Finnmarksvidda plain (Tenow and Holmgren 1987) in spite of the severe winter cli-

mate there. In a warmer climate with an inland extension of the maritime influence, a build up of the sub-populations on the Finnmarksvidda to mass-occurrence during outbreak periods will be possible. In fact, after the sequence of six unusually warm winters in 1987–1993, outbreaks during the current outbreak period occurred in 1993 and 1994 in several Finnmarksvidda localities where outbreaks of *E. autumnata* have seldom or never been reported previously (Tenow 1972). At the same time, outbreaks of *O. brumata* built up far inland, e.g. along the northern shore of Lake Torneträsk. *Operophtera brumata* eggs are less cold-hardy than those of *E. autumnata*. Therefore, as with *E. autumnata*, the extension of the outbreak area in 1994 could partly be a consequence of the preceeding warm winters.

In the same time period (1987–1994), *A. retinella* populations along the Norwegian coast grew to be a threat to the birch forest as far east as into Finnmark county (cf. Solberg et al. 1994). The distribution of the attacks is obviously maritime (Fig. 6). However, the biology and overwintering mode of this insect in the North is largely unknown and the role of climate for the emergence of the attacks is therefore uncertain.

If these changes in insect occurrence are all caused by an increase in oceanity and this increase is only transient, the zones of outbreaks will retreat to their former positions when climate reverses: the extension of the *E. autumnata* outbreak area into the Finnmarksvidda will be ephemeral, the *O. brumata* bridge-heads inland will be deserted and *A. retinella* attacks will disappear, leaving birch stands to recover. As an example of this, the winter of 1993/1994 was almost as cold as "normal" on the Finnmarksvidda plain and as a result the outbreak of *E. autumnata* was eliminated in the centre of the plain by a complete eradication of the above snow

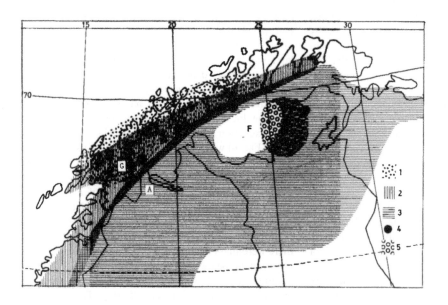

Fig. 6. Approximate distributions of insect outbreaks on birch forests in northern Fennoscandia. Numbered symbols:
1) *Argyresthia retinella*,
2) *Operophtera brumata*,
3) *Epirrita autumnata*,
4) *Eriocrania* spp.,
5) *Dineura virididorsata*.
A) the Abisko valley,
F) the Finnmarksvidda,
G) Gratangen Fjord (cf. Fig. 7).

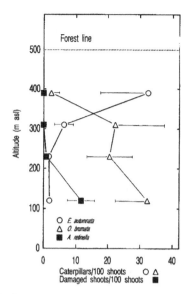

Fig. 7. Distribution of *Epirrita autumnata* and *Operophtera brumata* caterpillars and *Argyresthia retinella* damage along an upslope transect in a birch forest at Gratangen Fjord (G in Fig. 6), Troms, Norway, 1995. Horizontal bars: 2 × SE of the mean.

egg population. The beneath snow egg population survived by c. 50% (Tenow et al. unpubl.).

On the other hand, if the present trend of winter warming continues and escalates due to a predicted, anthropogenic greenhouse effect, then Finnmarksvidda will become established as a part of the outbreak area of *E. autumnata*, the northern shore of Lake Torneträsk will be permanently incorporated into the outbreak area of *O. brumata*, and *O. brumata* presence in the Abisko valley will increase, maybe causing significant changes in the forest/insect system of the valley. The *A. retinella* attacks may continue or repeat and penetrate higher upslope and deeper inland with potentially serious consequences for forest stands.

Epirrita autumnata caterpillars grow faster than birch leaves in warm summers and reach maturity before food quality deteriorates which should be positive for population growth (Ayres 1993). However, summers are anticipated to become not only warmer but also more rainy (Bennetts 1995). Warm winters and moist summers may increase the incidence of severe birch rust infestations which should reduce the quality of birch leaves as food in subsequent years (see above). Hypotheses on changes in inducible resistance of leaves against folivores due to climate change frequently agree but not always (Ayres 1993). These and other uncertainties and complications make it difficult to foresee effects of a changed summer climate on the population dynamics of *E. autumnata* and other insects.

Acknowledgements – Over the years, the staff of the Abisko Scientific Research Station and the former head of the station, the late G. Sandberg, and the present head, M. Sonesson, kindly supported the autumnal moth/mountain birch project. J. Bale, H. Bylund, T. Callaghan, B. Holmgren and S. Karlsson read the manuscript and suggested many improvements. The Royal Academy of Sciences provided living and working facilities. Financial support was provided by the Faculty of Mathematics and Natural Sciences of the University of Uppsala, the Swedish Natural Science Research Council, and the Swedish Environmental Protection Agency. All these supports are gratefully acknowledged.

References

Ayres, M. P. 1993. Plant defence, herbivory, and climate change. – In: Kareiva, P. M., Kingsolver, J. G. and Huey, R. B. (eds), Biotic interactions and global change. Sinauer, Sunderland, pp. 75–94.

Bennell, A. P. and Millar, C. S. 1984. Fungal pathogens of birch in Britain. – Proc. Royal Soc. Edinburgh 85B: 153–167.

Bennetts, D. A. 1995. The Hadley Centre transient climate experiment. – In: Harrington, R. and Stork, N. E. (eds), Insects in a changing environment. 17th Symp. Royal Entomol. Soc., Acad. Press, San Diego, pp. 49–58.

Bylund, H. 1995. Long-term interactions between the autumnal moth and mountain birch: the roles of resources, competitors, natural enemies, and weather. – Ph.D. thesis, Swedish Univ. of Agricultural Sci., Uppsala.

– and Tenow, O. 1994. Long-term dynamics of leaf miners, *Eriocrania* spp., on mountain birch: alternate year fluctuations and interaction with *Epirrita autumnata*. – Ecol. Entomol. 19: 310–318.

Dooley, H. L. 1984. Temperature effects on germination of uredospores of *Melampsoridium betulinum* and on rust development. – Plant Dis. 68: 686–688.

Edland, T. 1971. Wind dispersal of the winter moth larvae *Operophtera brumata* L. (Lep., Geometridae) and its relevance to control measures. – Norsk Entomol. Tidsskr. 18: 103–105.

Emanuelsson, U. 1987. Human influence on vegetation in the Torneträsk area during the last three centuries. – Ecol. Bull. (Stockholm) 38: 95–111.

Hämet-Ahti, L. 1963. Zonation of the mountain birch forest in northernmost Fennoscandia. – Ann. Bot. Soc. 'Vanamo' 34.

Haukioja, E. 1991. Cyclic fluctuations in density: interactions between a defoliator and its host tree. – Acta Oecol. 12: 77–88.

– and Heino, J. 1974. Birch consumption by reindeer (*Rangifer tarandus*) in Finnish Lapland. – Rep. Kevo Subarctic Res. Stat. 11: 22–25.

– and Iso-Iivari, L. 1976. Local and annual variation in secondary production by *Dineura virididorsata* (Hym., Tenthredinidae). – Rep. Kevo Subarctic Res. Stat. 13: 26–32.

–, Neuvonen, S., Hanhimäki, S. and Niemelä, P. 1988. The autumnal moth in Fennoscandia. – In: Berryman, A. A. (ed.), Dynamics of forest insect populations. Patterns, causes, implications. Plenum Press, New York, pp. 163–178.

Heath, J. 1976. The moths and butterflies of Great Britain and Ireland. Vol. I. Micropterigidae – Heliozelidae. – Blackwell Scientific Publ., Oxford.

Holmgren, B. and Tenow, O. 1987. Local extreme minima of winter air temperature in high-latitude mountainous terrain. – In: Alexandersson, H. and Holmgren, B. (eds), Climatological extremes in the mountains. Physical background, geomorphological and ecological consequences. Uppsala Univ., Dept of Physical Geography, UNGI Rapport 65: 25–41.

Jalkanen, R. and Närhi, P. 1993. Red belt phenomenon on the slopes of the Levi fell in Kittilä, western Lapland. – The Finnish For. Res. Inst. Res. Pap. 451: 55–60.
– and Nikula, A. 1993. Forest damage in northern Finland in 1992. – The Finnish For. Res. Inst. Res. Pap. 451: 30–35.
Johansen, T. J. and Kobro, S. 1996. Outbreaks of three lepidopterous species in Norway. – Fauna Norvegica, Ser. B, in press.
Kallio, P. 1984. Adaptation and evolution at the northern limits of life. – Acta Univ. Oul. F 1. Scripta Acad. pp. 131–150.
– and Lehtonen, J. 1973. Birch forest damage caused by Oporinia autumnata (Bkh.) in 1965–66, in Utsjoki, N Finland. – Rep. Kevo Subarctic Res. Stat. 10: 55–69.
Koponen, S. 1973. Herbivorous invertebrates of the mountain birch at Kevo, Finnish Lapland. – Rep. Kevo Subarctic Res. Stat. 10: 20–28.
– 1974. On the occurrence and ecology of Eriocrania spp. (Eriocraniidae) and other mining insects of the birch in northernmost Fennoscandia in 1973. – Rep. Kevo Subarctic Res. Stat. 11: 52–64.
– 1981. Outbreaks of Dineura virididorsata (Hymenoptera) and Eriocrania (Lepidoptera) on mountain birch in northernmost Norway. – Notulae Entomol. 61: 41–44.
Langlet, O. 1929. Några egendomliga frosthärjningar å tallskog jämte ett försök att klarlägga deras orsak. – Svenska Skogsvårdsför. Tidskr. 27: 423–461, in Swedish.
Lappalainen, J., Helander, M. J. and Palokangas, P. 1995. The performance of the autumnal moth is lower on trees infected by birch rust. – Mycol. Res., in press.
MacPhee, A. W. 1967. The winter moth, Operophtera brumata (Lepidoptera: Geometridae), a new pest attacking apple orchards in Nova Scotia, and its cold hardiness. – Can. Entomol. 99: 829–834.
Malliette, J. 1982. Structural dynamics of silver birch. I. The fates of buds. – J. Appl. Ecol. 19: 203–218.
Niemelä, P. 1980. Dependence of Oporinia autumnata (Lep., Geometridae) outbreaks on summer temperature. – Rep. Kevo Subarctic Res. Stat. 16: 27–30.
Nilssen, A. and Tenow, O. 1990. Diapause, embryo growth and supercooling capacity of Epirrita autumnata eggs from northern Fennoscandia. – Entomol. Exp. Appl. 57: 39–55.
Nilsson, N.-A. 1955. Studies on the feeding habits of trout and char in North Swedish lakes. – Rep. Inst. Freshw. Res. Drottnig. 36: 163–225.
Ruohomäki, K. 1994. Larval parasitism in outbreaking and non-outbreaking populations of Epirrita autumnata (Lepidoptera, Geometridae). – Entomol. Fennica 5: 27–34.
Palm, T. 1959. Följdverkningar av fjällbjörkmätarens härjning i Abiskodalen 1954–56. En koleopterologisk undersökning somrarna 1958 och 1959. – Ent. Tidskr. 34: 120–136, in Swedish with German summary.
Sandberg, G. 1963. Växtvärlden i Abisko nationalpark. – In: Curry-Lindahl, K. (ed.), Natur i Lappland. Almqvist and Wiksell, Uppsala, pp. 885–909, in Swedish.
– 1965. Abisko National Park. – National Parks of Sweden Series I.
Schöyen, W. M. 1897. Statsentomologens beretning. – Aarsberetn. off. Foranst. Landbr. Fremme i Aaret 1896: 61–116, in Norwegian.

Skjenneberg, S. and Slagsvold, L. 1968. Reindriften og dens naturgrunnlag. – Universitetsforlaget, Oslo, in Norwegian.
Skuncke, F. 1958. Renbeten och deras gradering. – Lappväsendet – Renforsk. Medd. 4, in Swedish.
Solberg, S., Venn, K., Solheim, H., Horntvedt, R., Austarå, Ö. and Aamlid, D. 1994. Cases of forest damage in Norway 1992 and 1993. – Res. Pap. Skogforsk 24, in Norwegian with English summary.
Sonesson, M. and Hoogesteger, J. 1983. Recent tree-line dynamics (Betula pubescens Ehrh. ssp. tortuosa (Ledeb.) Nyman) in northern Sweden. – Nordicana 47: 47–54.
Svensson, I. 1994. Lepidoptera-calendar. – H. Hallberg/I. Svensson, Kristianstad, Sweden, in Swedish with English summary.
Tenow, O. 1956. Fjällbjörkmätarens härjningar i abiskodalen sommaren 1955. – Sveriges Natur 6: 165–173, 184–187, in Swedish.
– 1963. Leaf-eating insects on the mountain birch at Abisko (Swedish Lapland). With notes on bionomics and parasites. – Zool. Bidr. Uppsala 35: 545–56.
– 1965. Fjällbjörkmätaren och frostmätaren. En presentation av den nordnorska björkskogens "lövmask". – Troms Skogeierforening, Årsmelding 1964, Tromsö, pp. 27–37, in Swedish.
– 1972. The outbreaks of Oporinia autumnata Bkh. and Operophthera spp. (Lep., Geometridae) in the Scandinavian mountain chain and northern Finland 1862–1968. – Zool. Bidr. Uppsala, Suppl. 2.
– 1975. Topographical dependence of an outbreak of Oporinia autumnata Bkh. (Lep., Geometridae) in a mountain birch forest in northern Sweden. – Zoon 3: 85–110.
– and Holmgren B. 1987. Low winter temperatures and an outbreak of Epirrita autumnata along a valley of Finnmarksvidda, the "Cold Pole" of northern Fennoscandia. – In: Alexandersson, H. and Holmgren, B. (eds), Climatological extremes in the mountains. Physical background, geomorphological and ecological consequences. Uppsala Univ., Dept of Physical Geography, UNGI Rapport 65: 203–216.
– and Bylund, H. 1989. A survey of winter cold in the mountain birch/Epirrita autumnata system. – Mem. Soc. Fauna Flora Fennica 65: 67–72.
– and Nilssen, A. 1990. Egg cold hardiness and topoclimatic limitations to outbreaks of Epirrita autumnata in northern Fennoscandia. – J. Appl. Ecol. 27: 723–734.
–, Holmgren, B. and Bylund, H. 1992. Mountain birch forests freeze in warm winters. – Disturbance related dynamics of birch and birch dominated ecosystems. A Nordic Symp., Illugastadir, Fnjoskadalur, Iceland, 18–22 September 1992, p. 12.
Tuhkanen, S. 1984. A circumboreal system of climatic- phytogeographical regions. – Acta Bot. Fennica 127: 1–50.
Vaarama, A. and Valanne, T. 1973. On the taxonomy, biology and origin of Betula tortuosa Ledeb. – Rep. Kevo Subarctic Stat. 10: 70–84.
Venn, K. 1962. Frostbelter. En sjelden frostskade i Krödsherad. – Norsk Skogbruk 18: 591–594, in Norwegian.
Wellington, W. G. 1976. Life at the cloud line. – Nat. Hist. 85: 100–105.

Ecological Bulletins 45: 115–120. Copenhagen 1996

The significance of carnivory for three Pinguicula species in a subarctic environment

P. Staffan Karlsson, Brita M. Svensson and Bengt Å. Carlsson

Karlsson, P. S., Svensson, B. M. and Carlsson, B. Å. 1996. The significance of carnivory for three *Pinguicula* species in a subarctic environment. – Ecol. Bull. 45: 115–120.

Studies on in situ prey capture and the influence of prey capture on nutrient economy and growth are summarised and related to demographic characteristics for three *Pinguicula* species growing in subarctic Scandinavia.
Based on results from experiments evaluating the effects of trapping rate on growth together with relationships between plant size, seed production, fecundity and mortality we estimated the increase in certain fitness components resulting from carnivory. Prey capture allowed *P. vulgaris* to increase its seed output, manifested as an increase in the amount of seeds (total mass) produced per reproductive plant and as a higher frequency of reproductive events, as well as its survival rate. The other two species only showed a marginal fitness increase. The difference in response between species can be ascribed to the higher rate of prey capture by *P. vulgaris*.

P. S. Karlsson, Abisko Scientific Research Station, S-981 07 Abisko, Sweden and Dept of Ecological Botany, Uppsala Univ., Villav. 14, S-752 36 Uppsala, Sweden. – B. M. Svensson and B. Å. Carlsson, Dept of Ecology, Plant Ecology, Lund Univ., Ecology Building, S-223 62 Lund, Sweden.

Plants inhabiting nutrient-poor environments often have special adaptations related to nutrient acquisition (Pate 1994). One such adaptation is carnivory, i.e. the ability to trap prey, digest them and assimilate nutrients from them (Chandler and Anderson 1976, Lüttge 1983). Carnivory, as well as other adaptations that enhance nutrient acquisition, could be expected to carry associated costs (cf. Givnish et al. 1984, Knight 1992). The benefits of carnivory have been suggested to be restricted to habitats where light and water are abundant but where the availability of mineral nutrients is restricted (Givnish et al. 1984).

In the subarctic parts of northern Scandinavia, three species in the carnivorous genus *Pinguicula* can be found. Although the species are similar in many respects, they show marked differences in their habitat preferences, ranging from the nutrient-rich calcareous soils inhabited by *P. alpina* to the very nutrient-poor *Sphagnum* bogs where *P. villosa* is found (cf. Karlsson 1986 and references therein). In view of this difference, these species lend themselves very well to analyses of the benefits of carnivory since carnivorous plants are most often found in nutrient-poor habitats, where, as stated above, carnivory is assumed to be most beneficial.

Although numerous studies have been made on the ecology and physiology of carnivorous plants (see reviews by, e.g. Lüttge 1983, Benzing 1987, Givnish 1989, Juniper et al. 1989), to our knowledge no one has tried to quantify the significance of the carnivorous habit for population performance. In this paper we make such an attempt, drawing from the results of a series of studies on the three *Pinguicula* species in subarctic Scandinavia to clarify the extent to which the natural range of prey capture increases fitness components of these species. The potential benefits can be expected to be inversely related to soil nutrient status. Thus, we hypothesise that the benefits are larger for the species growing in the most nutrient-poor habitats than for species growing on richer soils.

The species

The *Pinguicula* species of northern Scandinavia (*P. alpina* L., *P. villosa* L. and *P. vulgaris* L.) are all herbaceous, perennial geophytes. All have a basal leaf rosette, usually with 2–8 leaves, and one or a few flower stalks, each bearing one flower. *Pinguicula villosa* and *P. vulgaris* have small, annual root systems and survive the winter as a small bud situated c. 1 cm below the soil surface. *Pinguicula alpina* differs from the other two by having a relatively large, perennial root system and a smaller winter bud (Karlsson 1986).

In situ prey capture

In *Pinguicula*, large variations in prey capture exist among species, individuals, habitats and years (Hanslin 1994, Karlsson et al. 1994; cf. also Dixon et al. 1980, Watson et al. 1982). In the Abisko area, the catch of most *Pinguicula* individuals contributes little to their annual nutrient requirements; however some plants can obtain almost all the nitrogen and phosphorus that they need during a given year from prey (Karlsson et al. 1994).

Compared with the other two species, *P. vulgaris* is a more efficient carnivore and catches almost twice the biomass of prey per unit leaf area (Karlsson et al. 1987, 1994). Furthermore, flowering *P. vulgaris* individuals are almost twice as successful as carnivores compared with conspecific non-flowering individuals. Thus, increased prey capture during seed production contributes to the higher nutrient requirements of reproductive plants (Karlsson et al. 1994). For the other two species, no effect of reproductive status on prey capture was found.

Assimilation of nutrients from prey

It is commonly assumed that nitrogen is the most important limiting factor for carnivorous plants (Heslop-Harrison 1978, Thompson 1981). Under some conditions, however, such as at high soil pH, phosphorus may be the critical nutrient limiting plant growth. Thus, for *P. vulgaris* growing on calcareous soils in the Abisko area, the supply of artificial prey (tiny agar blocks) containing only phosphorus enhanced growth, whereas blocks containing nitrogen only did not (Karlsson and Carlsson 1984). The plants gained more N and P when fed blocks with a complete set of nutrients than when fed N or P only; i.e. the three-way interaction between N, P and "micronutrients" (all essential nutrients except N and P) was significant.

Feeding plants with prey increases plant nutrient gain from the soil (Hanslin and Karlsson 1996, cf. also Aldenius et al. 1983, Karlsson and Carlsson 1984). The mechanism responsible for this response is not known. It could be the indirect result of an increase in root growth, and thus an increase in uptake capacity per plant, or increased uptake capacity per unit root mass or length.

The efficiency of nitrogen assimilation from prey, measured as the transfer of ^{15}N from ^{15}N-enriched *Drosophila* flies to plants, was estimated to be 30–40% in the field and slightly higher (40–50%) for greenhouse-grown plants (Hanslin and Karlsson 1996). In a similar experiment, the Australian *Drosera erythrorhiza* assimilated 76% of the *Drosophila*-N (Dixon et al. 1980). The predominant types of natural prey for *Pinguicula* in subarctic Sweden, Nematocera and Collembola (Karlsson et al. 1994), are considerably smaller than the fruit flies. The smaller size of the natural prey may allow nutrients to be absorbed more efficiently from them than from *Drosophila*. Assimilation efficiency was found to be lower in situ than in greenhouse-grown plants. This reduced efficiency in situ may have been due to low temperature or to some prey being washed off the leaves by rain (cf. Karlsson et al. 1987, Hanslin and Karlsson 1996).

Effects of prey capture on nutrient economy and growth

Prey capture enhances growth and nutrient economy in subarctic *Pinguicula* (Aldenius et al. 1983, Karlsson and Carlsson 1984, Karlsson et al. 1991), as well as in a number of other carnivorous plants (see reviews in Givnish 1989, Juniper et al. 1989). The benefits of prey capture have been suggested to be limited to habitats that are sunny, nutrient-poor and moist (Givnish et al. 1984). However, this prediction could not be verified when testing the response to artificial feeding at different levels of soil nutrient supply (Karlsson et al. 1991). Therefore, the response model of Givnish et al. (1984) was modified so that the response to carnivory depends on the competitive ability of the carnivorous plants in relation to coexisting non-carnivorous plants, rather than on soil nutrient status (Fig. 1). In this model, the response to carnivory decreases as soil nutrient availability increases. The non-carnivorous species can be expected to be more responsive to increased nutrient availability, and thus increase their growth rate (and competitive ability) at a higher rate, compared with the carnivorous species.

In most cases, carnivory is assumed to supplement the supply of nutrients from the soil (Givnish 1989). To our knowledge, there is only one case described where prey seemed to be the exclusive source of nutrients (Karlsson and Pate 1992). In other cases the effects of natural levels of prey capture are marginal (e.g. Stewart and Nielsen 1992, 1993). Two different approaches have been used to quantify natural prey capture and its contribution to plant nitrogen economy: 1) Field monitoring of prey capture indicated an average nitrogen

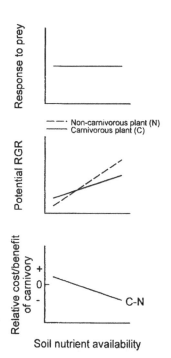

Fig. 1. Benefit-from-carnivory model (from Karlsson et al. 1991). Assume a carnivorous plant in which the gain from carnivory is independent of soil nutrient supply (top graph). Owing to the cost of carnivory the carnivorous plant (C) cannot increase its growth to the same degree as a similar non-carnivorous plant (N) (middle graph). Thus, the growth rate of the carnivorous plant relative to that of the non-carnivorous ones (C–N) decreases with increasing soil nutrient availability (lower graph).

contribution of c. 5–20% (Karlsson et al. 1994). 2) Using the fact that the natural abundance of ^{15}N in soil differs from that in prey, the relative contribution of nitrogen from prey to the plant nitrogen pool can be estimated by comparing the natural ^{15}N abundance of the carnivorous plant with that of its prey and non-carnivorous neighbours (Schulze et al. 1991). Prey importance estimates using this method are higher, viz. 34–49% (Hanslin 1994), than those based on prey counting (Karlsson et al. 1994). Results obtained with the ^{15}N natural abundance method also indicate that there is large variation among habitats. For example, in

P. alpina, prey dependence values ranging from 6% to 70% were obtained in four different habitats.

Demographic characteristics and resource investment into reproduction

In an eight-year study, between-year variation in plant density was higher in *Pinguicula villosa* than in the other two species, and *P. villosa* also had a considerably shorter life-span (Svensson et al. 1993, cf. Table 1), perhaps as a consequence of its more competitive habitat – the continuously growing *Sphagnum* carpet. Also, *P. villosa* invests comparatively more resources – i.e. dry matter, N and P – in flowering structures (Karlsson et al. 1990).

One can expect that in habitats with a harsh and unpredictable climate, such as the Subarctic, life-cycle events such as flowering and seedling establishment would be strongly dependent on current weather conditions. However, in many northern species flower primordia are formed up to three years before the plants actually flower (Sørensen 1941), thus indicating that flower production also depends on environmental conditions some years before. In either case, flowering and seedling emergence would be synchronised within a species, i.e. neighbouring study plots would behave similarly. *Pinguicula villosa* was found to differ from the other two species in this respect by showing less synchronisation in flowering and seedling establishment (Svensson et al. 1993). This, again, stresses the importance of its immediate neighbourhood: micro-environmental conditions influence the dynamics of *P. villosa* to a greater extent than they affect the dynamics of *P. alpina* and *P. vulgaris*.

Also, if a particular environmental factor (e.g. temperature or precipitation) is of equal importance for the three species, their flowering and seedling emergence behaviour would be similar and correlated to this factor. This was only partly the case: flowering was positively correlated with the current year's early-summer temperature in *P. alpina*, whereas the previous year's late-summer temperature showed a significant correlation with flowering in *P. villosa* and *P. vulgaris* (Svensson et al. 1993). This suggests that the latter two

Table 1. Comparison between success as carnivores (i.e. the dry weight of trapped prey per unit leaf area and season) and some demographic characteristics.

Species	Efficiency as carnivores (mg cm^{-2} yr^{-1})[A]	N gain from prey (% of annual turnover)[A]	Population			
			Reprod. effort[B] RE × P of flowering	P of flowering[C] (%)	Survival[C] (%)	Half-life[C] (years)
P. alpina	107	8–14	9.7	23	77	7.5
P. villosa	106	7–15	12.3	15	71	1.9
P. vulgaris	202	16–40	6.5	11	83	7.5

[A] Lower and upper value represent flowering and vegetative individuals (Karlsson et al. 1994). [B] Reproductive effort (in terms of biomass) from Karlsson (1988). [C] From Svensson et al. (1993).

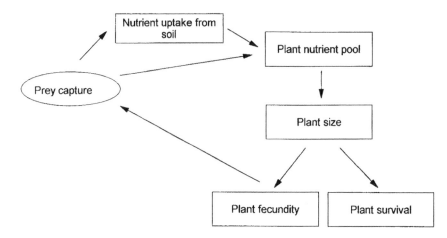

Fig. 2. Pathways through which prey capture can affect some plant fitness components (size, survival and reproduction).

species pre-form their flowers, whereas *P. alpina* does not. In *P. alpina*, seedling frequencies were positively correlated with late-summer temperatures the year before. This suggests that the seedlings of this species germinate in autumn.

There was large variation in the proportion of plants that flowered each year – from 8% to > 50%. In *P. villosa* and *P. vulgaris*, large vegetative individuals were the most likely to flower the following year, whereas in *P. alpina*, large flowering plants accounted for most of the flowering the year after (Svensson et al. 1993). This may be a reflection of the low relative somatic cost of reproduction (measured as the amount of dry matter, N and P in the winter bud) in this species (Karlsson et al. 1990), and of this species' perennial root system.

Seedling establishment also varied greatly between years and species: the proportion of seedlings varied from a few per cent to > 15% (Svensson et al. 1993). Seedling survival was generally very low. In *P. alpina*, some seedling cohorts had a higher survival compared with the other two species, thus compensating for the somewhat lower number of seedlings observed in this species.

Overall population growth rate (λ) measured over seven years, was close to unity for all three species, indicating stable populations (Svensson et al. 1993). However, there was large between-year variation in λ, especially for *P. villosa*. The most important transitions (based on elasticity analyses of size-class-based transition probability matrices) were small, vegetative individuals that either remained vegetative or flowered the following year. Generally, survival was more important than reproduction for the maintenance of population growth rate (Svensson et al. 1993).

Log-linear analyses revealed that for *P. alpina* and *P. vulgaris*, it was mainly size and reproductive status (i.e. flowering or vegetative) that determined their fate the following year (i.e. whether they would die, remain

vegetative or flower; Svensson et al. 1993). In *P. villosa*, however, the most important factor was environmental conditions in the current year rather than the status of the plant. Also, a significant interaction was found between size and reproductive status for *P. alpina* and *P. vulgaris* but not for *P. villosa*, again suggesting that this species is more dependent on current micro-environmental conditions than on former life-history events.

The significance of prey capture for population dynamics and fitness components

Individual *Pinguicula* plants can differ greatly in their success as carnivores; for example, we have observed over 50 nematocerans on one *P. vulgaris* individual, while another plant 20 cm away had trapped none. Such variability has also been observed between populations as well as between years in the same population. Indeed, most *Pinguicula* individuals were observed to trap only small amounts of prey, whereas a few successful individuals gained their whole annual nutrient requirement through trapped prey. Since we have not monitored the trapping success of the same plant individuals over several years, we do not know whether plants tend to be consistent in terms of their trapping success, or whether trapping success varies.

These characteristics make it difficult to generalise about benefits gained from the carnivorous habit at the population level. Nevertheless, we have attempted to summarise the effects of natural prey capture on some important components of plant fitness (Fig. 2). This was done by estimating plant size increase from the average natural prey capture levels, and comparing this with the average obtained from the 10% of the most efficient plants. Furthermore, we estimate the benefits of carnivory through a linear relationship between size

Table 2. Estimated increases in fitness components as a result of carnivory for the average plant (Mean) and for the 10% most efficient individuals of the population in trapping prey (Max.). The increase in fitness components are computed as the relative change in plant size (i.e. mass), the weight of the seeds produced, and in the likelihood that a plant will survive to, and reproduce during, the following season.

Species	Growth increase[A]	Seasonal catch[B]		Fitness increase (%)							
		Mean (mg)	Max. (mg)	Plant size		Seed mass		Survival		Flowering	
				Mean	Max.	Mean	Max.	Mean	Max.	Mean	Max.
P. alpina	14	0.21	0.71	3	10	1	4	0.4	1.2	5.1	17
P. villosa	73	0.05	0.21	4	15	0	0	0.8	2.8	5.1	19
P. vulgaris	47	0.61	1.39	29	65	12	23	6.9	15	48	110

[A] The growth response per unit of trapped prey (% [mg prey]$^{-1}$) is estimated from Aldenius et al. (1983), Karlsson and Carlsson (1984) and Karlsson et al. (1991). [B] From Karlsson et al. (1994).

gain and trapping success. Thus, using relationships between plant size, on the one hand, and in situ seed-output, mortality and fecundity on the other, we have estimated the fitness increase realised by trapping prey (Table 2). Generally, *P. vulgaris* showed a much stronger (6 to 17-fold) response to carnivory than the other two species.

We hypothesised that the benefits of carnivory should increase with decreasing soil nutrient availability (cf. Introduction). Thus, *P. villosa* could be expected to be the most dependent on prey, *P. vulgaris* intermediate, and *P. alpina* least dependent since it has the best prerequisites (a nutrient-rich substrate and a relatively large root system) for getting the nutrients it needs from the soil. The results presented here, however, show that this is not the case for the *Pinguicula* populations studied in the Abisko area. *Pinguicula alpina* and *P. villosa* are much less successful as carnivores compared to *P. vulgaris*. The higher trapping success of *P. vulgaris* is also the main reason for its larger fitness gain from carnivory (Table 2). However, it should be borne in mind that this benefit model is very simple, and that other mechanisms may exist that increase plant fitness through prey capture. Furthermore, our empirical database concerning the effect of prey capture on plant growth and size is limited, particularly for *P. alpina* and *P. villosa*. In a survey including several species with widely distributed populations, the hypothesised inverse relationship between the contribution of prey-derived nutrients and soil fertility (as indicated by soil pH) was, indeed, found to occur (Hanslin 1994).

The larger degree to which prey contributes to meeting the annual nutrient requirements in *P. vulgaris*, as compared with the other two species, does not seem to be mirrored in any of the demographic characteristics studied (Table 1). Neither their expected life span nor life-time seed output are correlated with their efficiency as carnivores. Thus, other differences, such as their habitat preferences and related factors, such as competition, could play such an important role that the effects of carnivory are not detected.

To be able to quantify more accurately the profits gained from carnivory, estimates of the associated costs that these plants have to pay for their carnivorous habit will be required. Here, we have focused on the contributions of prey derived nutrients to plant growth, nutrient economy and reproductive output. The direct costs associated with the ability to trap and digest prey (i.e. the costs of producing sticky mucopolysaccharides and enzymes necessary to digest the prey) and how these functions affect the photosynthetic capacity of the leaves are unknown.

Acknowledgements – We thank the Abisko Scientific Research Station, particularly its Director, M. Sonesson, for the support provided by him and the station staff during our work. M. Press and R. Zamora critically read the manuscript. D. Tilles corrected the English. Our studies have been supported by grants from the Swedish Natural Science Research Council (NFR).

References

Aldenius, J., Carlsson, B. and Karlsson, S. 1983. Effects of insect trapping on growth and nutrient content of *Pinguicula vulgaris* L. in relation to the nutrient content of the substrate. – New Phytol. 93: 53–59.

Benzing, D. H. 1987. The origin and rarity of botanical carnivory. – Trends Ecol. Evol. 2: 364–369.

Chandler, G. E. and Anderson, J. W. 1976. Studies on the nutrition and growth of *Drosera* species with special reference to the carnivorous habit. – New Phytol. 76: 129–141.

Dixon, K. W., Pate, J. S. and Bailey, W. J. 1980. Nitrogen nutrition of the tuberous sundew *Drosera erythrorhiza* Lindl. with special reference to catch of arthopod fauna by its glandular leaves. – Aust. J. Bot. 28: 283–297.

Givnish, T. J. 1989. Ecology and evolution of carnivorous plants. – In: Abrahmson, W. G. (ed.), Plant-animal interactions. McGraw-Hill, New York, pp. 243–290.

– , Burkhardt, E. L., Happel, R. E. and Weintraub, J. D. 1984. Carnivory of the bromeliad *Brocchinia reducta*, with a cost/benefit model for the general restriction of carnivorous plants to sunny, moist, nutrient-poor habitats. – Am. Nat. 124: 479–497.

Hanslin, H. M. 1994. Ecological aspects of the nitrogen nutrition and glandulation in some carnivorous *Pinguicula* and *Drosera* species. – Honour's thesis, Dept of Botany, Univ. of Trondheim, Norway.

– and Karlsson, P. S. 1996. Nitrogen uptake from soil and substrate as affected by prey capture level and plant reproductive status in four carnivorous plant species. – Oecologia, in press.

Heslop-Harrison, Y. 1978. Carnivorous plants. – Sci. Am. 238: 104–115.

Juniper, B. E., Robins, R. J. and Joel, D. M. 1989. The carnivorous plants. – Academic Press, London.

Karlsson, P. S. 1986. Seasonal pattern of biomass allocation in flowering and non-flowering specimens of three *Pinguicula* species. – Can. J. Bot. 64: 2872–2877.

– 1988. Seasonal patterns of nitrogen, phosphorus and potassium utilization by three *Pinguicula* species. – Funct. Ecol. 2: 203–209.

– and Carlsson, B. 1984. Why does *Pinguicula vulgaris* L. trap insects? – New Phytol. 97: 25–30.

– and Pate, J. S. 1992. Contrasting effects of supplementary feeding of insects or mineral nutrients on the growth and nitrogen and phosphorus economy of pygmy species of *Drosera*. – Oecologia 92: 8–13.

– , Nordell, K. O., Eirefelt, S. and Svensson, A. 1987. Trapping efficiency of three carnivorous *Pinguicula* species. – Oecologia 73: 518–521.

– , Svensson, B. M., Carlsson, B. Å. and Nordell, K. O. 1990. Resource investments in reproduction and their consequences for three *Pinguicula* species. – Oikos 59: 393–398.

– , Nordell, K. O., Carlsson, B. Å. and Svensson, B. M. 1991. The effect of soil nutrient status on prey utilization in four carnivorous plants. – Oecologia 86: 1–7.

– , Thorén, L. M. and Hanslin, H. M. 1994. Prey capture by three *Pinguicula* species in a subarctic environment. – Oecologia 99: 188–193.

Knight, S. E. 1992. Costs of carnivory in the common bladderwort *Utricularia macrorhiza*. – Oecologia 89: 348–355.

Lüttge, U. 1983. Ecophysiology of carnivorous plants. – In: Lange, O. L., Nobel, P. S., Osmond, C. B. and Ziegler, H. (eds), Encyclopedia of plant physiology NS 12C. Springer, Berlin, pp. 489–517.

Pate, J. S. 1994. The mycorrhizal association – just one of many nutrient acquiring specializations in natural ecosystems. – Plant Soil 159: 1–10.

Schulze, E.-D., Gebauer, G., Schulze, W. and Pate, J. S. 1991. The utilization of nitrogen from insect capture by different growth forms of *Drosera* from southwest Australia. – Oecologia 87: 240–246.

Sørensen, T. 1941. Temperature relations and phenology of the northeast Greenland flowering plants. – Medd. Grønl. 125: 1–305.

Stewart, C. N. and Nilsen, E. T. 1992. *Drosera rotundifolia* growth and nutrition in a natural population with special reference to the significance of insectivory. – Can. J. Bot. 70: 1409–1416.

– and Nilsen, E. T. 1993. Responses of *Drosera capensis* and *D. binata* var. *multifida* (Droseraceae) to manipulations of insect availability and soil nutrient levels. – N. Z. J. Bot. 31: 385–390.

Svensson, B. M., Carlsson, B. Å., Karlsson, P. S. and Nordell, K. O. 1993. Population dynamics of three *Pinguicula* species in a subarctic environment. – J. Ecol. 81: 635–645.

Thompson, J. N. 1981. Reversed animal-plant interactions: the evolution of insectivorous and ant-fed plants. – Biol. J. Linn. Soc. 16: 147–156.

Watson, A. P., Matthiessen, J. N. and Springett, B. P. 1982. Arthropod associates and macronutrient status of the red-ink sundew (*Drosera erythrorhiza* Lindl.). – Aust. J. Ecol. 7: 13–22.

Ecological Bulletins 45: 121–126. Copenhagen 1996

Photosynthetic characteristics of subarctic mosses and lichens

Barbara Schipperges and Carola Gehrke

Schipperges, B. and Gehrke, C. 1996. Photosynthetic characteristics of subarctic mosses and lichens. – Ecol. Bull. 45: 121–126.

This review deals with the photosynthetic characteristics of mosses and lichens in the Torneträsk area in northern Swedish Lapland. Most of the studied species have circumpolar distribution. It could be shown that the photosynthetic performance of cryptogams is well adapted to the special conditions of different subarctic microhabitats. Even different populations of the same species occur in different microhabitats. Probably in these cases there are not only plastic but also genetically fixed responses. Further, there were diurnal and seasonal variations in photosynthetic rates in mosses and lichens. Snow is an important mechanical and physiological factor for cryptogams in subarctic and arctic regions. The epiphytic lichen *Parmelia olivacea* could be used as indicator for the winter snow depth in the subarctic birch forest. Some species were able to photosynthesize even under a snow layer if light was not limiting (as in spring). Exposure of the heath moss *Hylocomium splendens* to ultraviolet-B radiation resulted in reduced biomass production. However, also stimulative effects of UV-B were found when UV-B was combined with additional irrigation. Interaction of UV-B and enhanced carbon dioxide concentrations reduced growth in *H. splendens* and in some lichen species. More ecophysiological work is needed to get a better understanding of the functioning of cryptogams in subarctic ecosystems.

B. Schipperges, Inst. for Polar Ecology, Wischhofstr. 1–3, Geb. 12, D-24148 Kiel, Germany (present address: Storvretsvägen 42, V., S-142 34 Skogås, Sweden). – C. Gehrke, Dept of Ecology, Plant Ecology, Ecology Building, S-223 62 Lund, Sweden.

Lichens and bryophytes form an increasing part of the vegetation towards the higher latitudes and altitudes. They partly dominate subarctic/arctic ecosystems, e.g. in the understorey of heathlands and forests, thus creating a specific temperature and humidity microenvironment for the underlaying root and soil system. Especially bryophytes can dominate whole ecosystems such as swamps, bogs and fens, accumulating carbon in deep peat layers. Many cryptogams have a high water holding capacity. They can stabilize the local water level supplying surrounding plants or even ecosystems with water. In addition, lichens (e.g. *Cladonia* and *Cetraria* species) are important as forage for reindeer in these regions, especially in winter.

Since many years several aspects of lichens and bryophytes are being studied in the Torneträsk area: their floristic composition and distribution (e.g. Fries 1907, Weimarck 1937, Magnusson 1952, Mårtensson 1955, 1956a, b, Klement 1959, Persson 1961, Sonesson 1970, Rumsey 1990, Alstrup 1991), systematics and taxonomy (e.g. Du Rietz 1926, Runemark 1956, Sones-son 1966) and ecology (e.g. Mårtensson and Berggren 1954, Rydén 1974, Sonesson and Johansson 1974, Sonesson et al. 1980, Baddeley 1991, Parsons 1991, Sonesson and Callaghan 1991, Malmer and Wallén 1996). During the last years ecophysiological studies in lichens and bryophytes (e.g. nitrogen-metabolism and CO_2 gas-exchange) on typical northern Scandinavian species have become important, supported by technical improvement of equipment for ecophysiological investigations.

The Torneträsk-area and the Abisko research station surroundings show a high diversity of geological and climatological conditions and thus offer a wide range of vegetation and habitat types. Detailed information is available about the flora and vegetation (e.g. Sylvén 1904, Fries 1913, 1919, Alm 1921, Gjaerevoll 1950, Persson 1961, Sandberg 1965, Sonesson 1970, Aronsson pers. comm.) and also about physical factors such as climate, geology and geomorphology (e.g. Sjögren 1909, Ångström and Tryselius 1934, Holdar 1960, Rapp 1981, Josefsson 1990, Nyberg and Lind 1990).

The aim of this paper is to summarize ecophysiological investigations, especially CO_2 gas-exchange, of lichens and bryophytes, carried out at Abisko and in the Torneträsk area. Such studies were encouraged by Mats Sonesson, to whose honour this volume is dedicated.

Special features of bryophytes and lichens

Bryophytes and lichens are poikilohydrous organisms, which means that they cannot control their tissue water content. Their water status is in balance with the humidity of the air. Cryptogams range from completely poikilohydrous (lichens, most bryophytes) to endòhydrous (few bryophytes, such as the genus *Polytrichum*). The gas exchange activity of cryptogams depends directly on the moisture content, light and temperature, which fluctuate with weather events. However, growth form (e.g. individual thalli, loose or tight cushions, mats) or morphology and anatomy (e.g. branching, development of water conduction elements and extended capillary networks in *Polytrichum*, and thickness of fungal layer in lichens) can influence their water status and gas exchange activity.

Growth strategies in different microhabitats

The first study concerning CO_2 exchange of cryptogams in the Abisko area was carried out in situ in 1974 in the Stordalen mire, c. 10 km south of Abisko, in connection with the International Biological Program, IBP, (Johansson and Linder 1980). The dominant bryophytes *Sphagnum fuscum* (in moderately dry hummock slopes) and *S. balticum* (on moderately wet depressions) showed different photosynthetic performances: *S. balticum* had higher photosynthetic rates than *S. fuscum* which was in agreement with biomass measurements in these species (Sonesson and Johansson 1974). Both species showed diurnal and seasonal variations in photosynthetic rates, with the highest rates around noon and in August. There were also different photosynthetic rates in different parts of the plants. In *S. fuscum* 98% of photosynthetic activity was found in the capitulum, whereas in *S. balticum* only 60% of the activity occured in the capitulum but 40% in the lower parts of the shoots. The limited number of data due to the destructive $^{14}CO_2$ method did not allow the authors to find a good relationship between the short-term photosynthetic performance and climatic variables.

The distribution pattern of two other common bryophyte species, *Polytrichum commune* and *Hylocomium splendens*, in the understorey of the subarctic birch forest was determined by their photosynthetic performance (Callaghan et al. 1978). *Hylocomium*

splendens, which has only aboveground tissue and shows photosynthetic activity only in the youngest segments (two years), grows in large carpets in cool, damp and undisturbed areas. The most limiting factor to its growth is water, and the seasons of maximum growth are spring and autumn. *Polytrichum commune* on the other hand has underground tissues and an internal water conduction system. This enables the species to colonize open unshaded and disturbed areas in the birch forest, because its photosynthesis can continue under dry and hot conditions. Growth of *P. commune* is limited mostly by low light, the maximum growth period is therefore the summer (Fig. 1).

The mosses *Polytrichum sexangulare* and *Anthelia juratzkana*, sampled at Låktatjåkka (in the mountains west to Lake Torneträsk; 68°24′N, 18°29′E) within a snowbed community, was determined by their capability to use low light intensities and by their low temperature optimum for photosynthesis (4–11°C) (Lösch et al. 1983). *Anthelia juratzkana* was found to be particularly adapted to the border zone along permanent snow patches. *Polytrichum sexangulare* showed higher photosynthetic rates and was more sucsessful at competing with higher plants at less extreme sites. However, both species can be photosynthetically active at temperatures

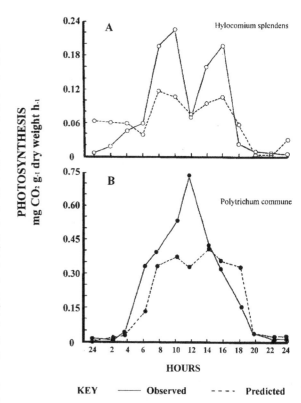

Fig. 1. Comparisons of observed diurnal photosynthesis (for 2 August 1975) with predictions made from multiple regression equations in (A) *Hylocomium splendens* and (B) *Polytrichum commune*. From Callaghan et al. (1978).

around 0°C and can probably also photosynthesize under the reduced light intensities under a snow cover (20–30 μmol m^{-2} s^{-1} PAR) which prolongs their time of productivity in spring.

In a field study the effect of microclimatic conditions on the primary production of the saxicolous lichens *Parmelia saxatilis* and *P. centrifuga* was investigated (Schulz and Schroeter 1992). Both species grew on the same boulder at different exposure, *P. centrifuga* on the top and southern side of the boulder and *P. saxatilis* at the northern slope of the boulder. Although *P. centrifuga* experienced higher temperatures overall than *P. saxatilis*, there was no difference in temperature and light conditions when the lichens were metabolically active. Water availability limited the range of temperature and irradiance usable for primary production in both *Parmelia* species.

Ecotypic differentiation

After the first gas exchange studies which were conducted with bryophytes, Sonesson (Sonesson 1986, Sonesson et al. 1992a) started CO_2 exchange measurements also with lichens. Of special interest was the question of ecotypic differentiation between lichen populations of the same species occuring at different altitudes. Lichens are symbiotic organisms consisting of an algal and a fungal partner. Therefore, ecotypic differentiation could exist in only one of the partners or in both. It is extremely difficult to measure the gas-exchange reactions of the two symbionts seperately. In general, lichens are considered to show high phenotypic plasticity which can be easily mistaken for genotypic variation (Kershaw 1985). Additional genetical experiments would be necessary to answer the question of ecotypic differentiation.

Two prominent subarctic lichen species, the chionophilous *Nephroma arcticum* and the chionophobous *Cetraria nivalis*, which occur both within the subalpine birch forest zone (at Abisko, c. 380 m a.s.l.) and in alpine fellfield sites above the treeline (at Slåttatjåkka, c. 1100 m a.s.l.) were the objects of the investigations. A combination of field ($^{14}CO_2$) and laboratory (IRGA) measurements showed two ecophysiological responses in each species (Sonesson 1986). The photosynthetic rates of the alpine population of *N. arcticum* were lower than those of the forest population (0.19 vs 0.36 mg CO_2 g^{-1} h^{-1}), whereas *C. nivalis* above the treeline showed higher photosynthetic rates (0.40 mg CO_2 g^{-1} h^{-1}) than on subalpine heaths below the treeline (0.25 mg CO_2 g^{-1} h^{-1}). Sonesson hypothesized that these differences were mainly due to ecotypic differentiation of the species.

In a following study with *N. arcticum* possible adaptations to the respective environments of the alpine and subalpine populations were investigated (Sonesson et al. 1992a). The total chlorophyll content (Chl a + b) in the subalpine population was about twice as high (from 0.6 to 1.0 mg Chl g^{-1}) as in the alpine population (from 0.4 to 0.6 mg Chl g^{-1}). Light compensation varied during the season between c. 40 and 100 μmol m^{-2} s^{-1} in the subalpine population and between c. 60 and 130 μmol m^{-2} s^{-1} in the alpine one. The photosynthetic performance of both populations in relation to light and also the differences in chlorophyll content reflected the differences in the light regimes at both altitudes. The subalpine and alpine populations were therefore referred to as "shade" and "sun" population respectively.

No differences in photosynthetic performance as related to light, temperature or thallus water content was found between the alpine and subalpine *C. nivalis* populations by Schipperges (unpubl.). However, transplantation experiments with *C. nivalis* populations from different latitudes and also measurements on specimen sampled along a latitudinal gradient from the Austrian Alps to Svalbard indicated that over long distances there is ecotypic differentiation in this species (Schipperges et al. 1995).

Influence of snow

The epiphytic lichen species *Parmelia olivacea* and *Parmeliopsis ambigua*, growing on birch tree trunks, are indicators of the limit of snow surface in winter. *Parmelia olivacea* grows only above the snow surface (this level is called the "Olivacea-limit"), whereas *P. ambigua* is mainly restricted to the bases of the tree trunks which are covered for a long period with snow (Sandberg 1958, Sonesson 1989). A combination of CO_2 exchange measurements in the laboratory and micrometeorological measurements in situ (Sonesson 1989) showed that *P. olivacea* can grow almost all year round. *Parmeliopsis ambigua* could only gain little of its annual carbon gain in late winter and early spring when snow covered. The shorter growing season of *P. ambigua* was not compensated by higher net photosynthetic or lower dark respiration rates. The lower saturation levels of light (300 vs 500 μmol m^{-2} s^{-1}), temperature (15 vs 20°C) and hydration (100% vs 200% of dry weight) for photosynthesis in *P. ambigua* were advantageous to its carbon gain. However, since the photosynthesis of *P. olivacea* in winter is mainly limited by light, even in this species most of the biomass must be formed during the snow free seasons.

Also along an 100 km long oceanic-continental gradient from Katterjåkka (68°25′N, 18°10′E; c. 20 km from the North Atlantic Ocean) to Rensjön (68°05′N, 19°45′E) the vertical distribution of *P. olivacea* and *P. ambigua* showed a positive correlation between the so called "*Parmelia olivacea*-limit" and the winter snow depth (Sonesson et al. 1994). The different degrees of sharpness of the limit at the different sites was probably due to mechanical factors such as windblown ice cristals rather than to physiological effects of snow. Other epiphytic lichens investigated (*Hypogymnia physodes* and *Parmelia sulcata*) showed only a weak correlation between their vertical distribution on trees and the snow depth. *Cetraria sepincula* showed no correlation with the height of the snow surface.

The influence of the spring snow cover (May–June) on the carbon balance of two epigeic (growing on soil) lichen species, *Cetraria nivalis* and *Cetraria delisei*, was investigated at Latnjajaure (68°21′N, 18°30′E; in the Abisko mountains at c. 1000 m a.s.l.), (Kappen et al. 1995). The chionophobous *C. nivalis* was covered only by a thin snow layer and became snow free early in the season (end of May). In contrast, the chionophilous *C. delisei* was most abundant at sites with a thick snow cover which became snow free about two weeks later. However, both lichen species experienced favourable conditions of light (up to c. 150 µmol m^{-2} s^{-1}), temperature (from -0.7 to $+1.2$°C) and thallus water content (140–250% of dry weight) for net photosynthesis under the snow. They were also able to utilize the higher CO_2 concentrations under the snow cover (mostly 450–500 ppm) compared to normal air for more efficient photosynthesis. This ability was stronger in *C. delisei* than in *C. nivalis* (Sommerkorn pers. comm.).

Global climate change

For some years the biological impact of changes in the composition of the earths' atmosphere has become a main item of research at Abisko. Factors of major concern are the depletion of the ozone layer, resulting in an enhancement of UV-B radiation, and the increasing atmospheric carbon dioxide concentration, probably resulting in a rise in global temperature and precipitation. The impacts of these changes are thought to be strongest at the high latitudes (Caldwell et al. 1980, Etkin 1990, Callaghan et al. 1992, Chapin et al. 1992, Cattle and Crosley 1995). High latitude ecosystems are already operating at their lower limit due to extreme climatic conditions, nutrient availability and short growing season. Despite the cryptogams' considerable contribution to ecosystem structure and functioning (Tenhunen et al. 1992), the knowledge of climate change impacts on these vulnerable organisms is just in the very beginning.

Impacts of enhanced UV-B radiation on bryophytes were studied mainly under field conditions. The moss *Hylocomium splendens*, growing in its natural heathland ecosystem, was exposed to enhanced UV-B radiation simulating a 15% ozone depletion (Gehrke et al. 1996). Biomass production was reduced after the second and third year of exposure. However, if additional water was supplied, growth of *H. splendens* was greatly stimulated under enhanced UV-B. This stimulative UV-B effects upon growth under optimal hydration conditions was confirmed in parallel greenhouse experiments. Further, stimulative effects upon physiology were shown in a short-term study (three months) during the unusually wet summer of 1992, when transplanted carpets of *H. splendens* were exposed to enhanced UV-B radiation, simulating a 10% ozone depletion in the atmosphere (Gehrke unpubl.). Photosynthetic efficiency, indicated by higher quantum yield, higher maximum net photosynthesis and lower light compensation point for photosynthesis, increased in young and old moss segments. Stimulation by UV-B radiation has been shown earlier for a few vascular plants (Tosserams and Rozema 1995) but so far mechanisms are unknown.

In another experiment, a bog dominated by the peat moss *Sphagnum fuscum* was exposed to enhanced UV-B, simulating a 15% ozone depletion. Height increment was reduced by 20% already after the first growing season of exposure (Gehrke et al. 1996).

The impact of different CO_2 concentrations on the gas exchange response of *Hylocomium splendens* was investigated in the laboratory (Sonesson et al. 1992b). Additional field measurements had shown that the ambient CO_2 concentrations, which *H. splendens* is normally exposed to at the forest floor, reached up to 1140 ppm for short periods, but were mostly between 400 and 450 ppm, a concentration that may occur as atmospheric CO_2 concentration over the next 50 yr. The data of the laboratory measurements suggested that CO_2 assimilation is limited by photosynthetic active radiation (PAR), temperature and water for most of the growing season. The higher CO_2 concentration in the immediate environment of the mosses counteracts some limitations of photosynthesis due to reduced PAR.

Interactive effects of enhanced UV-B and CO_2 concentrations on *H. splendens* were studied after five months of exposure in the laboratory (Sonesson et al. 1996). Enhanced CO_2 decreased photosynthetic efficiency, light compensation and maximum net photosynthesis, whereas no effect could be assigned to enhanced UV-B. The combination of CO_2 and UV-B resulted in decreased growth.

Three lichen species with different morphology, i.e. *Cladonia arbuscula*, *Cetraria islandica* and *Stereocaulon paschale*, were exposed to enhanced levels of UV-B radiation combined with normal and elevated CO_2 concentrations (Sonesson et al. 1995). Chlorophyll fluorescence measurements revealed an increase in photosystem II yield due to UV-B and no effect due to CO_2. At high CO_2 concentrations (1000 ppm) together with UV-B radiation the photosystem II yield was reduced. However, the authors point out that in both studies differences in the longwaved ultraviolet light

(UV-A, 320–400 nm)/UV-B/PAR ratios occurred between the UV-B treatments. Since UV-A is known to affect pigment production and to be involved in repair of UV-B damage (Middleton and Teramura 1994), further studies are needed to seperate the two counteractive effects.

Growth models

The calculation of productivity models for different species or ecosystems has become an important method to generalize and to simulate the performance of a species or an ecosystem under special conditions and to predict long-term reactions. Such models are quantitative descriptions of the relationship between a plant and the environmental influences on this plant. The growth model by Sommerkorn (pers. comm.) calculated for *Cetraria nivalis* and *Cetraria delisei* at the fellfield site Latnjajaure suggested that spring could be an important season for the annual carbon balance of both species. Although the lichens are still covered with snow, the light intensity is already high enough for photosynthesis. The low temperatures under the snow reduce dark repiration rate to a minimum during the nights. For *C. nivalis* about the same carbon gain was calculated during a short melting period as for *C. delisei* during a longer melting period (Kappen et al. 1995).

Schipperges (unpubl.) modelled productivity in *Cetraria nivalis* from Abisko for the snowfree months May to September using laboratory gas exchange data, results of field and laboratory water relation measurements and mesoclimatic data from the weather station at the research station. This model showed a considerable carbon gain in almost all of these months (between 0.73 and 5.72 g C kg C^{-1}), which will be necessary to balance respiration losses during the dark period in autumn and winter and for growth (Fig. 2).

Conclusions

The CO_2 gas-exchange studies carried out in lichens and bryophytes in the Torneträsk area during a period of c. 20 yr are reviewed. A lot of important results were obtained in these studies, which were briefly summarised here. Further research in the ecophysiology of lichens and bryophytes will be necessary with respect to the importance of these plant groups in subarctic vegetation. Growth models based on CO_2 gas-exchange data and data describing microhabitats, enable the prediction of cryptogam biomass production with respect to expected global climatic changes. Also the assessment of the carrying capacity for reindeer may be possible by means of growth models. More measurements of photosynthesis in lichens and bryophytes in situ even in winter is needed to assess the annual production of these poikilohydrous plants. Further, growth responses of the mosses *Polytrichum commune* and *Hylocomium splendens* to simulated environmental changes (increased temperature, water supply and nutrients) showed that nutrients play a larger role than expected (Potter et al. 1996). Therefore, nutrients and also the influence of higher plants on cryptogams (e.g. shading) must be considered in growth models, too. Up to date almost nothing was known about effects of ultraviolet-B radiation on mosses and lichens, and neither information existed about combined effects of UV-B and CO_2 on cryptogams. The experiments carried out at Abisko were therefore pioneer work in this field of research.

References

Alm, C. G. 1921. Floristiska anteckningar från Torneträskområdet. – Sv. Bot. Tidskr. 15: 224–227.

Alstrup, V. 1991. Lichens and lichenicolous fungi from the Torneträsk area. – Graphis Scripta 3: 54–67.

Ångström, A. and Tryselius, O. 1934. Total radiation from sun and sky at Abisko. – Geogr. Ann. 1: 53–69.

Baddeley, J. A. 1991. Effects of atmospheric nitrogen deposition on the ecophysiology of *Racomitrium lanuginosum* (Hedw.) Brid. – Ph. D. thesis, Univ. Manchester, Dept of Environ. Biol.

Caldwell, M. M., Robberecht, R. and Billings, W. D. 1980. A steep latitudinal gradient of solar ultraviolet-B radiation in the arctic-alpine life zone. – Ecology 61: 600–611.

Callaghan, T. V., Collins, N. J. and Callaghan, C. H. 1978. Photosynthesis, growth and reproduction of *Hylocomium splendens* and *Polytrichum commune* in Swedish Lapland. – Oikos 31: 73–78.

– , Sonesson, M. and Sömme, L. 1992. Responses of terrestrial plants and invertebrates to environmental change at high latitudes. – Phil. Trans. R. Soc. Lond. B. 338: 279–288.

Cattle, H. and Crosley, J. 1995. Modelling Arctic climate change. – Phil. Trans. R. Soc. Lond. A. 352: 201–213.

Chapin, F. S. III, Jefferies, R. L., Reynolds, J. F., Shaver, G. R. and Svoboda, J. (eds) 1992. Arctic ecosystems in a changing climate. An ecophysiological perspective. – Academic Press.

Du Rietz, G. E. 1926. Vorarbeiten zu einer "Synopsis Lichenum". I. Die Gattungen *Alectoria*, *Oropogon* und *Cornicularia*. – Ark. f. Botanik 20 A: 1–43.

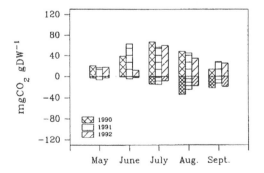

Fig. 2. Productivity balance (mgCO$_2$ gDW^{-1}) of *Cetraria nivalis* at Abisko 1990–1992 (May-September). According to Schipperges (unpubl.).

Etkin, D. A. 1990. Greenhouse warming: Consequences for arctic climate. – J. Cold Reg. Eng. 4: 54–66.

Fries, T. C. E. 1907. Om lavfloran i trakten af Torneträsk. – Sv. Bot. Tidskr. 1: 290.

– 1913. Botanische Untersuchungen im nördlichsten Schweden. Ein Beitrag zur Kenntnis der alpinen und subalpinen Vegetation in Torne Lappmark. – Vetenskapliga och praktiska undersökningar anordnade av LKAB.

– 1919. Floran inom Abisko nationalpark. – Arkiv f. Botanik 16: 1–48.

Gehrke, C., Johanson, U., Gwynn-Jones, D., Björn, L. O., Callaghan, T. V. and Lee, J. A. 1996. Effects of enhanced ultraviolet-B radiation on terrestrial subarctic ecosystems and implications for interactions with increased atmospheric CO_2. – Ecol. Bull. 45: 192–203.

Gjaerevoll, O. 1950: The snow-bed vegetation in the surroundings of lake Torneträsk, Swedish Lapland. – Sv. Bot. Tidskr. 44: 387–440.

Holdar, C.-G. 1960. The inland ice in the Abisko area. – Geogr. Ann. 41: 231–235.

Johansson, L.-G. and Linder, S. 1980. Photosynthesis of Sphagnum in different microhabitats on a subarctic mire. – Ecol. Bull. 30: 181–190.

Josefsson, M. 1990. The geoecology of subalpine heaths in the Abisko Valley, northern Sweden. A study of periglacial conditions. – UNGI Rapport, Diss. 78.

Kappen, L., Sommerkorn, M. and Schroeter, B. 1995. Carbon acquisition and water relations of lichens in polar regions – potentials and limitations. – Lichenologist 27: 531–546.

Kershaw, K. A. 1985. Physiological ecology of lichens. – Cambridge Univ. Press.

Klement, O. 1959. Zur Soziologie subarktischer Flechtengemeinschaften. – Nova Hedw. 1: 131–156.

Lösch, R., Kappen, L. and Wolf, A. 1983. Productivity and temperature biology of two snowbed bryophytes. – Polar Biol. 1: 243–248.

Magnusson, A. H. 1952. Lichens from Torne Lappmark. – Ark. f. Botanik Ser. 2, 2: 1–249.

Malmer, N. and Wallén, B. 1996. Peat formation and mass balance in subarctic ombrotrophic peatlands around Abisko, northern Scandinavia. – Ecol. Bull. 45: 79–92.

Mårtenson, O. 1955. Bryophytes of the Torneträsk area, Northern Swedish Lapland. I. Hepaticae. – KVA:s Avhandlingar i naturskyddsärenden 12.

– 1956a. Bryophytes of the Torneträsk area, Northern Swedish Lapland. II. Musci. – KVA:s Avhandlingar i naturskyddsärenden 14.

– 1956b. Bryophytes of the Torneträsk area, Northern Swedish Lapland. III. General part. – KVA:s Avhandlingar i naturskyddsärenden 15.

– and Berggren, A. 1954. Some notes on the ecology of the "copper mosses". – Oikos 5: 99–100.

Middleton, E. M. and Teramura, A. H. 1994. Understanding photosynthesis, pigment and growth responses induced by UV-B and UV-A irradiances. – Photochem. Photobiol. 60: 38–45.

Nyberg, R. and Lind, L. 1990. Geomorphic features as indicators of climatic fluctuations in a periglacial environment, northern Sweden. – Geogr. Ann. 72A: 203–210.

Parsons, A. N. 1991. Critical loads of nitrogen and sulphur for an ombrotrophic mire. – Ph. D. thesis, Univ. Manchester, Dept of Environ. Biol.

Persson, Å. 1961. Mire and spring vegetation in the area North of Lake Torneträsk. I. Description of vegetation. – Opera Bot. 6: 1–187.

Potter, J. A., Press, M. C., Callaghan, T. V. and Lee, J. A. 1996. Growth responses of Polytrichum commune (Hedw.) and Hylocomium splendens (Hedw.) Br. Eur. to simulated environmental change in the subarctic. – New Phytol., in press.

Rapp, A. 1981. Alpine debris flows in northern Scandinavia. – Geogr. Ann. 63 A: 183–196.

Rumsey, F. J. 1990. Additions to the bryophyte flora of the Torneträsk area, Swedish Lapland. – J. Bryol. 16: 199–208.

Runemark, H. 1956. Studies in Rhizocarpon. I. Taxonomy of the yellow species in Europe. – Opera Bot. 2.

Rydén, B. E. 1974. Bryophyte growth on a tundra mire in relation to radiant energy and water temperature. – Swedish IBP Tundra Biome Project, Tech. Rep. 16: 173–184.

Sandberg, G. 1958. Fjällens vegetationsregioner, vegetationsserier och viktigaste växtekologiska faktorer. – In: Skunke, F. (ed.), Renbeten och deras gradering. Lappväsendet – Renforskningen: Meddelande 4: 36–60.

Sandberg, G. 1965. Abisko Nationalpark. – K. Domänstyrelsen National parks of Sweden series 1965: 1–44.

Schulz, F. and Schroeter, B. 1992. The relevance of different microclimatic parameters on primary production in Parmelia saxatilis and Parmelia centrifuga. – In: Kärnefelt, I. (ed.), Second Int. Lichenol. Symp. Abstracts. Dept Syst. Bot. Univ. Lund 1992: 48–49.

Shipperges, B., Kappen, L. and Sonesson, M. 1995. Intraspecific variations of morphology and physiology of temperate to arctic populations of Cetraria nivalis. – Lichenologist 27: 517–529.

Sjögren, O. 1909. Der Torneträsk. Morphologie und Glazialgeologie. – GGF 31: 479–508.

Sonesson, M. 1966. On Drepanocladus trichophyllus in the Torneträsk area, northern Sweden. – Bot. Not. 119: 379–400.

– 1970. Studies on mire vegetation in the Torneträsk area, northern Sweden. III. Communities of the poor mires. – Opera Bot. 26.

– 1986. Photosynthesis in lichens populations from different altitudes in Swedish Lapland. – Polar Biol. 5: 113–124.

– 1989. Water, light and temperature relations of the epiphytic lichens Parmelia olivacea and Parmeliopsis ambigua in northern Swedish Lapland. – Oikos 56: 402–415.

– and Johansson, S. 1974. Bryophyte growth. Stordalen 1973. – Swedish IBP Tundra Biome Project, Tech. Rep. 16: 17–28.

– and Callaghan, T. V. 1991. Strategies of survival in plants of the Fennoscandian tundra. – Arctic 44: 95–105.

–, Persson, S., Basilier, K. and Stenström, T.-A. 1980. Growth of Sphagnum riparium Ångstr. in relation to some environmental factors in the Stordalen mire. – Ecol. Bull. 30: 191–207.

–, Schipperges, B. and Carlsson, B. Å. 1992a. Seasonal patterns of photosynthesis in alpine and subalpine populations of the lichen Nephroma arcticum. – Oikos 65: 3–12.

–, Gehrke, C. and Tjus, M. 1992b. CO_2 environment, microclimate and photosynthetic characteristics of the moss Hylocomium splendens in a subarctic habitat. – Oecologia 92: 23–29.

–, Osborn, C. and Sandberg, G. 1994. Epiphytic lichens as indicator of snow depths. – Arct. Alp. Res. 26: 159–165.

–, Callaghan, T. V. and Björn, L. O. 1995. Short-term effects of enhanced UV-B and CO_2 on lichens at different latitudes. – Lichenologist 27: 547–557.

–, Callaghan, T. V. and Carlsson, B. Å. 1996. Effects of enhanced ultraviolet radiation and carbon dioxide concentration on the moss Hylocomium splendens. – Global Change Biology 2: in press.

Sylvén, N. 1904. Studier över vegetationen i Torne lappmarks björkregion. – Ark. f. Botanik 3: 1–28.

Tenhunen, J. D., Lange, O. L., Hahn, S., Siegwolf, R. and Oberbauer, S. F. 1992. The ecosystem role of poikilohydrous tundra plants. – In: Chapin, F. S. III, Jefferies, R. L., Reynolds, J. F., Shaver, G. R. and Svoboda, J. (eds), Arctic ecosystems in a changing climate. An ecophysiological perspective. Academic Press, pp. 213–237.

Tosserams, M. and Rozema, J. 1995. Effects of ultraviolet-B radiation (UV-B) on growth and physiology of the dune grassland species Calamagrostis epigeios. – Environ. Poll. 89: 209–214.

Weimarck, H. 1937. Bryologiska undersökningar i nordligaste Sverige. – Sv. Bot. Tidskr. 31: 354–648.

Ecological Bulletins 45: 127–132. Copenhagen 1996

Resource investment in reproduction and its consequences for subarctic plants

L. Magnus Thorén, Åsa M. Hemborg and P. Staffan Karlsson

Thorén, L. M., Hemborg, Å. M. and Karlsson, P. S. 1996. Resource investment in reproduction and its consequences for subarctic plants. – Ecol. Bull. 45: 127–132.

The relative somatic cost of reproduction (RSC) is compared with reproductive effort (RE) for some subarctic plants in an attempt to determine whether the RSC is of the same magnitude as the RE. The relationship between RSC and RE varied between species, between habitats within species and to some extent between resources. In some cases RSC was found to be substantially less than RE. Possible physiological mechanisms enhancing resource acquisition in reproductive plants are discussed. We suggest that detailed analyses of resource economics during reproduction may help to identify the conditions under which plants can compensate for their reproductive resource investments and thus avoid reproductive costs. It is also discussed whether plants in cold environments i.e. at high altitudes or latitudes, should differ from plants in other environments in their relative reproductive investments.

L. M. Thorén, Dept of Ecology, Plant Ecology, Lund Univ., Ecology Building, S-223 62 Lund, Sweden. – Å. M. Hemborg, Dept of Ecological Botany, Uppsala Univ., Villavägen 14, S-752 36 Uppsala, Sweden. – P. S. Karlsson, Abisko Scientific Research Station, S-981 07 Abisko, Sweden and Dept of Ecological Botany, Uppsala Univ., Villavägen 14, S-752 36 Uppsala, Sweden.

It is generally accepted that various functions in an organism, such as growth, reproduction and defence, compete for a limited set of resources (cf. Calow and Townsend 1981, Bazzaz et al. 1987, de Jong and van Noordwijk 1992). An investment in one function should hence occur at the expense of the other(s). Therefore, the pattern of resource partitioning between functions is expected to reflect important life-history characteristics. This has also led to the assumption that a resource investment in reproduction should be associated with a cost (Stearns 1989). A cost that follows as a consequence of a resource investment in reproduction should initially be expressed as a resource depletion in the reproductive individual, i.e. as a somatic cost of reproduction. This somatic cost has implications for future growth, survival and/or reproduction. Costs mediated by resource investments have been called physiological costs. There may, however, also be other ways in which an individual experiences costs for reproduction; for example, ecological costs occur if reproduction increases the risk of encountering herbivores, parasites or diseases, etc. (Reznick 1992, Roff 1992). Still another type of cost is that reproduction may reduce the number of meristems available for growth (Maillette 1982, Tuomi et al. 1982, Watson 1984).

Although reproduction is generally expected to have associated costs, such costs have not always been found (Horvitz and Schemske 1988, Jennersten 1991, Syrjänen and Lehtilä 1993, Lehtilä et al. 1994, Wikberg et al. 1994, cf. also Reekie and Bazzaz 1987a). An inability to find evidence for reproductive costs could be due to the use of inadequate methods (Reznick 1985, Roff 1992) Alternatively, the organisms in question may have means of avoiding or decreasing the costs (Tuomi et al. 1983).

In this paper we will focus on the consequences of resource investments in reproduction by subarctic plants: Is there a somatic cost of reproduction that is similar in magnitude to the reproductive effort made by a plant, or do plants have the ability to compensate for the resources invested in reproduction? If so, by which mechanisms can this be achieved? Also, we will discuss whether plants in cold environments should be expected to differ from those in other environments, i.e. lower latitudes or altitudes, with respect to resource investment in reproduction and its consequences.

Resource investment in reproduction and its proximate consequences

Conceptual framework

A relative somatic cost of reproduction (RSC) has been defined as (Tuomi et al. 1983, cf. also Calow 1979):

$$RSC = (I_n - I_s)/I_n \qquad (1)$$

where I_n is the somatic investment in a non-reproductive individual and I_s the somatic investment in a reproducing one. The reproductive and non-reproductive individuals used for the comparison are assumed to be identical in all respects except for some being reproductive. This measure is closely related to the relative investment in reproduction i.e. reproductive effort (RE), defined as:

$$RE = I_r/(I_r + I_s) \qquad (2)$$

(Tuomi et al. 1983) where I_r is resource investment in reproduction.

Since RSC and RE are closely related and can be quantified in the same currency (i.e. the same resource) these two expressions provide a potential tool for use in analysing the resource economics of reproduction. A comparison between the two can reveal whether or not the cost of reproduction is equal to the reproductive investment. In the simplest case, if the resources invested in reproduction are allocated from somatic tissues to reproductive parts without any interaction with other plant functions, then $I_r + I_s = I_n$, and RSC = RE. If resource investment in reproduction decreases the foraging capacity of the reproductive individual, e.g. through a decreased investment in photosynthetic machinery or in roots, RSC should be larger than RE. On the other hand, if resource acquisition is enhanced by reproduction RSC should be less than RE.

Although the concept of relative somatic cost was presented more than 10 yr ago (Tuomi et al. 1983), to our knowledge, it has only been used rarely to analyse the consequences of plant reproduction. Reekie and Bazzaz (1987b) used a reproductive effort term (their RE6) that is identical to RSC as defined by Tuomi et al. (1983). Caesar and Macdonald (1984) used RSC to quantify costs of male reproduction for *Betula papyrifera* in terms of long shoot leaf mass and area. In addition, we have used RSC in studies of subarctic plants (Karlsson et al. 1990, Hemborg 1992, Thorén et al. 1996).

Relationships between RE and RSC

A number of estimations of RE and RSC for the three *Pinguicula* species in the Abisko area indicate a substantial variation in the relationship between RE and RSC (Fig. 1). In some cases RSC is substantially lower than the corresponding reproductive investment (RE). On the other hand, a survey of 9 populations comprising 6 species indicates that RSC, in most cases, relatively closely matches RE (Fig. 2).

These results indicate that plants sometimes have the ability to compensate for resource investment in reproduction. More importantly, at this functionally more narrow scale, there are considerable variations at all levels: between species, between populations of the same species, between years and between different currencies (resources). A similar pattern was obtained by Reekie and Bazzaz (1987b) when comparing RE (their RE3) and RSC (their RE6) for three genotypes of *Agropyron repens*. In general, their RSC values (for dry matter) were c. 0.15–0.20 lower than RE. However, the RSC varied between genotypes and even more strikingly, within genotypes between treatments (cf. Fig. 1 and 2 in Reekie and Bazzaz 1987b).

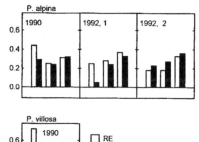

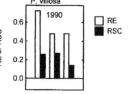

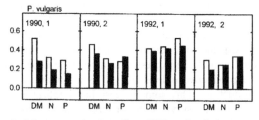

Fig. 1. Mean reproductive effort (RE) and relative somatic cost (RSC), in terms of dry matter, nitrogen and phosphorus, for three *Pinguicula* species growing in the Abisko area (data from Hemborg 1992 and Thorén et al. 1996). Numbers (1 or 2) after the year indicate that two populations were studied that year.

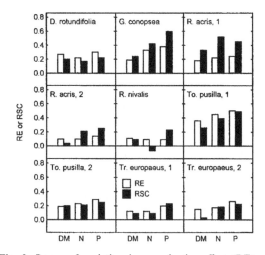

Fig. 2. Range of variation in reproductive effort (RE) and relative somatic cost (RSC) in terms of dry matter, nitrogen and phosphorus, for 15 populations comprising 9 species in the Abisko area (data from Hemborg 1992). Key to names of genera: D, *Drosera*; G, *Gymnadenia*; R, *Ranunclulus*; To, *Tofieldia* and Tr, *Trollius* (names according to Lid 1987). Numbers after the species names indicate that two different populations were studied.

Mechanism of resource compensation

The empirical results discussed above show that the somatic cost of reproduction may be considerably less than the reproductive effort. Thus, some species, under certain conditions, apparently can compensate for resource investments in reproduction. There are many reports of reproductive structures on green plants supplying a substantial part of their own carbon requirements through photosynthesis (e.g. Bazzaz et al. 1979). In some species foliage close to reproductive structures show an increased photosynthetic capacity (Luxmore 1991) which further enhances the CO_2-assimilation capacity of the reproductive individual. However, there can also be a cost of reproduction expressed as a decreased ability of the foliage to assimilate CO_2. In *Rhododendron lapponicum* leaves, photosynthetic capacity and photosynthetic nitrogen use efficiency are lower after reproduction than before (Karlsson 1994).

In many cases, photosynthesis in the reproductive organs contribute to their carbohydrates requirements. Thus, one could expect that the difference between RE and RSC commonly should be greater when measured in terms of dry matter (as an approximation for carbon) than when measured in terms of mineral nutrients. However, our results on the *Pinguicula* species as well as other species in the Abisko area do not support this prediction. There may be at least two possible reasons for this: 1) many species have mechanisms allowing for an enhanced nutrient gain in reproductive individuals, 2) carbon costs (as assessed in terms of biomass) are affected more strongly by the mineral nutrient depletion

during reproduction than by the actual carbon investment in reproduction. Biomass would then be reduced in proportion to nutrient allocation.

To our knowledge few studies have addressed the question of whether the capacity to acquire other resources, most importantly nitrogen and mineral nutrients, can be enhanced by reproductive activity. One example of such an enhancement has, however, been found in a carnivorous plant species, *Pinguicula vulgaris* (Karlsson et al. 1994). In this species the amount of trapped prey was found to be about twice as high in reproductive individuals as compared with naturally non-reproductive ones. Furthermore, there seems to be an interaction between trapping success and nitrogen uptake from soil; i.e. a high catch apparently stimulates N uptake (Hanslin and Karlsson 1996). This interaction between trapping success and nitrogen uptake form soil seems to be independent of reproductive status.

Discussion

Plant strategies concerning resource investments in reproduction in cold environments

Cold environments, i.e. those at high latitudes or altitudes, differ from other environments in a number of characteristics (e.g. Chapin and Shaver 1985, Bliss 1985, Billings 1987, Sonesson and Callaghan 1991). For plant growth and reproduction two of these characteristics could be expected to be the most important, viz. the low temperatures and short growing season. Low temperatures affect plants for at least two reasons: First, physiological processes within the plants are temperature dependent. Second, low soil temperatures result in low microbial activity and a slow release of soil nutrients. However, conditions experienced by plants may be less severe than those indicated by standard air temperature measurements (cf. Körner and Larcher 1988). Furthermore, at high latitudes the short season may partly be compensated for by long days during the growing season. Nevertheless, the low temperatures and short summer may apply some special restrictions on plant resource acquisition and allocation patterns and their combined effects creates an environment where sexual reproduction often fails (Bliss 1971, Callaghan and Emanuelsson 1985).

Tundra plants often rely on vegetative rather than sexual reproduction (Billings and Mooney 1968, Bliss 1971, Callaghan et al. 1996). To our knowledge no attempts have been made to analyse explicitly how resource investments in sexual reproduction should change as the Arctic or alpine regions are approached. Several attempts have, however, been made to relate life-history strategies to habitat conditions (e.g. MacArthur and Wilson 1967, Southwood 1977, Grime 1979, Taylor et al. 1990). Although life-histories cannot be predicted based on habitat conditions in a simple manner (Stearns 1992) the frequencies of different life-history types varies between habitats (Hart 1977). Assuming that arctic and alpine environments are characterised by higher stress levels (sensu Grime 1979) and lower productivity than habitats at lower altitudes or latitudes, a prediction can be made: Reproductive allocation should decrease as the stress level increases (Grime 1979) and decrease with decreasing environmental carrying capacity (Taylor et al. 1990). Both these suggest that plants in cold environments should have lower relative reproductive investments than plants growing under more favourable conditions.

This is in agreement with the conclusion of Callaghan and Emanuelsson (1985) that annual seed output is lower for arctic plants than for their counterparts in warmer regions. Further they point out that life-time reproductive output is higher due to much longer life-spans and that "actual costs of sexual reproduction may be very high because of low survival probabilities" of seeds and seedlings. Also Chester and Shaver (1982) state that reproductive effort of arctic plants tend to be lower than RE of temperate species. However, all species in their study do not support this conclusion.

Another line of arguments results in a different prediction: The reproductive effort is commonly size dependent (Samson and Werk 1986). Perennial plants relatively often show an inverse relationship between size and reproductive effort (Table 1). Thus, since plants tend to increase in size when approaching alpine and arctic environments, reproductive effort can be expected to decrease. This pattern has been described for *Heloniopsis orientalis* along an alpine transect (Kawano and Masuda 1980). However, it should be born in mind that several of the reports cited in Table 1 point out that for a particular species there are marked variations the relationship between RE and size, both in space and time (e.g. Ohlson 1988).

Many reproductive characteristics of tundra plants are highly variable (Molau 1993), and no consistent pattern

Table 1. Relationship between reproductive effort (RE) and plants size for some perennial plants. Negative relationships are those where RE decreases with plant size or the intercept (b) in the linear regression (y = b + mx) of reproductive allocation (y) versus plant size (usually above ground biomass as x), is significantly greater than zero (cf. Samson and Werk 1986). For annual species positive relationships seems to be more common than negative (Samson and Werk 1986).

Type of relationship	No. of species	Sources
Negative	12	1
No. significant rel.	7	2
Positive	3	3

Sources: 1: Kawano and Masuda 1980, Soule and Werner 1981, Waite and Hutchings 1982, Kawano et al. 1982, Aarsen and Taylor 1992, Karlsson and Lindström unpubl. 2: Pitelka et al. 1980, Douglas 1981, Hume and Cavers 1983, Aarsen and Taylor 1992. 3: Aarsen and Taylor 1992.

has emerged from empirical studies on reproductive investments in arctic environments. It seems clear, however, that almost all plants in these environments depend much more on vegetative reproduction than on sexual reproduction for their proliferation. Nevertheless, more or less all species in these environments do reproduce sexually. For clonal plants, the strain put on sexually reproducing individuals, i.e. the cost of reproduction, may be a more important factor for their life-history characteristics than the number of sexually produced offspring emerging. When populations are maintained mainly by asexual propagation a genotype whose costs for sexual reproduction are low may be more successful than a genotype having high costs. It may, finally, be hypothesised that due to the environmental constraints (cf. above) low reproductive investments may have substantial costs associated with them.

Cost of reproduction

Since the concept of reproductive cost was introduced by Williams (1966), this field has developed fast, both theoretically and empirically (Jönsson and Tuomi 1994). Because the theories usually have taken it for granted that there should be costs associated with resource investments in reproduction (e.g. Roff 1992, Stearns 1992) the inability to produce empirical evidence supporting theories of reproductive costs has received attention (Reznick 1985). There have also been theoretical analyses showing that costs need not always be expected (Tuomi et al. 1983).

We suggest that two lines of approaches are needed to resolve the extent to which plants may avoid somatic costs of reproduction and if so how this is accomplished: First, reliable methods must be identified. For example, there has been a debate over whether reproductive costs can be assessed without having control of the genetic constitution of the organisms studied (Reznick 1985, Partridge 1992). Furthermore, all measures of reproductive costs require some kind of non-reproductive control. The use of phenotypic correlations between naturally reproductive and non-reproductive individuals may not produce reliable results (Antonovics 1980, Roff 1992). Several types of experimental methods have been used to create non-reproductive controls. One is surgery, i.e. removal of reproductive parts (e.g. Montalvo and Ackerman 1987, Horovitz and Schemske 1988, and our work). Another approach is to control the initiation of reproduction by growing plants at different daylengths (but similar diurnal photon flux rates, Reekie and Bazzaz 1987a,b). A third method is to use gibberellic acid to induce reproduction (Reekie and Reekie 1991). Do these methods produce plants that are identical in all respects except for reproductive activity?

Second, to understand the resource economics of reproduction, detailed analyses of resource acquisition

130

and allocation patterns are needed. For example, do roots of reproductive and non-reproductive plants have similar nutrient-uptake efficiencies? Is the proportional allocation to roots similar in reproductive and non-reproductive plants? Although the increased carbon acquisition capacity of reproductive plants is well documented, very little is known about how the capacity to acquire other resources changes during reproduction. Detailed analyses of growth and resource allocation during reproduction, that were made under controlled conditions by Reekie and Bazzaz (1987a,b) and Reekie and Reekie (1991), need to be repeated on more species and combined with field studies to gain a more complete understanding of the resource economics of reproduction in plants.

Conclusions

Only scattered investigations are available of resource dynamics during reproduction – in cold environments as well as in other environments. Presently no pattern has emerged on whether arctic or alpine plants differ from e.g. temperate plants with respect to their reproductive investments. Neither do we know if a specific investment in reproduction leads to similar costs in different environments. However, the results presented here indicate that there are no simple relationships between investment and cost. We thus conclude that further studies of plant resource economy during reproduction are required before any general patterns regarding investments and their potential costs can be detected.

Acknowledgements – We thank the staff of the Abisko station especially its Director M. Sonesson, for their support. E. Reekie and J. Tuomi provided constructive comments on the manuscript. Financial support for this work has been obtained from the Swedish Natural Sciences Research Council, the Abisko Scientific Research Station and the Royal Swedish Academy of Sciences. D. A. Tilles corrected the English.

References

Aarssen, L. W. and Taylor, D. R. 1992. Fecundity allocation in herbaceous plants. – Oikos 65: 225–232.

Antonovics, J. 1980. Concepts of resource allocation and partitioning in plants. – In: Staddon, J. E. R. (ed.), The allocation of individual behavior. Academic Press, pp. 1–35.

Bazzaz, F. A., Carlson, R. W. and Harper, J. L. 1979. Contribution to reproductive effort by photosynthesis of flowers and fruits. – Nature 279: 554–555.

– , Chiariello, N. R., Coley, P. D. and Pitelka, L. F. 1987. Allocating resources to reproduction and defence. – BioScience 37: 58–67.

Billings, W. D. 1987. Constraints to plant growth, reproduction and establishment in arctic environments. – Arct. Alp. Res. 19: 357–365.

– and Mooney, H. A. 1968. The ecology of arctic and alpine plants. – Biol. Rev. 43: 481–529.

Bliss, L. C. 1971. Arctic and alpine plant life cycles. – Ann. Rev. Ecol. Syst. 2: 405–437.

– 1985. Alpine. – In: Chabot, B. F. and Mooney, H. A. (eds), Physiological ecology of North American plant communities. Chapman and Hall, pp. 41–65.

Caesar, J. C. and Macdonald, A. D. 1984. Shoot development in *Betula papyrifera*. V. Effect of male inflorescence formation and flowering on long shoot development. – Can. J. Bot. 62: 1708–1713.

Callaghan, T. V. and Emanuelsson, U. 1985. Population structure and processes of tundra plants and vegetation. – In: White, J. (ed.), The population structure of vegetation. W. Junk Publ. pp. 399–439.

– , Svensson, B. M. and Carlsson, B. Å. 1996. Some apparently paradoxical aspects of the life cycles, demography and population dynamics of plants from the subarctic Abisko area. – Ecol. Bull 45: 133–143.

Calow, P. 1979. The cost of reproduction – a physiological approach. – Biol. Rev. 54: 23–40.

– and Townsend, C. R. 1981. Energetics, ecology and evolution. – In: Townsend, C. R. and Calow, P. (eds), Physiological ecology. Blackwell, pp. 3–19.

Chapin, F. S. III and Shaver, G. R. 1985. Arctic. – In: Chabot, B. F. and Mooney, H. A. (eds), Physiological ecology of North American plant communities. Chapman and Hall, pp. 16–40.

Chester, A. L. and Shaver, G. R. 1982. Reproductive effort in cottongrass tussock tundra. – Holarct. Ecol. 5: 200–206.

de Jong, G. and van Noordwijk, A. J. 1992. Acquisition and allocation of resources: genetic (co)variances, selection, and life histories. – Am. Nat. 139: 749–770.

Douglas, D. 1981. The balance between vegetative and sexual reproduction of *Mimulus primuloides* (Scorphulariaceae) at different altitudes in California. – J. Ecol. 69: 295–310.

Grime, J. P. 1979. Plant strategies and vegetation processes. – John Wiley, Chichester, England.

Hanslin, H. M. and Karlsson, P. S. 1996. Nitrogen uptake from prey and substrate as affected by prey capture level and plant reproductive status in four carnivorous plant species. – Oecologia, in press.

Hart, R. 1977. Why are biennials so few? – Am. Nat. 111: 792–799.

Hemborg, Å. M. 1992. Costs of reproduction in herbs in a subarctic environment. – Honours thesis, Dept Ecol. Botany, Uppsala Univ.

Horvitz, C. C. and Schemske, D. W. 1988. Demographic cost of reproduction in a neotropical herb: An experimental field study.– Ecology 69: 1741–1745.

Hume, L. and Cavers, P. B. 1983. Resource allocation and reproductive and life-history strategies in widespread populations *Rumex crispus*. – Can. J. Bot. 61: 1276–1282.

Jennersten, O. 1991. Cost of reproduction in *Viscaria vulgaris* (Caryophyllaceae): a field experiment. – Oikos 61: 197–204.

Jönsson, K. I. and Tuomi, J. 1994. Costs of reproduction in a historical perspective. – Trends Ecol. Evol. 9: 304–307.

Karlsson, P. S. 1994. Photosynthetic capacity and photosynthetic nutrient use efficiency of *Rhododendron lapponicum* leaves as related to leaf nutrient status, leaf age and branch reproductive status. – Funct. Ecol. 8: 694–700.

– , Svensson, B. M., Carlsson, B. Å. and Nordell, K. O. 1990. Resource investment in reproduction and its consequences in three *Pinguicula* species. – Oikos 59: 393–398.

– , Thorén, L. M. and Hanslin, H. M. 1994. Prey capture by three *Pinguicula* species in a subarctic environment.– Oecologia 99: 188–193.

Kawano, S. and Masuda, J. 1980. The productive and reproductive biology of flowering plants. VII. Resource allocation and reproductive capacity in wild populations of *Heloniopsis orientalis* (Thumb.) C. Tanaka (Liliaceae). – Oecologia 45: 307–317.

– , Hiratsuka, A. and Hayashi, K. 1982. Life history characteristics and survivorship of *Erythronium japonicum*. – Oikos 38: 129–149.

Körner, C. and Larcher, W. 1988. Plant life in cold climates. – In: Long S.-F. and Woodward, F.-I. (eds), Plants and temperature. Symp. Exp. Biol. Vol. 42: 25–57, Cambridge.

Lehtilä, K., Tuomi, J. and Sulkinoja, M. 1994. Bud demography of the mountain birch Betula pubescens ssp. tortuosa near tree line. – Ecology 75: 945–955.

Lid, J. 1987. Norsk og Svensk flora. – Det Norske Samlaget, Oslo.

Luxmoore, R. J. 1991. A source-sink framework for coupling water, carbon, and nutrient dynamics of vegetation. – Tree Physiol. 9: 267–280.

MacArthur, R. H. and Wilson, E. O. 1967. The theory of island biogeography. – Princeton Univ. Press, Princeton, New Jersey.

Mailette, L. 1982. Structural dynamics of silver birch. I. The fate of buds. – J. Appl. Ecol. 19: 203–218.

Molau, U. 1993. Relationships between flowering phenology and life history strategies in tundra plants. – Arct. Alp. Res. 25: 391–402.

Montalvo, A. M. and Ackerman, J. D. 1987. Limitations to fruit production in Ionopsis utricflariodes (Orchidaeae). – Biotropica 19: 24–31.

Ohlson, M. 1988. Size-dependent reproductive effort in three populations of Saxifraga hirculus in Sweden. – J. Ecol. 76: 1007–1016.

Partridge, L. 1992. Measuring reproductive costs. – Trends Ecol. Evol. 7: 99–100.

Pitelka, L. F., Stanton, S. and Peckenham, M. O. 1980. Effects of light and density on resource allocation in a forest herb, Aster acuminatus (Compositae). – Am. J. Bot. 67: 942–948.

Reekie, E. G. and Bazzaz, F. A. 1987a. Reproductive effort in plants. 1. Carbon allocation to reproduction. – Am. Nat. 129: 867–896.

– and Bazzaz, F. A. 1987b. Reproductive effort in plants. 3. Effects of reproduction on vegetative activity. – Am. Nat. 129: 907–919.

– and Reekie, J. Y. C. 1991. The effect of reproduction on canopy structure, allocation and growth in Oenothera biennis. – J. Ecol. 79: 1061–1071.

Reznick, D. 1985. Cost of reproduction: an evaluation of the empirical evidence. – Oikos 44: 257–267.

– 1992. Measuring the costs of reproduction. – Trends Ecol. Evol. 7: 42–45.

Roff, D. A. 1992. The evolution of life histories; theory and analysis. – Chapman and Hall.

Samson, D. A. and Werk, K. S. 1986. Size-dependent effects in the analysis of reproductive effort in plants. – Am. Nat. 127: 667–680.

Sonesson, M. and Callaghan, T. V. 1991. Strategies of survival in plants of the Fennoscandian tundra. – Arctic 44: 95–105.

Soule, J. D. and Werner, P. A. 1981. Patterns of resource allocation in plants, with special reference to Potentilla recta L. – Bull. Torr. Bot Club 108: 311–319.

Southwood, T. R. E. 1977. Habitat, the templet for ecological strategies? – J. Anim. Ecol. 46: 337–365

Stearns, S. C. 1989. Trade-offs in life-history evolution. – Funct. Ecol. 3: 259–268.

– 1992. The evolution of life histories. – Oxford Univ. Press, Oxford, England.

Syrjänen, K. and Lehtilä, K. 1993. The cost of reproduction in Primula veris: differences between two adjacent populations. – Oikos 67: 465–472.

Taylor, D. R., Aarssen, L. W. and Loehle, C. 1990. On the relationship between r/K selection and environmental carrying capacity: a new habitat templet for plant life history strategies. – Oikos 58: 239–250.

Thorén, L. M., Karlsson, P. S. and Tuomi, J. 1996. Somatic cost of reproduction in three carnivorous plant species (Pinguicula) – Oikos, in press.

Tuomi, J., Niemelä, P. and Mannila, R. 1982. Resource allocation on dwarf shoots of birch (Betula pendula): reproduction and leaf growth. – New Phytol. 91: 483–487.

– , Hakala, T. and Haukioja, E. 1983. Alternative concepts of reproductive effort, cost of reproduction and selection in life-history evolution. – Am. Zool. 23: 25–34.

Waite, S. and Hutchings, M. J. 1982. Plastic energy allocation patterns in Plantago coronopus. – Oikos 38: 333–342.

Watson, M. 1984. Developmental constraints: effect on population growth and patterns of resource allocation in a clonal plant. – Am. Nat. 123: 411–426.

Williams, G. C. 1966. Natural selection, the cost of reproduction, and a refinement of Lack's principle. – Am. Nat. 100: 687–690.

Wikberg, S., Svensson, B. M. and Carlsson, B. Å. 1994. Fitness, population growth rate and flowering in Carex bigelowii, a clonal sedge. – Oikos 70: 57–64.

Ecological Bulletins 45: 133–143. Copenhagen 1996

Some apparently paradoxical aspects of the life cycles, demography and population dynamics of plants from the subarctic Abisko area

Terry V. Callaghan, Bengt Å. Carlsson and Brita M. Svensson

Callaghan, T. V., Carlsson, B. Å. and Svensson, B. M. 1996. Some apparently paradoxical aspects of the life cycles, demography and population dynamics of plants from the subarctic Abisko area. – Ecol. Bull. 45: 133–143.

Although plant species of the Arctic and Subarctic have evolved a range of adaptations to survive the harsh abiotic environment there, such adaptations are not always successful, and the population dynamics of many arctic plant species are characterised by infrequent and/or intermittent recruitment from sexual reproduction. Plant traits and population structures which we observe today are the result of selection pressures exerted in past environmental conditions often different from those occurring at present. We use specific examples from research based at the Abisko Scientific Research Station in the Subarctic, a critical ecotone in the context of future climate change, to speculate and provoke discussion on some implications of various forms of life cycles and population dynamics in the Subarctic.

We specifically address the potential immortality of genets in the context of environmental stability and we consider the implications for plants of high stress and disturbance in some habitats. We present and discuss apparent paradoxes in the between-year variability of flowering and in the longer-term population dynamics in the Subarctic. Deterministic, non-plastic plant growth and great mobility of clonal plants are presented as alternative strategies to survive subarctic environments. We show how these strategies can be facilitated by density dependent survival, in contrast to density dependent mortality associated with temperate latitudes, and complex physiological controls within clones to reduce intraclonal and interspecific competition. Finally, we infer likely responses of some subarctic plant populations to predicted future climatic changes suggesting that slow clonal growth and slow migration of boreal species compared with the predicted rate of climate change will lead to a situation in which the ecological memory in these plant populations is even greater than at present.

T. V. Callaghan, Sheffield Centre for Arctic Ecology, Univ. of Sheffield, 26 Taptonville Road, Sheffield, UK S10 5BR. – B. Å. Carlsson and B. M. Svensson, Dept of Ecology, Plant Ecology, Lund Univ., Ecology Building, S-223 62 Lund, Sweden.

The arctic environment imposes severe abiotic constraints on plant growth, seedling development, survival and reproduction (Warming 1909, Sørensen 1941, Porsild 1951, Billings and Mooney 1968, Bliss 1971, Savile 1972). Long winters and persistent snow cover reduce the period available for plant growth (Bliss 1956), while low temperatures and low availability of soil nutrients (Chapin 1980) in summer further restrict growth and reproduction by limiting growth and development rates (Warren Wilson 1966, Callaghan 1974). The environmental constraints of the Arctic represent strong selective forces which have resulted in the development of a

flora in which many plants exhibit patterns of growth, development and reproduction characterised by slow rates over long periods; annuals are rare (Bliss 1971) and long life cycles are common (Callaghan and Emanuelsson 1985). Processes such as flowering can be prolonged over two or even more years (Fries 1913, Sørensen 1941, Bliss 1971), and plant development can be interrupted by events such as freezing and ice encapsulation only to continue successfully on the return of favourable conditions (Bliss 1971, Crawford et al. 1994). Such adaptations are not always successful, however, and the population dynamics of many arctic plant species are characterised by infrequent and/or intermittent recruitment from sexual reproduction (Wager 1938, Callaghan 1976, Carlsson and Callaghan 1990a). This leads to age class distributions in which some age classes are missing (Wager 1938, Callaghan 1987, Philipp et al. 1990) and often a strong reliance on clonal growth (Carlsson et al. 1990) (where clonal may be defined as growth producing descendants vegetatively or by apomixis from a single plant). In some plant populations, the interannual fluctuations in reproduction, recruitment or survival appear to be non-random and cyclical patterns are evident (Makarova 1993, Carlsson and Callaghan 1994, Laine et al. 1995).

Plant traits which we observe today are the result of selection pressures exerted in past environmental conditions often different from those occurring at present. For example, the occasional large investment of plant resources into sexual reproduction appears to be inappropriate in the many cases where seedlings are rare or absent. One can then ask the questions why these traits are preserved when they appear to be non-adaptive, or whether they are adaptive under conditions beyond those which we normally focus on. The answer may lie in the extended generation times of arctic plants: time is needed for genetic changes to appear in the population. Indeed, the life history traits of many arctic plant populations discussed above, present barriers against genetic change (Wikberg et al. 1994). Under conditions of rapid change, environmental shifts may have occurred faster than genetic change and we may now have "obsolete" populations carrying genetic memories of past environmental conditions. Vavrek et al. (1991) showed just such an example when a seed bank of *Carex bigelowii* formed in cooler conditions some 200–300 yr ago had lower temperature optima for growth and germination than current populations.

This contribution is not intended to be an exhaustive review of research into plant population dynamics in the Arctic/Subarctic, nor is it intended to be a generalisation of processes there. Instead, we use specific examples from research based at the Abisko Scientific Research Station in the Subarctic, a critical ecotone in the context of future climate change (Bonan et al. 1992), to speculate and provoke discussion on some implications of various forms of life cycles and population dynamics for plant–environment interactions in the longer term, and in the Subarctic. We focus on aspects which are particularly problematic in terms of lack of data and on the interpretation of existing data which seem counter-intuitive or paradoxical from our perspective of ecological principles developed in temperate latitudes. We finally apply our current understanding to infer population responses to future environmental changes predicted for arctic regions (Cattle and Crossley 1995).

Subarctic environments and vegetation of the Abisko area

The Abisko station is situated just over 200 km north of the Arctic Circle in the northern part of the Fennoscandian mountain range (68°21′N, 18°49′E). The climate is subarctic with relatively short, cool summers and long winters characterised by cold spells alternating with short periods of thawing conditions. The Abisko valley is relatively dry compared with the surrounding regions, having a mean annual precipitation of 304 mm, of which c. 50% falls as rain between June and September (Andersson et al. 1996). Mean daily air temperatures are above 0°C from early May to October, with mean monthly temperature exceeding 10°C in July only.

The subalpine vegetation of the Abisko area is dominated by mountain birch *Betula pubescens* ssp. *tortuosa* (Ledeb.) Nyman woodland interspersed with small mires (Sonesson and Lundberg 1974). Most of this woodland is of a heath character with a field layer dominated by dwarf shrubs. In areas with moving ground water, herb-rich, meadow-type forests occur, often dominated by tall-growing herbs and broad-leaved grasses.

Above treeline, at c. 600 m a.s.l., the alpine vegetation is characterised by a strongly heterogeneous mixture of very diverse plant communities, ranging from exposed hilltops with a hardy flora of compact cushion plants and lichens to various dwarf shrub and grass heath communities below.

Potential immortality and environmental stability

The subarctic environment at Abisko includes gradients of the three major plant strategies sensu Grime (1977, 1979), i.e., competition, stress and disturbance. Competitive environments are found, for example, in the completely vegetated birch meadow communities with relatively high productivity, multi-layered canopy and high diversity. Stressed environments can be found from the birch heath vegetation, through the treeline heaths to the fellfields where stress is exerted through low nutrient status and/or harsh climate (Havström et al. 1993, Jonasson et al. 1993). Disturbed environments are best exemplified by patterned ground with freeze-thaw cycles (Jonasson 1986) and unstable mountain slopes subject to solifluction, mud slides, rock falls, avalanches, etc. (Rapp 1992). In these environments, the high stress and high disturbance environments can be found for which Grime found no plant strategy.

In the forest habitats, large-scale disturbances such as forest fires (Zackrisson 1977) seem to play only a minor role and perhaps the greatest disturbance is limited to defoliation of the birch trees about every 10–11 yr (Tenow 1972, Bylund 1995) which is insufficient to kill the dominant ground vegetation. Such stability seems to have favoured the existence of extremely long-lived vegetatively reproducing clonal plants which, paradoxically, have the potential for immortality. This growth form dominates the vegetation of the Torneträsk region (Jónsdóttir et al. 1996). In the clonal growth form, fitness has to be interpreted within the context of great longevity (Eriksson and Jerling 1990, Fagerström 1992). Three examples of clonal and/or long-lived growth forms are presented below.

Lycopodium annotinum – a stress-tolerant, foraging species

Lycopodium annotinum L. is a clonal stoloniferous/ rhizomatous pteridophyte. Horizontally growing modules are composed of a series of annually produced segments on which vertical modules and roots are attached. On average, horizontal modules grow 60 mm yr^{-1} and form a system of interconnected modules which eventually get separated from each other by the proximal decay of the oldest parts after c. 18–25 yr. In this way one clone splits into a multitude of clonal fragments or phenets (Callaghan et al. 1990, Klimeš et al. unpubl.). Apical growth and distal senescence is a rejuvenating process which produces phenets which are all <c. 25 yr, but the clone of which the phenets are part may, of course, be much older. Fairy rings (Komissarova et al. 1992) can be formed by centrifugal growth and have been observed in areas which have been burnt (man-made or accidentally). Knowing the date of fire, one can relate the size of the ring to age, and conclude that clones initiated 250 yr ago can still be represented by living and growing phenets (Oinonen 1968). The success of this species and the relative stability of the habitat, despite natural climate-change events such as the Little Ice Age and the Hypsithermal, can be inferred from the presence of *Lycopodium* spores in peat and sediment profiles for approximately the last 9500 calendar ^{14}C yr BP (Berglund et al. 1996).

Apical meristems can die in unfavourable microsites but this releases apical suppression (Svensson and Callaghan 1988) and, so long as at least one meristem survives, the phenet and clone can continue to grow (Fig. 1). Such a response to temporary and microscale disturbances buffers clones against adversity in both space and time and contributes to the phenomenon of potential immortality, a somewhat paradoxical or counter-intuitive aspect of harsh environments where survival would be expected to be low.

Betula pubescens ssp. tortuosa – a sedentary competitor

Lycopodium annotinum's longevity is associated with constant mobility; either centrifugal growth (Oinonen

1968) or multidirectional foraging (Callaghan et al. 1986a, 1990) and the ability to continually produce new root systems. In contrast, mountain birch trees *Betula pubescens* ssp. *tortuosa* in the Subarctic have great longevity through continual rejuvenation while occupying the same site and sharing the same root system in a similar way to unitary organisms.

The mountain birch woodland is characterised by an open tree canopy and widely spaced trees of generally low stature (see Ovhed and Holmgren 1996). However, the trees are strong competitors compared with co-occurring species which they over-top. On dry soils and south-facing aspects, the trees are taller and mono-cormic (single-stemmed); on wetter soils and cooler sites, the polycormic form is dominant. The polycormic form proliferates with basal sprouts which constantly rejuvenates the tree in a natural process similar to the management of coppiced trees. Ages of monocormic trees range up to 151 yr, the ages of branches of polycormic trees are similar, ranging up to 156 yr (Sonesson and Hoogesteger 1983). However, the longevity of polycormic "clones" is unknown but presumably far greater (cf. the often high ages of coppiced trees compared with ordinary trees, Rackham 1980).

Insect defoliation of the birch trees controls tree mortality and should be more effective on terminating monocormic trees, whereas the sprouting ability of polycormic trees facilitates recovery and, possibly, potential immortality. However, extensive mortality of

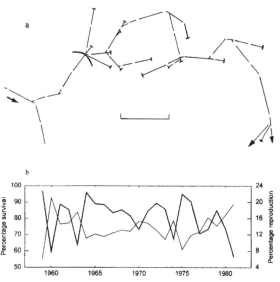

Fig. 1. Rejuvenation mechanisms in clonal *Lycopodium annotinum* which confer potential immortality. a) simulated phenet showing no net increase in number of apical meristems (arrows) after 25 yr, but persistence due to turnover of meristems (dead meristems denoted by bars). Scale equals 20 cm. From Callaghan et al. (1990). b) buffering of temporary and small-scale spatial adversity through release of apical suppression resulting in meristem proliferation following death of apical growing points. Thin line indicates survival, thick line reproduction. From Callaghan et al. (1986b).

polycormic mountain birch has been described in sub-arctic Finland and attributed to sequential impacts of defoliating insects and reindeer which browsed the new sprouts (Kallio and Lehtonen 1975); this was termed àn ecocatastrophy and was suggested as an alternative explanation to direct climatic controls for many now treeless mountain tops in northern Finland.

In contrast to *Lycopodium annotinum* which produces new roots on the phenet and occupies new horizontal space as it grows and in which modules can survive more or less independently of each other (Cook 1979, Tuomi and Vuorisalo 1989, Eriksson and Jerling 1990), birch stems depend on a root system shared with other stems. Thus, potential immortality of birch shoot systems is constrained by the mortality of their roots (Groff and Kaplan 1988, Lehtonen and Heikkinen 1995).

Hylocomium splendens – a sedentary stress-tolerator

The pleurocarpic moss *Hylocomium splendens* (Hedw.) Br. Eur. forms carpets which can be extensive in the Subarctic and further south (Tamm 1953, Callaghan et al. 1978, Økland and Økland 1996). It has a particularly effective method of rejuvenation: young modules are produced at the top of the moss carpet while older, shaded modules at the base die. Unlike *Lycopodium annotinum*, in which modules interchange resources in a horizontal dimension, the moss modules are separate and the shoots can be thought of as a "colony" of modules (Callaghan 1988). Analyses of samples collected from throughout the Arctic/Subarctic show that the growth period of a module can last between 2 and 4 yr and the degeneration phase can last between 9 and 17 yr (Callaghan et al. unpubl.). In the Abisko area (Stordalen and Latnjajaure), these periods are 3 and 9 yr, respectively, but the period over which the moss has occupied the site as clones is unknown. However, the infrequency of sporophyte production together with low rates of branching of shoots implies that the moss carpets of the Abisko area are very old. Also, the high density (Økland and Økland 1996) and vertical integrity of shoots required for optimum water balance within moss carpets suggests that prolonged site stability is required for extensive moss stand development.

Species of highly stressed and disturbed habitats

In the triangle of three fundamental strategies developed by Grime (1977, 1979), the combination of high disturbance and high stress is not included and strategies to combat this combination are also considered to be lacking, basically because of the paradox between longevity associated with stress-tolerators and the ruderal (short life-cycle) strategy associated with species of disturbed habitats. In practice, of course, highly stressed and disturbed habitats exist in the Arctic as noted above. In an attempt to identify possible plant adaptations to a habitat with quite high levels of both stress and disturbance – actively sorted polygons – Jonasson and Callaghan (1992) used a biomechanical approach arguing that plant roots must be specifically adapted to withstand the upward forces and displacement of frost heave (Perfect et al. 1988). Jonasson and Callaghan hypothesised that roots must be either elastic, extendible, annual or rudimentary to avoid damage during displacement. Measurements of the same species on actively sorted polygons and stable ground, showed that smaller and weaker roots were produced in the disturbed areas, but that they were the strongest per cross sectional area, i.e., they had the greatest breaking resistance. Overall, it was found that forbs and graminoids characterised the polygons, dwarf shrubs characterised the stable ground. The polygons also selected for species such as *Pinguicula vulgaris* with particularly small and annual roots.

The active polygons selected against horizontally spreading clonal plants typical of the vegetation of the area as a whole, presumably because the modules within independent physiological units would be broken. However, seedlings of some of these species were present. Seedlings of many subarctic species such as *Carex bigelowii* (but see Jónsdóttir 1995) and *Cassiope tetragona* (Havström 1995) are rare or absent where the mature clones occur (Callaghan and Emanuelsson 1985), but seedlings can be found in disturbed areas such as the polygons (Jonasson 1992). As mature plants are not found here, there is a paradox with a missing transitional habitat – or conditions – where all age classes can be found from seedlings to mature individuals. How, then, is the transition from seedling to mature clone achieved? Perhaps by a change in the habitat with an increase in stability following seedling establishment allowing mature clones to survive; so far, the process has not, apparently, been described.

Episodic recruitment, population structure and environmental memory in populations

Mechanisms enabling the potential immortality discussed above can be selected for in stable environments. However, changing environments and instability select for life history traits which rely on recruitment to populations through repeated flowering (Stearns 1992). Three important differences exist between flowering in the Arctic and elsewhere. Firstly, the flowering process is prolonged (Fries 1913) with the flowering cycle sometime extending over three years (Bliss 1971). This long cycle reduces the impacts of any particular growing

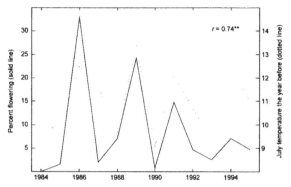

Fig. 2. Between-year variability in flowering of *Carex bigelowii* showing strong correlation between flowering intensity and July mean temperature in the year before flowering occurs. From Carlsson and Callaghan (1994).

season and, unlike the situation at more southerly latitudes, the degree of flowering integrates and represents the conditions over several years rather than just one year. Although flowering initiation is often related to plant/ramet size (Carlsson and Callaghan 1990b, Svensson et al. 1993) as in other regions (e.g., Silvertown 1985, Cooper and McGraw 1988), arctic plants grow slowly and flowering occurs after longer periods of development than in temperate plants. For example, tillers of *Phleum alpinum* may flower after 9 yr (Callaghan 1977), whereas temperate grass species may have several generations of tillers flowering within one year (Law 1979). Thus, the monocarpic flowering event in an arctic graminoid represents an integration of a multi-year environmental history. A third difference between flowering at arctic and temperate latitudes relates to flowering intensity and between-year variability. Often, strong positive correlations exist between flowering intensity and temperature in the Arctic (Svensson et al. 1993, Carlsson and Callaghan 1994, Havström et al. 1995a; Fig. 2). However, in the few cases examined in detail, it is apparent that a low mean flowering rate in colder sites (Fig. 3) or in colder climatic periods (Havström et al. 1995a) is associated with variability between years which ranges from zero flowering to high flowering intensities. This "all or nothing effect" seems to contrast with variability in warmer sites, warmer climatic periods, or at temperate latitudes where between-year variation exists, but a higher mean flowering intensity ensures some degree of flowering, and recruitment, in most years, as, e.g., in salt marsh communities (Torstensson 1987) or chalk grasslands (Grubb 1986).

Zero flowering in some years is the first constraint which imposes an episodic nature on recruitment of individuals in the Arctic, although the possibility for annual recruitment from seed banks exists (e.g., McGraw 1980, Lévesque and Svoboda 1995). Seedling recruitment may require specific climatic conditions,

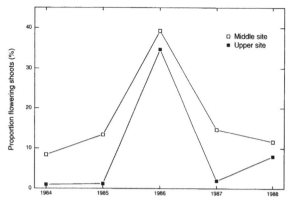

Fig. 3. Between-year variability in flowering of two *Carex bigelowii* populations at two sites on Mt Slåttatjåkka, Swedish Lapland. The climatically harsher upper site is characterised by generally very low flowering rates interrupted by years of heavy flowering. In contrast, the more favourable middle site shows less between-year variability and manages to flower in all years. Modified after Carlsson and Callaghan (1990a).

unavailable in every growing season (Wager 1938), but little seems to have been published on this aspect. These two constraints on recruitment lead to age class distributions in which all age classes are not represented (e.g., Wager 1938, Philipp et al. 1990; Fig. 4). In extreme cases, current populations may have been recruited during periods some decades or even a hundred or more years ago when warming, and presumably disturbance occurred as the Little Ice Age event came to an end (Havström 1995, Molau 1996). Calculated age class distributions from Svalbard and Abisko were used by Havström (1995) to estimate establishment

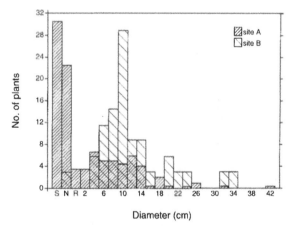

Fig. 4. Diameter (more or less equal to age) class distributions of two Greenlandic populations of *Silene acaulis* showing intermittent and low recent recruitment of seedlings at site B, and recent prolific seedling production at the more recently disturbed site A. In order to achieve the high frequency of 6–12-cm diameter individuals at site B, the population structure must have resembled that at site A at some time in the past. S: seedlings, N: one rosette, R: 2–3 rosettes. From Philipp et al. (1990).

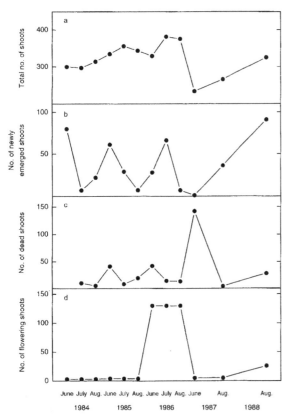

Fig. 5. Determination of tiller population dynamics in *Carex bigelowii*. In this species, fluctuations in flowering controls tiller population dynamics by a physiological mechanism, rather than by the recruitment of seedlings. The proportion of tillers flowering varies enormously between years (d). Since tillers show monocarpic growth, the growth of a tiller is terminated with the production of a flowering stem, and its aboveground part will die. This means that the amount of flowering one year will be reflected in a corresponding decrease in tiller population size in the next year (a) and (c). This suggests a cyclical behaviour of the tiller population: the tiller systems pass through an accumulation phase, when biomass and number of tillers in the system increase. A climatically triggered outburst in flowering will follow, leading to a large decrease in tiller numbers, thus initiating a new cycle. After Carlsson and Callaghan (1990a).

indices which were positively correlated with summer temperatures. Havström used these results to argue that the establishment of new genets of *Cassiope tetragona* at least since 1915 has been associated with periods with warmer-than-average summers when snow banks recede further than normal, releasing relatively competitor-free ground suited to colonisation by *C. tetragona*. Thus, the current population structure we see in some arctic plant populations may include a long memory of past periods unfavourable (Wager 1938, Philipp et al. 1990) or favourable (Havström 1995, Molau 1996) for recruitment.

Such episodic events in the establishment of populations may also be interpreted by considering the morphological and physiological differences between

individuals of different age classes within the same species. In many arctic plant populations, particularly those of clonal plants in long-established communities with complete plant cover, seedlings are rarely observed (e.g., Callaghan 1976, Carlsson and Callaghan 1990a). This is not surprising as modules of clonal plants have specific attributes, such as physiological integration (see Jónsdóttir et al. 1996), which increase survival probabilities of vegetative offspring compared with seedlings (Callaghan and Emanuelsson 1985) which lack subsidised growth. It may be assumed, therefore, that seedling form and function have been selected by environmental conditions in a locality different from that where mature plants are found (see above) or, in environmental conditions associated with a different climatic/environmental period. In perhaps the most extreme cases of the cryptogams, the individuals of the two phases of the life cycle are so different that they must have been selected by totally different environmental conditions (Keddy 1981). Gametophytes of *Lycopodium annotinum*, for example, are heterotrophic, underground structures some few cm long but have never been observed in the Abisko area. However, as noted above, studies by Oinonen (1968) in northern Finland and Kommissarova et al. (1992) in Russia suggest that recruitment from gametophytes takes place under specific conditions associated with forest fires. As clones of sporophytes may be several hundred yr old, we can assume that the environmental conditions favouring recruitment are brief but that the memory of these conditions is maintained, amplified ecologically in the growth of clones, but hidden by the current relationships between current clonal interactions with current environmental conditions.

Another example of memory in episodic flowering is exemplified in *Carex bigelowii*. In this species, apparently cyclical flowering (Carlsson and Callaghan 1990a, 1994) controls tiller population dynamics by a physiological mechanism, rather than by the recruitment of seedlings (Fig. 5). The proportion of tillers flowering varies enormously between years (between 1% and 32%), and is quite closely correlated to summer temperature in the previous year (Carlsson and Callaghan 1994). Since tillers show monocarpic growth, the growth of a tiller is terminated with the production of a flowering stem, and its aboveground part will die. This means that the amount of flowering one year will be reflected in a corresponding decrease in population size in the next year (Carlsson and Callaghan 1990a). This suggests a cyclical behaviour of the population: the tiller systems pass through an accumulation phase, when biomass and number of tillers in the system increase. A climatically triggered outburst in flowering will follow, leading to a large decrease in tiller numbers, thus initiating a new cycle.

In the same species, disjunction between seed production and potential seedling recruitment 200–300 yr later due to burial of a seed bank under a solifluction lobe showed that modern genotypes had higher temperature optima for growth/germination than the old population growing during Little Ice Age temperatures (Vavrek et al. 1991). Thus, a memory of past climatic conditions was maintained in this case showing maladjustment to present conditions.

Plant plasticity, mobility, density-dependent survival and competition

Bradshaw (1965) suggested that the lack of mobility in plants, which prevented them from exhibiting behaviour found in animals in responses to foraging for resources and escaping adverse conditions was compensated for by a high degree of plasticity. Such plasticity could give a competitive advantage to individuals in "benign" environments where resource limitation (e.g., light) limits growth more than the physical environment (e.g., wind, desiccation) and high density leads to self-thinning and competitive displacement. In the Subarctic, there is the apparent paradox with these generalisations in that some plants are relatively non-plastic and high shoot densities can enhance survival whereas other plants are quite mobile.

Deterministic, non-plastic growth ensures that plants in the harsher arctic environments (fell-fields, polar semi-deserts) can 1) avoid wind, ice crystal abrasion, desiccation and grazing by growing close to the ground surface in depressions and between stones (Savile 1972), 2) maintain aerodynamic shapes which produce large temperature differentials between vegetative and flowering meristems, leaves and the air above (Mølgaard 1982), and 3) remain below the insulating snow cover during winter thereby avoiding winter desiccation and reducing grazing damage (Sonesson and Callaghan 1991). In such conditions, plastic, opportunistic responses to short-term increases in temperature or reduction in light (etiolation) would be fatal over the longer term. Such deterministic growth contrasts with the often plastic, opportunistic growth of plants from more southerly regions where vertical growth is often essential to compete for light, pollinators, pollen/seed dispersal, etc., and is perhaps without the same risks to survival as in the Arctic.

Deterministic growth in the Arctic sometimes involves the tight-packing of individual shoots within individual plants (e.g., the cushion growth form) and even between individuals of different species within communities. This is associated with a phenomenon of density dependent survival (Økland and Økland 1996) which contrasts with the density dependent mortality most often considered in other latitudes (Westoby 1981). Close-packing of shoots of different species can enhance performance (Carlsson and Callaghan 1991)

and, we assume, survival due to a shelter effect. This has been interpreted as a positive interaction between plants in response to harsh abiotic conditions and has been contrasted with negative, competitive interactions in closed vegetation further south. However, competition for resources has been claimed to be greatest where resources are most limiting (Oksanen 1980, Moen 1993). As the harsh abiotic environment is apparently moderated by positive responses between individuals in some arctic plants, it must be assumed that these abiotic conditions have a greater selective force than competition for resources. Evidence for this claim comes from species removal experiments in which increased biomass of remaining species was not observed even though competitors were removed (Jonasson 1992, Shevstova et al. 1995).

In contrast to Bradshaw's (1965) view of the lack of mobility in plants, those such as *Lycopodium annotinum* may be highly mobile extending some 60 mm yr^{-1} and they have sophisticated physiological and morphological controls to forage for resources, escape competition from other species, and reduce competition between modules within individual clones.

By modelling the horizontal growth architecture of *L. annotinum*, it has been shown that the most common branching angle between mother branch and daughter branch (53°) and the most common expression of apical dominance, i.e., a sequential reduction in length of successive branch generations, together resulted in a pattern of spatially separated modules where the overlaps between segments were minimised (Callaghan et al. 1990). This means that the vertical branches, being the important source for carbon, are free to develop and grow without competition from each other. Simulating the vegetative spread, using field data on morphology and population dynamics as input values, the rate of spread of one clone was 134 cm times 90 cm per 28 yr. Overall, the mobility of *L. annotinum* allows flexible responses to spatial heterogeneity and to directional changes over time in one place, e.g., the establishment and growth of a competitor. The flexible response is facilitated by variation in branching patterns, length increments, and directional changes in the apical growing point (Svensson and Callaghan 1988, Callaghan et al. 1990, Klimeš et al. unpubl.). However, the degree of flexibility is limited and once a module of *L. annotinum* locates a favourable patch in terms of below-ground resources and light, residence time is short and the plant grows out of the patch due to its fixed mobility. As *L. annotinum* invests only 5% of its biomass in below-ground tissues yet grows in vegetation characterised by biomass investments into below-ground tissues an order of magnitude greater, this constant mobility has been interpreted as an escape strategy from below-ground competitors.

In contrast, *Carex bigelowii* shows a flexibility in its mobility and, typical of arctic plants, invests a large component of biomass (c. 20%) in below-ground tissues. A division of labour between sibling tillers (Carlsson and Callaghan 1990b, Jónsdóttir et al. 1996) enables it to efficiently locate and exploit the sparse and patchy occurrences of nutrients available in its habitat. The first daughter tiller is always initiated ventrally on the parent's rhizome/stem and has a long-rhizome (or guerrilla) tiller type. This maximises the forward "search" for favourable patches. Should these be encountered, as was shown in a fertiliser experiment (Carlsson and Callaghan 1990b), the increased nutrient availability releases apical dominance enabling the production of more than one daughter tiller, which then will be mainly of a short-rhizome or phalanx type, thus efficiently exploiting the favourable micro-site. Physiological integration between the tiller types and generations (Jónsdóttir et al. 1996) allows the maintenance of a particularly flexible system of soil resource location, exploitation and competition.

Conclusions and inferences for future environmental changes

This contribution has raised some apparently paradoxical and counter-intuitive aspects of the relationship between plant performance in the Arctic and the environment there. Although it is unwise to generalise, some lessons can be applied to understanding likely plant responses to future environmental changes, particularly climatic change, which might differ from conventional perceptions.

Climatic change associated with the greenhouse effect is predicted to be particularly great in the Arctic and Subarctic (Mitchell et al. 1990, Maxwell 1992, Cattle and Crossley 1995) and arctic ecosystems are suggested to be particularly sensitive to these changes (Callaghan et al. 1992, Chapin et al. 1992, Callaghan and Jonasson 1995a, b). Loss of biodiversity in a warming Arctic is a particular concern (Chapin and Körner 1995), yet the positive responses of flowering to increased temperatures demonstrated above and increased recruitment in infrequent favourable climatic periods suggest an increase in biodiversity at least at the sub-specific level, which cannot be predicted by relating current recruitment to current environmental conditions. Many of the long-lived arctic species have already experienced considerable variability in climate in their life spans and many can apparently buffer fluctuations in climate (Fig. 1b) and can survive warmer conditions, despite predictions that warming might lead to adverse carbon balances within plants (Diemer 1996). If arctic species can survive climatic warming, how will the predicted vegetation change (Bonan et al. 1992) occur? It has been shown that some arctic species are poor competitors due to their deterministic growth pattern or evidenced by

escape strategies as in *L. annotinum*. Thus, it is competition (e.g., shade in *C. tetragona* – Havström et al. 1995b) which is likely to displace current arctic species during future warming. However, competition between existing ecotypes, as has been suggested by Crawford et al. (1995), may be slow to occur in practice, particularly in species such as *Cassiope tetragona* (Havström 1995) and *Diapensia lapponica* (Molau 1996) in which centrifugal growth of clones is only $1.5-5$ mm yr^{-1}. Also, it is doubtful if southern species can migrate at the same rate as climatic zones shift (c. $400-500$ km by the year 2020), but data, apart from tree line migration (up to 300 m yr^{-1} for conifers and up to 1000 m yr^{-1} for alder and birch; see references in Callaghan and Jonasson 1995a), are lacking. Such a scenario will lead to a situation in which the ecological memory in arctic plant populations is even greater than at present, with a wider separation of those climatic conditions which allowed establishment of clones/populations and selected for particular biotypes, and the future structure and behaviour of the populations. This will impose even greater challenges on our interpretations of future plant behaviour, and will increase the need for interpretations to be made in wider spatial and temporal contexts.

Acknowledgements – We are greatly indebted to Mats Sonesson, Director of the Abisko Scientific Research Station for encouraging and facilitating many of the studies described here. We are also grateful to the staff of the Abisko Scientific Research Station for help. TVC wishes to thank the U.K. Natural Environment Research Council for support. BÅC and BMS wish to thank the Swedish Natural Science Research Council for support. We are grateful to P. S. Karlsson and J. Tuomi for constructive comments on the manuscript.

References

Andersson, N. Å., Callaghan, T. V. and Karlsson, P. S. 1996. The Abisko Scientific Reseach Station. – Ecol. Bull. 45: 11–14.

Berglund, B., Barnekow, L., Hammarlund, D., Sandgren, P. and Snowball, I. F. 1996. Holocene forest dynamics and climatic changes in the Abisko area, northern Sweden – the Sonesson model of vegetation history reconsidered and confirmed. – Ecol. Bull. 45: 15–30.

Billings, W. D. and Mooney, H. A. 1968. The ecology of arctic and alpine plants. – Biol. Rev. 43: 481–529.

Bliss, L. C. 1956. A comparison of plant development in microenvironments of arctic and alpine tundras. – Ecol. Monogr. 26: 303–337.

– 1971. Arctic and alpine life cycles. – Annu. Rev. Ecol. Syst. 2: 405–438.

Bonan, G. B., Pollard, D. and Thompson, S. L. 1992. Effects of boreal forest vegetation on global climate. – Nature 359: 716–718.

Bradshaw, A. D. 1965. Evolutionary significance of phenotypic plasticity in plants. – Adv. Genet. 13: 115–155.

Bylund, H. 1995. Long-term interactions between the autumnal moth and mountain birch: the roles of resources, competitors, natural enemies, and weather. – Ph.D. thesis, Swedish Univ. of Agricultural Sciences, Uppsala.

Callaghan, T. V. 1974. Intraspecific variation in *Phleum alpinum* L. with special reference to polar populations. – Arct. Alp. Res. 6: 361–401.

– 1976. Growth and population dynamics of *Carex bigelowii* in an alpine environment. – Oikos 27: 402–413.

– 1977. Adaptive strategies in the life cycles of South Georgian graminoid species. – In: Llano, G. A. (ed.), Adaptations within antarctic ecosystems. Smithsonian Inst., Washington, pp. 981–1002.

– 1987. Population processes in arctic and boreal plants. – Ecol. Bull. 38: 58–68.

– 1988. Physiological and demographic implications of modular construction in cold environments. – In: Davy, A. J., Hutchings, M. J. and Watkinson, A. R. (eds), Plant population ecology. Blackwell, Oxford, pp. 111–135.

– and Emanuelsson, U. 1985. Population structure and processes of tundra plants and vegetation. – In: White, J. (ed.), The population structure of vegetation. Junk, Dordrecht, pp. 399–439.

– and Jonasson, S. 1995a. Arctic terrestrial ecosystems and environmental change. – Philos. Trans. R. Soc. Lond. B. 352: 259–276.

– and Jonasson, S. 1995b. Implications for changes in arctic plant biodiversity from environmental manipulation experiments. – In: Chapin, F. S. III and Körner, C. (eds), Arctic and alpine biodiversity: patterns, causes and ecosystem consequences. Springer, Berlin, pp. 151–164.

–, Collins, N. J. and Callaghan, C. H. 1978. Photosynthesis, growth and reproduction of *Hylocomium splendens* and *Polytrichum commune* in Swedish Lapland. – Oikos 31: 73–88.

–, Headley, A. D., Svensson, B. M., Lixian, L., Lee, J. A. and Lindley, D. K. 1986a. Modular growth and function in the vascular cryptogam *Lycopodium annotinum*. – Proc. R. Soc. Lond. B 228: 195–206.

–, Svensson, B. M. and Headley, A. D. 1986b. The modular growth of *Lycopodium annotinum*. – Fern Gaz. 13: 65–76.

–, Svensson, B. M., Bowman, H., Lindley, D. K. and Carlsson, B. Å. 1990. Models of clonal plant growth based on population dynamics and architecture. – Oikos 57: 257–269.

–, Sonesson, M. and Sømme, L. 1992. Responses of terrestrial plants and invertebrates to environmental change at high latitudes. – Philos. Trans. R. Soc. Lond. B. 338: 279–288.

Carlsson, B. Å. and Callaghan, T. V. 1990a. Effects of flowering on the shoot dynamics of *Carex bigelowii* along an altitudinal gradient in Swedish Lapland. – J. Ecol. 78: 152–165.

– and Callaghan, T. V. 1990b. Programmed tiller differentiation, intraclonal density regulation, and nutrient dynamics in *Carex bigelowii*. – Oikos 58: 219–230.

– and Callaghan, T. V. 1991. Positive plant interactions in tundra vegetation and the importance of shelter. – J. Ecol. 79: 973–983.

– and Callaghan, T. V. 1994. Impact of climate change factors on *Carex bigelowii*: implications for population growth and spread. – Ecography 17: 321–330.

–, Jónsdóttir, I. S., Svensson, B. M. and Callaghan, T. V. 1990. Aspects of clonality in the Arctic: a comparison between *Lycopodium annotinum* and *Carex bigelowii*. – In: van Groenendael, J. and de Kroon, H. (eds), Clonal growth in plants: regulation and function. SPB Acad. Publ., the Hague, pp. 131–151.

Cattle, H. and Crossley, J. 1995. Modelling arctic climate change. – Phil. Trans. R. Soc. Lond. A 352: 201–213.

Chapin, F. S. III 1980. The mineral nutrition of wild plants. – Annu. Rev. Ecol. Syst. 11: 233–260.

– and Körner, C. (eds) 1995. Arctic and alpine biodiversity: patterns, causes and ecosystem consequences. – Springer, Berlin.

–, Jefferies, R. L., Reynolds, J. F., Shaver, G. R. and Svoboda, J. (eds) 1992. Arctic ecosystems in a changing climate. An ecophysiological perspective. – Academic Press, San Diego.

Cook, R. E. 1979. Asexual reproduction: a further consideration. – Am. Nat. 113: 769–772.

Cooper, S. D. and McGraw, J. B. 1988. Constraints on reproductive potential at the level of the shoot module in three ericaceous shrubs. – Funct. Ecol. 2: 96–108.

Crawford, R. M. M., Chapman, H. M. and Hodge, H. 1994. Anoxia tolerance in High Arctic vegetation. – Arct. Alp. Res. 26: 308–312.

–, Chapman, H. M. and Smith, L. C. 1995. Adaptation to variation in growing season length in arctic populations of *Saxifraga oppositifolia* L. – Bot. J. Scotl. 47: 177–192.

Diemer, M. 1996. A comparison of photosynthetic performance and leaf carbon gain of temperate and subarctic genotypes of *Geum rivale* and *Ranunculus acris* in northern Sweden. – Ecol. Bull. 45: 144–150.

Eriksson, O. and Jerling, L. 1990. Hierarchical selection and risk spreading in clonal plants. – In: van Groenendael, J. and de Kroon, H. (eds), Clonal growth in plants: regulation and function. SPB Acad. Publ., the Hague, pp. 79–94.

Fagerström, T. 1992. The meristem-meristem cycle as a basis for defining fitness in clonal plants. – Oikos 63: 449–453.

Fries, T. C. 1913. Botanische Untersuchungen im nördlichsten Schweden. – Vetensk. och prakt. unders. i Lappland, Flora och Fauna 2: 1–361.

Grime, J. P. 1977. Evidence for the existence of 3 primary strategies in plants and its relevance to ecological and evolutionary theory. – Am. Nat 111: 1169–1194.

– 1979. Plant strategies and vegetation processes. – Wiley, Chichester.

Groff, P. A. and Kaplan, D. R. 1988. The relation of root systems to shoot systems in vascular plants. – Bot. Rev. 54: 387–422.

Grubb, P. J. 1986. Problems posed by sparse and patchily distributed species in species-rich plant communities. – In: Diamond, J. and Case, T. J. (eds), Community ecology. Harper and Row, New York, pp. 207–225.

Havström, M. 1995. The establishment of new genets of *Cassiope tetragona* may increase if the arctic climate gets warmer. – In: Havström, M. Arctic plants and climate change – experimental and retrospective studies of *Cassiope tetragona*. Ph.D. thesis, Univ. of Göteborg, Sweden.

–, Callaghan, T. V. and Jonasson, S. 1993. Differential growth responses of *Cassiope tetragona*, an arctic dwarf shrub, to environmental perturbations among three contrasting high- and subarctic sites. – Oikos 66: 389–402.

–, Callaghan, T. V., Svoboda, J. and Jonasson, S. 1995a. Little ice age temperature estimated by growth and flowering differences between subfossil and extant shoots of *Cassiope tetragona*, an arctic heather. – Funct. Ecol. 9: 650–654.

–, Callaghan, T. V. and Jonasson, S. 1995b. Effects of simulated climate change on the sexual reproductive effort of *Cassiope tetragona*. – In: Callaghan, T. V., Oechel, W. C., Gilmanov, T., Holten, J. I., Maxwell, B., Molau, U., Sveinbjörnsson, B. and Tyson, M. (eds), Global change and arctic terrestrial ecosystems. Proc. contributed and poster papers from the int. conf., 21–26 August 1993, Oppdal, Norway. Commission of the European Comm. Ecosyst. Res. Report, Brussels, pp. 109–114.

Jonasson, S. 1986. Influence of frost heaving on soil chemistry and the distribution of plant growth forms. – Geogr. Ann. 68A: 185–195.

– 1992. Growth responses to fertilisation and species removal in tundra related to community structure and clonality. – Oikos 63: 420–429.

– and Callaghan, T. V. 1992. Root mechanical properties related to disturbed and stressed habitats in the Arctic. – New Phytol. 122: 179–186.

–, Havström, M., Jensen, M. and Callaghan, T. V. 1993. In situ mineralization of nitrogen and phosphorus of arctic soils after perturbations simulating climate change. – Oecologia 95: 179–186.

Jónsdóttir, I. S. 1995. The importance of sexual reproduction in arctic clonal plants and their evolutionary potential. – In: Callaghan, T. V., Oechel, W. C., Gilmanov, T., Holten,

J. I., Maxwell, B., Molau, U., Sveinbjörnsson, B. and Tyson, M. (eds), Global change and arctic terrestrial ecosystems. Proc. contributed and poster papers from the int. conf., 21–26 August 1993, Oppdal, Norway. Commission of the European Comm. Ecosyst. Res. Report, Brussels, pp. 81–88.
– , Callaghan, T. V. and Headley, A. D. 1996. Resource dynamics within arctic clonal plants. – Ecol. Bull. 45: 53–64.
Kallio, P. and Lehtonen, J. 1975. On the ecocatastrophy of birch forests caused by Oporinia autumnata (Bkh) and the problems of reforestation. – Rep. Kevo Subarct. Res. Stn 10: 70–84.
Keddy, P. A. 1981. Why gametophytes and sporophytes are different: form and function in a terrestrial environment. – Am. Nat. 118: 452–454.
Komissarova, I. F., Filin, V. R., Alekhina, N. A. and Nauyalis, I. I. 1992. Changes in certain soil properties under the influence of Lycopodium fairy rings. – Eurasian Soil Sci. 24: 45–59.
Laine, K., Malila, E. and Siuruainen, M. 1995. How is annual climatic variation reflected in the production of germinable seeds of arctic and alpine plants in the Northern Scandes? – In: Callaghan, T. V., Oechel, W. C., Gilmanov, T., Holten, J. I., Maxwell, B., Molau, U., Sveinbjörnsson, B. and Tyson, M. (eds), Global change and arctic terrestrial ecosystems. Proc. of contributed and poster papers from the int. conf., 21–26 August 1993, Oppdal, Norway. Commission of the European Comm. Ecosyst. Res. Report, Brussels, pp. 89–95.
Law, R. 1979. The cost of reproduction in annual meadow grass. – Am. Nat. 113: 3–16.
Lehtonen, J. and Heikkinen, R. K. 1995. On the recovery of mountain birch after Epirrita damage in Finnish Lapland, with a particular emphasis on reindeer grazing. – Écoscience 2: 349–356.
Lévesque, E. and Svoboda, J. 1995. Germinable seed bank from polar desert stands, Central Ellesmere Island, Canada. – In: Callaghan, T. V., Oechel, W. C., Gilmanov, T., Holten, J. I., Maxwell, B., Molau, U., Sveinbjörnsson, B. and Tyson, M. (eds), Global change and arctic terrestrial ecosystems. Proc. of contributed and poster papers from the int. conf., 21–26 August 1993, Oppdal, Norway. Commission of the European Comm. Ecosyst. Res. Report, Brussels, pp. 97–107.
Makarova, O. A. 1993. The growing wild berries purchase in the Murmansk Region. – In: Kalabin, G. V. (ed.), The vegetable resources of the European north: productivity, rational use, protection. Kola Science Centre, Apatity, pp. 24–25, in Russian.
Maxwell, B. 1992. Arctic climate: potential for change under global warming. – In: Chapin, F. S. III, Jefferies, R. L., Reynolds, J. F., Shaver, G. R. and Svoboda, J. (eds), Arctic ecosystems in a changing climate. An ecophysiological perspective. Academic Press, San Diego, pp. 11–34.
McGraw, J. B. 1980. Seed bank size and distribution of seeds in cottongrass tussock tundra, Eagle Creek, Alaska. – Can. J. Bot. 58: 1607–1611.
Mitchell, J. F. B., Manabe, S., Meleshko, V. and Tokioka, T. 1990. Equilibrium climate change – and its implications for the future. – In: Houghton, J. T., Jenkins, G. J. and Ephraums, J. J. (eds), Climate change. The IPCC scientific assessment. Cambridge Univ. Press, Cambridge, pp. 131–172.
Moen, J. 1993. Positive versus negative interactions in a high alpine block field: germination of Oxyria digyna seeds in a Ranunculus glacialis community. – Arct. Alp. Res. 25: 201–206.
Molau, U. 1996. Climatic impacts on flowering, growth, and vigour in an arctic-alpine cushion plant, Diapensia lapponica, under different snow cover regimes. – Ecol. Bull. 45: 209–219.
Mølgaard, P. 1982. Temperature observations in high Arctic

plants in relation to microclimate in the vegetation of Peary Land, North Greenland. – Arct. Alp. Res. 14: 105–115.
Oinonen, E. 1968. The size of Lycopodium clavatum and L. annotinum L. stands as compared to that of L. complanatum L. and Pteridium aquilinum (L.) Kuhn. stands, the age of the tree stand and the dates of fire, on the site. – Acta Forest. Fenn. 87: 1–53.
Økland, R. H. and Økland, T. 1996. Population biology of the clonal moss Hylocomium splendens in Norwegian boreal spruce forests. II. Effects of density. – J. Ecol. 84: 63–69.
Oksanen, L. 1980. Abundance relationships between competitive and grazing tolerant plants in productivity gradients on Fennoscandian mountains. – Ann. Bot. Fennici 17: 410–429.
Ovhed, M. and Holmgren, B. 1996. Modelling and measuring evapotranspiration in a mountain birch forest. – Ecol. Bull. 45: 31–44.
Perfect, E., Miller, R. D. and Burton, B. 1988. Frost upheaval of overwintering plants: a qualitative study of displacement processes. – Arct. Alp. Res. 20: 70–75.
Philipp, M., Böcher, J., Mattsson, O. and Woodell, S. R. J. 1990. A quantitative approach to the sexual reproductive biology and population structure in some arctic flowering plants: Dryas integrifolia, Silene acaulis and Ranunculus nivalis. – Medd. Grønl. Biosci. 34.
Porsild, A. E. 1951. Plant life in the Arctic. – Can. Geogr. J. 42: 120–145.
Rackham, O. 1980. Ancient woodland. Its history, vegetation and uses in England. – Arnold, London.
Rapp, A. 1992. Kärkevagge revisited. Field excursions on geomorphology and environmental history in the Abisko mountains, Sweden. – Sver. Geol. Unders. Ser. Ca 81: 269–276.
Savile, D. B. O. 1972. Arctic adaptations in plants. – Canada Dept Agriculture, Monogr. 6: 1–81.
Shevstova, A., Ojalala, A., Neuvonen, S., Vieno, M. and Haukioja, E. 1995. Growth and reproduction of dwarf shrubs in a subarctic plant community: annual variations and above-ground interactions with neighbours. – J. Ecol. 83: 263–275.
Silvertown, J. 1985. Survival, fecundity and growth of wild cucumber Echinocystis lobata. – J. Ecol. 73: 841–850.
Sonesson, M. and Lundberg, B. 1974. Late quaternary forest development of the Torneträsk area, north Sweden. 1. Structure of modern forest ecosystems. – Oikos 25: 121–133.
– and Hoogesteger, J. 1983. Recent tree-line dynamics (Betula pubescens Ehrh. ssp tortuosa (Ledeb.) Nyman) in northern Sweden. – Nordicana 47: 47–54.
– and Callaghan, T. V. 1991. Life in the polar winter – strategies of survival. – Arctic 44: 95–105.
Sørensen, T. 1941. Temperature relations and phenology of the northeast Greenland flowering plants. – Medd. Grønl. 125: 1–305.
Stearns, S. C. 1992. The evolution of life histories. – Oxford Univ. Press, Oxford.
Svensson, B. M. and Callaghan, T. V. 1988. Apical dominance and the simulation of metapopulation dynamics in Lycopodium annotinum. – Oikos 51: 331–342.
– , Carlsson, B. Å., Karlsson, P. S. and Nordell, K. O. 1993. Comparative long-term demography of three species of Pinguicula. – J. Ecol. 81: 635–645.
Tamm, C. O. 1953. Growth, yield and nutrition in carpets of a forest moss (Hylocomium splendens). – Medd. Statens Skogsforskningsinst. 43: 1–140.
Tenow, O. 1972. The outbreaks of Oporinia autumnata Bkh. and Operophthera ssp. (Lep. Geometridae) in the Scandinavian mountain chain and northern Finland 1862–1968. – Zool. Bidr. Upps. Suppl. 2: 1–107.
Torstensson, P. 1987. Population dynamics of the annual halophyte Spergularia marina on a Baltic seashore meadow. – Vegetatio 68: 169–172.

Tuomi, J. and Vuorisalo, T. 1989. What are the units of selection in modular organisms? – Oikos 54: 227–233.

Vavrek, M. C., McGraw, J. B. and Bennington, C. C. 1991. Ecological genetic variation in seed banks. III. Phenotypic and genetic differences between young and old seed populations of *Carex bigelowii*. – J. Ecol. 79: 645–662.

Wager, H. G. 1938. Growth and survival of plants in the Arctic. – J. Ecol. 26: 390–410.

Warming, E. 1909. Oecology of plants. – Oxford Univ. Press, Oxford.

Warren Wilson, J. 1966. An analysis of plant growth and its control in Arctic environments. – Ann. Bot. N.S. 30: 383–402.

Westoby, M. 1981. The place of the self-thinning rule in population dynamics. – Am. Nat. 118: 581–587.

Wikberg, S., Svensson, B. M. and Carlsson, B. Å. 1994. Fitness, population growth rate and flowering in *Carex bigelowii*, a clonal sedge. – Oikos 70: 57–64.

Zackrisson, O. 1977. Influence of forest fires on the north Swedish boreal forest. – Oikos 29: 22–32.

Ecological Bulletins 45: 144–150. Copenhagen 1996

A comparison of photosynthetic performance and leaf carbon gain of temperate and subarctic genotypes of Geum rivale and Ranunculus acris in northern Sweden

Matthias Diemer

Diemer, M. 1996. A comparison of photosynthetic performance and leaf carbon gain of temperate and subarctic genotypes of *Geum rivale* and *Ranunculus acris* in northern Sweden. – Ecol. Bull. 45: 144–150.

Gas exchange characteristics and carbon gain of transplanted temperate zone and native subarctic *Geum rivale* and *Ranunculus acris* were evaluated in a common garden under a subarctic growth regime. Photosynthetic capacity (A_{cap}) and rates of dark respiration of temperate zone genotypes grown in the subarctic did not differ significantly from subarctic genotypes, indicating a significant amount of phenotypic adjustment, since A_{cap} of temperate zone genotypes were c. 40% higher in central Europe. Modelled daily leaf carbon balances in the subarctic summer were highest in subarctic genotypes, due to more efficient utilization of low quantum flux densities. Furthermore, temperate zone genotypes transplanted to the subarctic tended to have higher daily leaf carbon balances, than at their latitudinal origin in central Europe. Since neither functional leaf lifespans nor the ratio of assimilatory to non-assimilatory tissues differed appreciably among subarctic and transplanted temperate zone genotypes, it is unlikely that southern provenances attain greater plant carbon balances in the subarctic, than native genotypes. Based on these data it is unlikely that subarctic provenances will be replaced by temperate zone forb genotypes, in the course of proposed poleward migrations of plants associated with climatic warming.

M. Diemer, Inst. für Botanik, Univ. Innsbruck, Sternwartestr. 15A, A-6020 Innsbruck, Austria (present address: Botanisches Inst. der Univ. Basel, Schönbeinstr. 6, CH-4056 Basel, Switzerland).

The factor thought to be most limiting to annual biomass production and carbon gain in arctic plants is the duration of the growth period (Chapin 1983). Plants of cold environments are well adapted to low temperatures, with respect to net photosynthesis and respiration (Billings 1974, Körner and Larcher 1988). Furthermore relative growth rates of arctic graminoids are comparable to temperate zone counterparts (Chapin 1983). Yet total plant biomass of forbs tends to decline with increasing latitude (Prock and Körner 1996).

Proposed changes associated with climate warming at high latitudes include increased plant productivity via increased photosynthesis and/or extended duration of the growth period (Callaghan et al. 1992). On the other hand, potential gains in productivity could be compensated by deleterious effects of higher levels of tissue respiration, increased cloudiness and/or higher UV-B levels. Another potential consequence of global warm-

ing is the poleward migration of species (cf. Peters and Darling 1985, Callaghan et al. 1992). Hence in the long-run, as subarctic species migrate northward they could be replaced by more competitive temperate zone ecotypes in their present subarctic habitat. In fact, Holten (1993) predicts a northward expansion of temperate zone lowland species in Norway in response to climate warming, which could threaten arctic/alpine species.

Northward migration of temperate zone ecotypes also entails acclimation or adaptation to differing photoperiods in the north. Hence the phenotypic plasticity of genotypes, which are adapted to warmer climatic conditions, with respect to the photoperiod and duration of active growth in the subarctic, will ultimately determine the success of a northward migration. A possible approach to evaluate the roles of climate on growth responses of plants are reciprocal transplant

(=common garden) experiments. This approach has been widely used to study altitudinal ecotypes (cf. Clausen et al. 1940, Mooney and Billings 1961). Yet field studies involving latitudinal ecotypes have, aside from Turesson's pioneening studies (Turesson 1930), been carried out merely on *Carex aquatilis* (Chapin and Chapin 1981, Chapin and Oechel 1983), *Phleum alpinum* (Callaghan 1974) and herbaceous perennials (Prock and Körner 1996). Results of these field studies suggest that growth rate underlies strong species-specific genotypic controls and that populations tend to be most successful at their latitudinal origin, in particular if steep environmental gradients are involved.

On the other hand, processes associated with carbon balance tended to lack ecotypic variation in *C. aquatilis* (Chapin and Oechel 1983, but see Mooney and Billings 1961). If so, extended photoperiods during the arctic summer should result in extensive daily carbon surplusses for migrating temperate zone genotypes. Laboratory experiments, in which confounding effects of light intensity and dose as well as temperature can be controlled, have shown that increased photoperiods do in fact result in dramatic dry matter gains (Hay 1990, Solhaug 1991). The consensus of these investigations is that increasing photoperiod (or latitude) tend to increase leaf area ratio (LAR), primarily through an increase in leaf area and SLA. In a field transplant experiment temperate zone low elevation genotypes transplanted to the subarctic had significantly higher LWR (from +17 to +29%), but lower total biomass and SLA, compared to central Europe (Prock and Körner 1996).

Since temperature plays a subordinate role in influencing gas exchange (Chapin 1983, Körner and Diemer 1987), daily carbon gain and growth of herbaceous perennial plants should be equivalent or greater in the subarctic summer, compared to the temperate zone. A likely means of achieving higher carbon balances is continuous photosynthesis of leaves during the 24 h photoperiod of the subarctic summer (see Warren-Wilson 1954, Johansson and Linder 1975, Mayo et al. 1977). In the absence of nighttime dark respiration (R_D) daily carbon gains should be higher in the subarctic, provided daily quantum flux and photosynthetic characteristics of subarctic species are equivalent to conditions in the temperate zone. Leaf carbon balances have been developed for herbaceous perennial plants from the temperate zone (Diemer and Körner 1996), which can serve as a basis for comparison. Thus the objectives of this study are twofold:

1) In a comparison of photosynthetic characteristics of latitudinal genotypes of two common European herbaceous perennial meadow species in situ effects of acclimation to growing conditions in the subarctic were examined.

2) I want to test whether daily leaf carbon gain of herbaceous perennial plants in the subarctic summer is significantly different from the temperate zone, and whether or not differences exist among latitudinal genotypes. If daily sums of net carbon gain among subarctic and transplanted temperate zone genotypes are similar, then it is unlikely that temperate zone genotypes will have a competitive advantage in a warmer subarctic.

Materials and methods

The species utilized in this study were two herbaceous perennial species common throughout Europe. *Ranunculus acris* L. (Ranunculaceae) is a pasture species in most of central Europe, but occupies stream banks and moist open woodlands in its subarctic habitat near Abisko, Sweden. *Geum rivale* L. (Rosaceae) also grows near stream banks in the subarctic, but extends its habitat to include wet meadows in central Europe. All measurements were carried out in a common transplant garden at Abisko Scientific Research Station (68°1′N, 18°49′E), which was established in June-July 1987 by Körner (Prock 1994, see also Prock and Körner 1996). Subarctic genotypes were obtained from the vicinity of Abisko, while temperate zone material originated from a locality near Innsbruck, Austria (47°15′N, 11°25′E). At the time of gas exchange measurements all experimental plants were well established and temperate season transplants had experienced their second growing season in the subarctic.

Gas exchange measurements were carried out on mature, fully developed leaves during a two-week period in July 1989. A portable gas exchange system was utilized (LCA-2). Gas exchange rates were calculated on a projected leaf area basis and corrected for the depletion of CO_2 in the cuvette as described in Diemer and Körner (1996). Diurnal measurements of net photosynthesis and dark respiration determined under a variety of weather conditions were used to model gas exchange characteristics. In addition, photosynthetic capacity (A_{cap}) defined as net photosynthetic rates at saturating quantum flux density (QFD), optimal leaf temperatures and ambient CO_2 partial pressures was extracted from the data set. Based on laboratory light response curves determined on temperate zone genotypes in Innsbruck a QFD > 1200 μmol m^{-2} s^{-1} was used to delimit light saturation (cf. Körner and Diemer 1987, Diemer and Körner 1996). Nighttime dark respiration (R_D) was determined in situ during 01:30–13:00 when QFD declined <5 μmol m^{-2} s^{-1}.

Photosynthetic light response was modelled with the equation derived by Küppers and Schulze (1985), according to the procedure described by Diemer and Körner (1996). The equation utilized was:

$$A_{QFD} = A_{cap} * (1 - e^{-a_1 * (QFD - \Gamma_Q)}) \tag{1}$$

where a_1 represents a coefficient determined iteratively. Based on measurements in the twilight I utilized 20 $\mu mol \ m^{-2} \ s^{-1}$ as the light compensation point (Γ_Q). Since measurements were carried out under a range of weather conditions (clear to overcast days) I assumed that the modelled light response curves incorporated the effect of temperature on leaf photosynthesis as well.

Temperature response of dark respiration (R_D) was modelled with a second-order polynome, in the form

$$R_D = a + b_1 * T_{air} + b_2 * T_{air}^2 \qquad (2)$$

where T_{air} represents air temperature and a, b_1 and b_2 are regression coefficents. For fitting temperature response curves of R_D I assumed a Q_{10} of 2.0 and a cessation of R_D at $-10°C$ (see Diemer and Körner 1996). Gas exchange rates obtained in Abisko, were compared to published data for *R. acris* and *G. rivale* from central Europe (Diemer and Körner 1996), which comprise the same genotypes that were transplanted to the subarctic and which were maintained in the Innsbruck transplant garden (see Prock and Körner 1996).

Daily leaf CO_2 balances were calculated for mature, fully developed leaves during a clear (20 July) and overcast day (28 July) during 1989. Since QFD at the meteorological station was higher than inside the transplant garden, particularily during the night (shade cast by surrounding vegetation), a third scenario involving a 20 h photoperiod during a clear day (20 July 1989) was used (Fig. 2A). I utilized hourly means of QFD (for daylight hours) and air temperature (for the subarctic night at QFD $< 10 \ \mu mol \ m^{-2} \ s^{-1}$) obtained from the nearby Abisko meteorological station to calculate daily CO_2 gains and losses of leaves, utilizing the equations above (for methodology see Diemer 1990, Diemer and Körner 1996). In this simulation effects of vapor pressure deficits on photosynthesis via stomatal closure were disregarded, since air temperatures did not exceed $11.2°C$ during the days utilized for my calculations.

Results and discussion

Photosynthetic capacity and photosynthetic light response

A_{cap} of the differing genotypes growing in the Abisko transplant garden were not significantly different (Table 1). Mean overall A_{cap} of *R. acris* was roughly 20% higher than in *G. rivale*. Values obtained for temperate zone genotypes transplanted to the subarctic were significantly lower than in central Europe (*R. acris* in central European transplant garden: 13.6 $\mu mol \ CO_2$ $m^{-2} \ s^{-1}$, *G. rivale* 10.5 $\mu mol \ CO_2 \ m^{-2} \ s^{-1}$, Diemer 1990). This reduction appears to reflect an acclimation to the subarctic light climate. In the Norwegian subarctic Skre (1975) obtained mean A_{cap} quite similar (*R.*

Table 1. Photosynthetic capacity ($\mu mol \ CO_2 \ m^{-2} \ s^{-1}$) of subarctic and temperate zone genotypes in the Abisko transplant garden, expressed as means and standard errors of 2–6 leaves of different plants. NS: indicates no significant differences between genotypes (two-tailed t-test).

Species	Latitudinal origin		t-test
	Subarctic	Temperate zone	
Ranunculus acris	9.1 ± 1.5	10.0 ± 1.5	NS
Geum rivale	8.0 ± 0.6	7.4 ± 0.1	NS

acris: 9.5 $\mu mol \ CO_2 \ m^{-2} \ s^{-1}$, *G. rivale* 7.6 $\mu mol \ CO_2$ $m^{-2} \ s^{-1}$) to our values. Rates were however determined at QFD of merely 400 $\mu mol \ m^{-2} \ s^{-1}$ (Skre 1975), at which none of the Abisko species attained photosynthetic light saturation.

Modelled photosynthetic responses to QFD are depicted in Fig. 1. Initial slopes of photosynthetic light response of both subarctic genotypes were steeper compared to transplanted temperate zone genotypes. Hence potential carbon gain of leaves of subarctic genotypes was higher at lower QFD, compared to transplanted temperate zone genotypes. In general, net photosynthe-

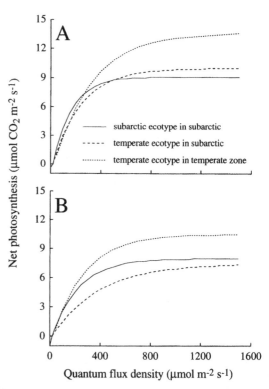

Fig. 1. Photosynthetic response to QFD for subarctic and temperate zone genotypes of A) *Ranunculus acris* and B) *Geum rivale* determined in the subarctic (Abisko), and for temperate zone genotypes at central Europe (data from Diemer 1990). Fitted curves represent means of 2–7 leaves. See text for the methodology utilized.

Table 2. Nighttime dark respiration (μmol CO_2 m^{-2} s^{-1} at QFD < 10 μmol m^{-2} s^{-1}) of subarctic and temperate zone genotypes in the Abisko transplant garden, expressed as means and standard errors of 6–7 leaves of different plants. Nighttime air temperatures ranged from 12 to 14°C. NS: indicates no significant differences between genotypes (two-tailed t-test).

| Species | Latitudinal origin | | t-test |
	Subarctic	Temperate zone	
Ranunculus acris	0.4 ± 0.2	0.4 ± 0.1	NS
Geum rivale	1.0 ± 0.2	0.9 ± 0.4	NS

sis was saturated at lower QFD in subarctic genotypes (500–700 μmol m^{-2} s^{-1}, Fig. 1). In temperate zone genotypes of *G. rivale* transplanted to the subarctic both the slope and the level of the photosynthetic light response curve were drastically lower, compared to plants growing in central Europe (Fig. 1).

Prock and Körner (1996) summarized pertinent climatic parameters of both temperate zone and subarctic research sites. In Abisko mean daily air temperatures during the growth period were 4.5 K lower (mean: 8.1°C) and mean daytime QFD (>30 μmol m^{-2} s^{-1}) was 42% lower (mean: 446 μmol m^{-2} s^{-1}), than at Innsbruck. On the other hand the photoperiod was 8 h longer at Abisko when averaged over the entire growth period. Hence photoperiod extension in the subarctic is associated with an overall lower mean QFD (Nilsen 1985). Subarctic genotypes appear to have responded with higher rates of net photosynthesis at low QFD.

Nighttime dark respiration

Measurements of nighttime dark respiration (R_D) were carried out during two nights with cuvette temperatures ranging from 12 to 14°C. No significant differences existed between subarctic and transplanted temperate zone genotypes (Table 2). In general, *Geum* had roughly twofold higher rates of R_D than *Ranunculus*. Furthermore the R_D of temperate zone *R. acris* determined at 15°C in central Europe (0.7 μmol CO_2 m^{-2} s^{-1}) were somewhat higher than estimates obtained on either genotypes in the subarctic. The reverse

was observed in *G. rivale*: nighttime R_D of temperate zone genotypes determined at 15°C at the geographical origin in central Europe was 33% lower (Diemer 1990), than in the subarctic (Table 2). As with A_{cap} transplanted temperate zone genotypes adjusted their R_D to rates exhibited by subarctic genotypes, supporting the conclusions of Billings (1974) and Chapin and Oechel (1983), i.e. pronounced phenotypic plasticity of gas exchange traits.

Daily carbon gain

Rates of calculated daily carbon gain of mature leaves are summarized in Table 3 and Fig. 2B. Irrespective of weather conditions values were highest for the two *R. acris* genotypes. Differences between provenances were negligible in *R. acris* (p = 0.24, paired t-test), compared to *G. rivale* genotypes (p = 0.001). Leaves of transplanted temperate zone *G. rivale* had on the average 30% lower daily carbon gains than subarctic *G. rivale*. Differences tended to be most pronounced during the low QFD conditions of the overcast day, reflecting higher rates of net photosynthesis of subarctic genotypes in low light (Fig. 1).

Comparable data for mature leaves of temperate zone genotypes obtained in central Europe extracted from the data of Diemer (1990) yielded daily means of 306 ± 20 mmol CO_2 m^{-2} d^{-1} for *R. acris* (range: 60–469 mmol CO_2 m^{-2} d^{-1}, n = 53 days of two leaf cohorts) and 264 ± 15 mmol CO_2 m^{-2} d^{-1} for *G. rivale* (range: 31–401 mmol CO_2 m^{-2} d^{-1}, n = 47 days) during April and May. Hence daily net carbon gain of leaves in the subarctic summer, particularly of subarctic genotypes, was appreciably higher than in the temperate zone, although some overlap exists. It is surprising that lower A_{cap} of temperate zone genotypes transplanted to the subarctic does not automatically translate into lower daily carbon gains, compared to central Europe. These latitudinal differences can be explained in part by the virtual absence of dark respiration in the subarctic, because R_D amounts to roughly 5–10% of daily carbon fixation in mature leaves of temperate zone genotypes in central Europe (Diemer 1990).

Table 3. Estimates of daily net CO_2 uptake of subarctic and temperate zone genotypes during clear sky (20 July 1989) and overcast (28 July 1989) weather conditions at Abisko, as well as clear sky conditions (20 July) with reduced 20 h photoperiod (see methodology). Daily sums are expressed as mmol CO_2 m^{-2} d^{-1}.

Weather conditions	Species			
	Ranunculus acris		*Geum rivale*	
	Latitudinal origin			
	Subarctic	Temperate	Subarctic	Temperate
Clear sky 24 h photoperiod	532	517	421	313
Overcast 24 h photoperiod	347	289	240	146
Clear sky 20 h photoperiod	491	484	381	280

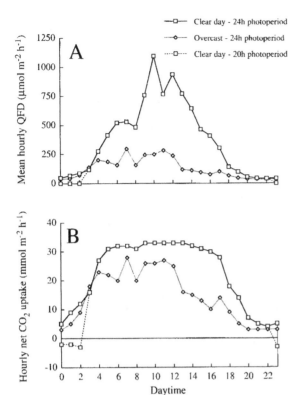

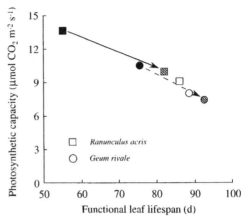

Fig. 3. Relationships between photosynthetic capacity and functional leaf lifespan for subarctic and temperate zone genotypes of *R. acris* and *G. rivale*. The arrows indicate the extent and direction of phenotypic adjustment of temperate zone genotypes in the subarctic. Solid symbols represent temperate zone genotypes in central Europe, hatched ones temperate zone genotypes transplanted to the subarctic, while open symbols depict subarctic genotypes of the subarctic. Leaf lifespan data were obtained from Prock and Körner (1996) and Diemer and Körner (1996).

Fig. 2. Diel changes in A) QFD and B) corresponding hourly CO_2 gain of mature leaves of the subarctic genotype of *Ranunculus acris* for the days utilized in carbon balance calculations.

Plant carbon status and migration potential

Obviously the migration potential of a particular species depends upon demographic traits, such as vegetative growth, survivorship, seed production and viability, as well as dispersal. However, these traits are closely linked to the carbon status of a plant, which is dependent upon the ratio of gains to losses (cf. Bloom et al. 1985). Carbon gains can be described by the lifetime carbon surplus of leaves, while leaf weight ratio (LWR) represents the ratio of assimilatory to non-assimilatory tissues.

The lifetime net carbon surplus of leaves is dependent on gas exchange characteristics, leaf construction cost and lifespan. Diemer and Körner (1996) have shown for herbaceous plants, that lifespan plays a greater role than A_{cap} in determining the leaf carbon balance. According to Prock and Körner (1996) functional leaf lifespans of temperate zone transplants of *R. acris* and *G. rivale* at Abisko were virtually identical to subarctic genotypes, but 22 to 49% greater than at their latitudinal origin in central Europe. They attribute increased longevity of temperate zone genotypes transplanted to Abisko to photoperiod extension in the subarctic. Hence latitudinal displacement of temperate zone geno-

types causes a reduction of A_{cap} and an increase in functional leaf lifespan (Fig. 3).

In the subarctic leaf construction costs, which can be estimated via leaf weight per area (LWA, = 1/SLA), were identical (*R. acris*) or 25% higher (*G. rivale*) in subarctic genotypes compared to transplanted temperate zone genotypes (Prock and Körner 1996). With respect to lifetime leaf carbon surplus, which consists of net daily carbon gain, leaf costs and functional lifespan (Table 3, Fig. 3), transplanted temperate zone genotypes of *R. acris* should be equally successful in the subarctic, as subarctic genotypes. Since daily carbon gain of temperate zone transplants of *G. rivale* was lower in the subarctic, and lifespan was equal, lifetime leaf carbon surplus must be higher in subarctic genotypes. The discrepancy could be somewhat mediated by the higher leaf construction costs of subarctic genotypes of *G. rivale*. Irrespective, leaf lifetime carbon surplus of transplanted temperate zone genotypes of both species is higher in the subarctic, than in central Europe.

Although biomass fractionation appears to be a rather conservative trait (Körner 1994), in this case the LWR of transplanted temperate zone *R. acris* and *G. rivale* tended to increase in the subarctic (Körner and Prock 1996). Unfortunately data of the biomass fractionation of transplanted species is incomplete, no data is available for *R. acris*. For *G. rivale*, temperate zone genotypes transplanted to the subarctic partitioned nearly 30% of total biomass to leaves, which corresponds closer to patterns observed in subarctic genotypes (29%), than temperate zone genotypes in central Europe (20%). Since the proportions of belowground fractions also did not differ appreciably between subarctic and transplanted temperate zone genotypes, potential biomass carbon gain : loss ratios are equivalent, assuming that belowground fractions are at an equilibrium. If transplants of *R. acris* exhibit similar trends, then plant carbon balances of transplanted temperate zone genotypes for the growth period are equal (*R. acris*) or lower (*G. rivale*) than for subarctic genotypes. Hence based on carbon balance considerations there is no evidence that temperate zone genotypes will presently outcompete subarctic genotypes.

It is unlikely that projected temperature increases (+2 K in summer and +3 to +4 K during winter, Holten 1993) will favor temperate zone genotypes of the species examined here in the subarctic, unless the frequency distribution of QFD shifts towards higher light intensities. Furthermore, a simulation of a 2 K increase in canopy temperature had a negligible effect on lifetime leaf carbon gain in central European *R. acris* and *G. rivale* amounting to −2 to +4% (Diemer and Körner 1996).

Long-term persistence of latitudinal genotypes may well depend on their behaviour during periods in which light levels are too low for photosynthetic carbon gain. Chapin and Chapin (1981) demonstrated high autumn and winter mortality in transplanted *C. aquatilis* shoots and suggested that mortality may be related to a depletion of (carbon) storage reserves. Thus appreciable portions of carbon surplusses accrued during the extended photoperiod of the subarctic summer are subsequently respired during the subarctic autumn and winter. Based on estimates of potential carbon gain subarctic genotypes (particularly *G. rivale*) are favoured over transplanted temperate zone genotypes in this respect.

Acknowledgements – C. Körner, P. S. Karlsson and T. Callaghan commented on the manuscript. S. Prock supplied data on plant traits and growth conditions. Meteorological data were provided by the staff of the Abisko Naturvetenskapliga Station. Travel funds were made available by the Swedish Academy of Sciences. Further funding was provided by the Austrian Fonds zur Förderung der wiss. Forschung (to C. Körner).

References

Billings, W. D. 1974. Arctic and alpine vegetation: plant adaptations to cold climates. – In: Ives, J. D. and Barry, R. G. (eds), Arctic and alpine environments. – Methuen, London, pp. 403–443.

Bloom, A. J., Chapin, F. S. III and Mooney, H. A. 1985. Resource limitation in plants – an economic analogy. – Annu. Rev. Ecol. Syst. 16: 363–392.

Callaghan, T. V. 1974. Intraspecific variation in *Phleum alpinum* L. with special reference to polar populations. – Arct. Alp. Res. 6: 361–401.

– , Sonesson, M. and Sømme, L. 1992. Responses of terrestrial plants and invertebrates to environmental change at high latitudes. – Phil. Trans. R. Soc. Lond. B. 338: 279–288.

Chapin, F. S. III 1983. Direct and indirect effects of temperature on arctic plants. – Polar Biol. 2: 47–52.

– and Chapin, M. C. 1981. Ecotypic differentiation of growth processes in *Carex aquatilis* along latitudinal and local gradients. – Ecology 62: 1000–1009.

– and Oechel, W. C. 1983. Photosynthesis, respiration, and phosphate absorption by *Carex aquatilis* ecotypes along latitudinal and local environmental gradients. – Ecology 64: 743–751.

Clausen, J., Keck, D. D. and Hiesey, W. M. 1940. Experimental studies on the nature of species. I. Effect of varied environments on western North American plants. – Carnegie Inst. Publ. 520.

Diemer, M. 1990. Die Kohlenstoffbilanz von Blättern krautiger Pflanzen aus dem Hochgebirge und der Niederung. – Ph. D. thesis, Univ. Innsbruck.

– and Körner, C. 1996. Lifetime leaf carbon balances of herbaceous perennial plants from low and high altitudes in the central Alps. – Funct. Ecol. 10, in press.

Hay, R. K. M. 1990. The influence of photoperiod on the dry-matter production of grasses and cereals. – New Phytol. 116: 233–254.

Holten, J. I. 1993. Potential effects of climatic change on distribution of plant species, with emphasis on Norway. – In: Holten, J. I., Paulsen, G. and Oechel, W. C. (eds), Impacts of climatic change on natural ecosystems, with emphasis on boreal and arctic/alpine areas. NINA, Trondheim, pp. 84–104.

Johansson, L. G. and Linder, S. 1975. The seasonal pattern of photosynthesis of some vascular plants on a subarctic mire. – In: Wiegolaski, F. E. (ed.), Fennoscandian tundra ecosystems 1. Plants and microorganisms. Springer, pp. 194–200.

Körner, C. 1994. Biomass fractionation in plants: a reconsideration of definitions based on plant functions. – In: Roy, J. and Garnier, E. (eds), A whole plant perspective on carbon-nitrogen interactions. SBP, The Hague, pp. 141–157.

– and Diemer, M. 1987. In situ photosynthetic responses to light, temperature and carbon dioxide in herbaceous plants from low and high altitude. – Funct. Ecol. 1: 179–194.

– and Larcher, W. 1988. Plant life in cold climates. – In: Long, S. F. and Woodward, F. I. (eds), Plants and temperature. Symp. Soc. Exp. Biol. 42. CBL, Cambridge, pp. 25–57.

Küppers, M. and Schulze, E. D. 1985. An empirical model of net photosynthesis and leaf conductance for the simulation of diurnal courses of CO_2 and H_2O exchange. – Austr. J. Plant Physiol. 12: 513–526.

Mayo, J. M., Hartgerink, A. P., Despain D. G., Thompson, R. G., van Zinderen Bakker, E. M. and Nelson, S. D. 1977. Gas exchange studies of *Carex* and *Dryas*, Truelove Lowland. – In: Bliss, L. C. (ed.), Truelove Lowland, Devon Island, Canada: A high arctic ecosystem. Univ. of Alberta Press, Alberta, pp. 265–279.

Mooney, H. A. and Billings, W. D. 1961. Comparative physiological ecology of arctic and alpine populations of *Oxyria digyna*. – Ecol. Monogr. 31: 1–29.

Nilsen, J. 1985. Light climate in northern areas. – In: Kaurin, A., Junttila, O. and Nilsen, J. (eds), Plant production in the north. Norwegian Univ. Press, Tromsø, pp. 62–72.

Peters, R. L. and Darling, J. D.S. 1985. The greenhouse effect and nature reserves. – Bioscience 35: 707–717.

Prock, S. 1994. Vergleichende ökologische Untersuchungen zur Phänologie, Blattlebensdauer und Biomassallokation von krautigen Tal- und Gebirgspflanzen aus der Alpenregion, der Subarktis und der Arktis. – Ph. D. thesis, Univ. Innsbruck.

– and Körner, C. 1996. A cross-continental comparison of phenology, leaf dynamics and dry matter allocation in arctic and temperate zone herbaceous plants from contrasting altitudes. – Ecol. Bull. 45: 93–103.

Skre, O. 1975. CO_2 exchange in Norwegian tundra plants studies by infrared gas analysis. – In: Wiegolaski, F. E.

(ed.), Fennoscandian tundra ecosystems 1. Plants and microorganisms. Springer, pp. 168–183.

Solhaug, K. A. 1991. Long day stimulation of dry matter production in *Poa alpina* along a latitudinal gradient in Norway. – Holarct. Ecol. 14: 161–168.

Turesson, G. 1930. The selective effect of climate upon the plant species. – Hereditas 14: 99–152.

Warren-Wilson, J. 1954. The influence of 'midnight sun' conditions on certain diurnal rhythms in *Oxyria digyna*. – J. Ecol. 42: 81–94.

Ecological Bulletins 45: 151–158. Copenhagen 1996

Interactions between hemiparasitic angiosperms and their hosts in the subarctic

Malcolm C. Press and Wendy E. Seel

Press, M. C. and Seel, W. E. 1996. Interactions between hemiparasitic angiosperms and their hosts in the subarctic. – Ecol. Bull. 45: 151–158.

Aspects of the growth, nutrient and carbon relations of some of the common hemiparasites of the Lake Torneträsk region of subarctic Sweden are described with respect to the ways in which the plants utilise resources and the consequences of this method of nutrition for both the parasite and the host. Some species are facultative parasites and can grow in the absence of a host plant. Under these circumstances their growth is severely limited by an inability to both absorb and assimilate inorganic solutes. Attachment to a host plant greatly stimulates growth, with the nature of the host determining the extent to which such stimulation occurs. The most vigorous performance of the parasites occurs on nitrogen-fixing legumes, when the highest rates of growth and photosynthesis are observed. Thus heterotrophic organic nitrogen supply may play a major role in determining the success of these hemiparasites. Less is known about the influence of these hemiparasitic angiosperms on the performance of their host. A study of the *Rhinanthus minor-Poa alpina* ssp. *vivipara* association revealed that infected *P. alpina* plants accumulated less biomass and had lower shoot:root ratios than uninfected controls, and invested less material in reproductive tissue. However, the parasite did not influence rates of photosynthesis in the host.

M. C. Press, Dept of Animal and Plant Sciences, Univ. of Sheffield, Sheffield, U.K. S10 2TN. – W. E. Steel, Dept of Plant and Soil Science, Univ. of Aberdeen, Aberdeen, U.K. AB9 2UD.

Low nutrient availability is amongst the major factors which limit plant growth in the Arctic. Many arctic species, along with those plants living in other nutrient poor habitats, have thus evolved strategies which facilitate nutrient acquisition by either enhancing or replacing direct solute uptake via the root system. In this chapter we consider one of these strategies, the parasitism of neighbouring plants by angiosperms. Other examples of facilitated nutrient acquisition systems in arctic plants, such as carnivory and mycorrhizas, are reviewed by Karlsson et al. (1996) and Jonasson and Michelsen (1996).

Worldwide there are between 3000 and 4000 species of flowering plants which derive some or all of their water, inorganic and organic solutes from a neighbouring angiosperm(s) (see e.g. Kuijt 1969, Press and Graves 1995). Parasitic angiosperms are widely distributed both geographically and taxonomically, occurring in c. 17 plant families, and ranging in distribution from arctic to tropical environments. They may be sub-divided morphologically, depending on whether they are attached to the host root or shoot, and functionally depending on the presence or absence of chlorophyll (with species being referred to as hemi- and holoparasites, respectively).

The subarctic is particularly rich in root hemiparasitics within the Scrophulariaceae. In the Abisko area of Lake Torneträsk, Lewejohann and Lorenzen (1983) report ten species of hemiparasitic Scrophulariaceae from five genera, with five of the species being described as common (Table 1). Some of the annual parasites in this group, such as *Euphrasia frigida* and *Rhinanthus minor*, are described as facultative because they are capable of completing their life cycle in the absence of a host. This, however, is an event which rarely occurs in nature because of the density of neighbours. The growth of unattached, potentially parasitic plants is poor compared to that of attached individuals. Moreover, for genera such as *Euphrasia* and *Rhinanthus* which are capable of attaching to a range of plants,

Table 1. Parasitic plants present in the Abisko region of subarctic Sweden, together with their life-form and a note on their distribution (from Lewekohann and Lorenzen 1983).
Vassijaure, Vassitjåkka, Kärkevagge (Va); Kärketjårro, Låktatjåkka (Kå); Koppårasen to Vadvetjåkka (Ko); Vadvetjåkka national park (Vv); Njunjes (Nj); Jebrentjåkke, Lairevare, Pessisvare (Je); Vakketjåkka, Vaivantjåkka (Vk); Torneham, Björkliden (To); Njulla, Slåttatjåkka (Na); Abisko, from Abiskojåkka to Tjuonovaggejåkka (Ab); Abiskojaure, Pallentjåkka (Aj); Lapp-porten, Tjuonavagge (Lp); Bergfors, Rensjön (Br).

Species	Life form	Location
Bartsia alpina	perennial	common
Euphrasia arctica ssp. *tenuis*	annual	Je, To, Ab, Aj
E. frigida	annual	common
E. lapponica	annual	Kä, Je, To, Ab
Melampyrum sylvaticum	annual	common
M. pratense	annual	common
Pedicularis lapponica	perennial	common
P. flammea	perennial	Je
P. hirsuta	perennial	Va, Kä, Vv, Nj, Vk, To, Na, Ab, Aj, Lp
P. sceptrum-carolinum	perennial	Vk, Na, Ab, Br

growth of the parasite is variable depending on the nature of the host to which it attaches (Yeo 1964, Seel et al. 1993a). Other parasitic angiosperms are highly host specific, such as *Pedicularis sceptum-carolinum* which only parasitises one species of willow. The performance of these parasites is presumably governed at least in part by effects of the environment on host growth and productivity, although the relationship between host and parasite performance in these associations has not been well investigated.

In this paper we examine aspects of the growth, nutrient and carbon relations of some of the most common hemiparasites found in the Swedish subarctic. The data are discussed with respect to i) the way in which this group of plants utilise resources and ii) the consequences of this method of nutrition for both the parasite and the host.

Growth of unattached hemiparasites

It is possible to use facultative parasite species to examine the extent to which the growth of unattached plants is limited by the supply of either inorganic or organic solutes. *Rhinanthus minor* is a facultative annual root hemiparasite which is present in large numbers as an alien along road-sides and in disturbed areas of the Lake Torneträsk region. The response of unattached *R. minor* to inorganic solutes was investigated by supplying plants with nitrogen, phosphorus and potassium, both individually in a pot study and in combination in a field study (Seel et al. 1993b). Both height-growth and dry matter accumulation were stimulated in the field by the addition of nutrients, but only to a limited extent compared to the response following attachment to a host. The pot study showed that the response of *R. minor* to fertilisation in the field was

mainly attributable to the addition of phosphate (Fig. 1), with no growth stimulation in response to the addition of either nitrogen or potassium. For non-parasitic plants, the extent to which growth is stimulated by nutrient supply has been used as an indication of the extent to which the nutrient or nutrients involved are limiting plant growth. Since the addition of nutrients did not stimulate the growth of *R. minor*, either the supply of inorganic solutes was not limiting or else the parasite was not able to absorb and/or assimilate the nutrients. Measurements of the concentrations of nitrogen and phosphorus in the leaves of *R. minor* (and the related Mediterranean species *Bartsia trixago*) suggest that unattached parasites have at least some capacity to absorb inorganic solutes from the soil (Seel et al. 1993b, Press et al. 1993). In addition Seel and Woo (unpubl.) have measured high concentrations of nitrate in *R. minor* leaves, especially when the anion was supplied at a high rate. Thus it appears that an inability to assimilate inorganic ions may explain the absence of any large response to nutrients by unattached facultative hemiparasites, and low activities of the nitrate assimilating enzyme nitrate reductase in root hemiparasites support this conclusion (Gebauer et al. 1988).

Given that when attached to a host hemiparasites are more likely to be in receipt of organic, rather than inorganic solutes, we might expect the plants to show greater growth responses to amino acids than to inorganic nitrogen. However, a subsequent pot study in which *R. minor*, *Melampyrum sylvaticum* and *Euphrasia frigida* were watered with a solution containing a mixture of amino acids demonstrated that this was not the case either (Seel and Press unpubl.). We therefore conclude that a combination of both poor uptake of solutes and a low capacity for their assimilation are major limitations to the autotrophic growth of these hemiparasites.

152

The influence of host type on parasite growth

Like other facultative root hemiparasites, the growth of *R. minor* and *E. frigida* is greatly stimulated following attachment to a host plant (Seel and Press 1993). However the extent to which growth is stimulated is dependent on the nature of the host to which the parasite becomes attached, and biomass accumulation is much greater when the plants are parasitising a legume rather than a grass (Seel and Press 1993; see also Gibson and Watkinson 1991). Dry matter accumulation by unattached *E. frigida* plants, and those attached to *Festuca ovina* differed by a factor of approximately five, but the difference in dry matter accumulation between unattached plants and those attached to the legume *Astragalus alpinus* was a factor of almost 70. The corresponding values for *R. minor*

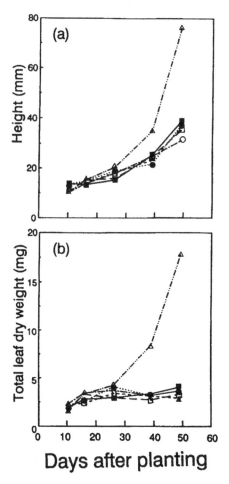

Fig. 1. Height (a) and total leaf dry weight (b) of *Rhinanthus minor* grown in sand in the absence of a host and supplied with different inorganic solutes, over a 50 day period: water only (—■—); 2.5 mM NaNO₃ (- - ▲ - -); 2.5 mM NH₄Cl (··· ● ···); 2.5 mM NH₄NO₃ (—□—); 1 mM Na₂HPO₄ (·· —△— ··); 2.5 mM KCl (—○—). Means of 12 plants are plotted and standard errors (not shown, for clarity) are <10% of the mean value. (Taken from Seel et al. 1993b).

growing on *Festuca rubra* and *Vicia cracca* were 7 and 377, respectively. In addition to gross differences in biomass accumulation, the host type also influences both the allometry and the architecture of the parasites (Seel and Press 1993). Plants parasitic on a legume are considerably more branched and leafy than those on a graminiferous host. This change in branching pattern and leafiness may have important consequences for resource capture (carbon dioxide and light), although a cost-benefit analysis of the different architectures of the plants parasitic on the different host types has not yet been conducted.

These findings suggest that the performance of at least some hemiparasites may be largely determined by the host, and in order to investigate this phenomenon further, *R. minor* was grown on 11 different hosts in a greenhouse study (Seel et al. 1993a). *Rhinanthus minor* was selected for this study because of its wide host range. From observations at just five sites Gibson and Watkinson (1989) concluded that the plant could parasitise at least 50 species from 18 families, with the Leguminosae and Gramineae being the preferred hosts. Seel et al. (1993a) observed large differences in the growth of the parasite in relation to host type (Fig. 2), with parasites growing on the legume *Trifolium repens* reaching more than ten times the height of unattached individuals. Variation in parasite growth was also observed with different grass hosts, and may relate to the relative growth rate of the grass and the ability of the grass to supply the parasite with resources.

The nature of the host can therefore have a profound effect on the growth and allometry of the hemiparasite, for those species with low host specificity. The most likely explanation for the variation in parasite growth in relation to host type is the different ability of hosts to supply the parasite with resources in a form which the parasite can use for growth.

Foliar nutrient content of attached hemiparasites

Despite the importance of phosphorus in stimulating the growth of unattached *R. minor* seedlings, in this study there was no clear pattern in the phosphorus content of leaves of the hemiparasites. With the exception of *R. minor* parasitising *F. rubra*, the parasites did not accumulate phosphate to a greater degree than the hosts.

Foliar nutrient analyses in the study of *R. minor* and *E. frigida* growing on either legume or grass hosts (Seel and Press 1993) showed that the highest nitrogen concentrations were in parasites attached to legumes. Unattached plants had concentrations either similar to, in the case of *R. minor*, or lower than, in the case of *E. frigida*, those found in plants parasitising grasses. The foliar nitrogen content of the parasites exceeded that of

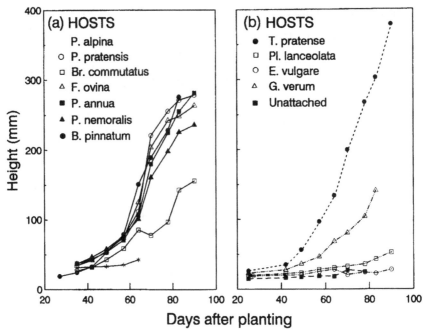

Fig. 2. Height of *Rhinanthus minor* over a 90 day period on either grass hosts (a) or dicotyledonous hosts (b). Data for unattached plants (also shown in [b]) are presented for comparison. Means of 10–36 measurements are shown, and standard errors (not shown, for clarity) are <10% of the mean value. (Taken from Seel et al. 1993a).

the host in all cases, and such accumulation of nitrogen as well as other elements is a characteristic feature of both root and shoot hemiparasites and is attributable at least in part to their high rates of transpiration (see below).

Photosynthesis

A strong positive relationship has been reported between foliar nitrogen concentration and rates of photosynthesis in many C_3 plants (see e.g. Evans 1989) because of the importance of nitrogen for the construction of photosynthetic pigments and proteins. Very low rates of photosynthesis were measured in *R. minor* plants grown in the absence of a host, ranging from 0.6 to 2.3 μmol m^{-2} s^{-1}, depending on the nutrients supplied (Seel et al. 1993b). The supply of neither nitrogen nor phosphate stimulated photosynthesis in the plants, although attachment to a host did, with more than a ten-fold increase in the rate of CO_2 fixation being observed when the host was a legume (*Trifolium repens*) (Seel et al. 1993a). There was a strong positive relationship between the rates of photosynthesis in *R. minor* and its growth on different hosts (Fig. 3). This relationship is at least partly a result of differences in the supply of organic nitrogen compounds from host to parasite, which may be used for a number of processes including the construction of photosynthetic machinery, as illustrated by higher concentrations of chlorophyll in

the leaves of *R. minor* associated with nitrogen-rich hosts (Seel et al. 1993a). Differences in the concentration of organic nitrogen supplied to hemiparasites by different hosts can be illustrated by data from a study of the SW Australian root hemiparasitic shrub *Olax phyllanthi*, in which xylem sap nitrogen concentrations for legume hosts exceeded those of non-legume hosts by a factor of more than three (Tennakoon and Pate 1996).

In addition to receiving nitrogenous compounds from the host which can be used to fix carbon autotrophically, the parasites are probably also in receipt of host carbon which may be directly used for growth. There is no direct evidence to support this supposition for the species studied here, but significant transfer of carbon from hosts has been reported for other associations involving root hemiparasitic Scrophulariaceae (e.g. Rogers and Nelson 1962, Govier et al. 1967, Cechin and Press 1993), and also for other groups of parasitic plants. In the mistletoes, for example, estimates of heterotrophic carbon gain range from 5 to 62% of the total leaf carbon pool (see references in Press and Whittaker 1993, Tennakoon and Pate 1996). Greater growth on legume hosts may also reflect a higher input of carbon as well as nitrogen because of the carbon associated with the nitrogenous solutes, in addition to carbon supplied in the form of non-nitrogen based compounds (Tennakoon and Pate 1996). In their classic study of the root hemiparasite *Odontites verna* parasitising either barley or white clover, Govier et al. (1967) examined the form in which the parasite received

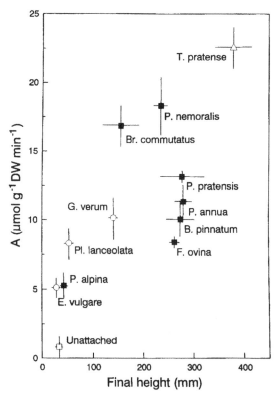

Fig. 3. Relationship between mean light saturated rates of photosynthesis (A) and mean height of *Rhinanthus minor* after 90 days growth, on 11 different host species or unattached. Symbols indicate different types of host: grasses (■); legume (△); non-legume dicotyledons (○); unattached (□). (Taken from Seel et al. 1993a).

those parasites with a wide host range, the costs of processing incoming solutes in the haustorium are also likely to vary between one host and the next. In nature some parasitic angiosperms obtain water and solutes from more than one host, as has been demonstrated for *Olax phyllanthi* (Pate et al. 1990). Although in the case of *Olax* it seems that water relations determine the temporal pattern of host dependency, there are likely to be associated differences in the nature of solutes received from the host, which may have implications for parasite growth. Some of the hemiparasites studied here are likely to form attachments to more than one host simultaneously, for example, *Pedicularis sylvatica* and *Bartsia alpina*, although studying the nutrient relations of these perennial species has proved to be rather difficult.

Transpiration-photosynthesis relationship

Rates of transpiration and stomatal conductances of hemiparasitic angiosperms are characteristically high (see references in Press and Whittaker 1993), and the hemiparasites studies in the subarctic conform with this pattern of behaviour (Table 2; see also Press et al. 1988, Seel and Press 1994). Since these rates are greater than those of the host plant, and the photosynthetic capacity of even attached parasites is generally less than that of the host, hemiparasites tend to have significantly lower water use efficiencies (WUE, defined as the molar ratio of photosynthesis to transpiration) than their hosts.

There has been much interest in attempting to interpret the functional significance of high rates of transpiration in hemiparasites. Based on studies of leafy mistletoes, Schulze et al. (1984) proposed a nitrogen parasitism hypothesis, in which they suggested that high transpiration rates were primarily necessary to maintain a high flux of nitrogen to the parasite in order to stimulate photosynthesis and growth (see also Ehleringer et al. 1985). Although a high rate of transpiration will ensure a large flux of nitrogen, the theory goes on to suggest that when hemiparasites are in receipt of nitrogen-rich sap, the difference between host and parasite water use efficiency is smaller, since the parasite would not need to expend so much water in order to obtain sufficient nitrogen. In other words, nitrogen may

radiolabelled carbon fed to the host. When parasitising white clover, the parasite received 87% of the total ^{14}C in the form of amino acids and amides, >90% of which was recovered as aspartic acid and aspargine. In contrast, with barley as a host, monosaccharides and other non-nitrogenous compounds accounted for most of the heterotrophic carbon. Pate and co-workers also provide some evidence for the metabolism of host-derived solutes within the haustorium (Govier et al. 1967, Pate et al. 1991), and high levels of metabolic activity will carry associated energetic costs for the parasite. Thus there are both qualitative and quantitative differences in the nature of both organic and inorganic solutes available to parasitic angiosperms, depending on the host. For

Table 2. Transpiration rates and water use efficiency for hemiparasitic angiosperms measured in the Abisko area (from Press et al. 1988).

Species	Transpiration rate (mmol m^{-2} s^{-1})	Water use efficiency (mmol CO$_2$ mol^{-1} H$_2$O)
Euphrasia frigida	7.1	0.5
Rhinanthus minor	9.2	0.7
Pedicularis lapponica	7.0	0.3
Bartsia alpina	7.0	0.5

play a role in modulating parasite water relations. The validity of this theory has subsequently been questioned by a number of studies (Press and Whittaker 1993, Panvini and Eickmeier 1993, see also Marshall et al. 1994), and we have found no evidence that nitrogen modulates the transpiration rates of the subarctic root hemiparasites that we have studied (Seel et al. 1993a). Moreover, there is evidence that under conditions where the plants are supplied with additional nitrogen, hemiparasites will readily accumulate nitrogen in excess of that which is used for growth without any effect on stomatal conductance (Seel and Woo unpubl.). It would thus seem unlikely that there is an operative feedback system by which nitrogen could act as a signal to induce a reduction in the rate of transpiration.

Effects of parasitism on the host

A small number of parasitic genera are weeds of agricultural importance, and the effects that they have on their hosts have been well characterised. Both *Striga* and *Orobanche*, the two most important root parasitic weeds of herbaceous crops, not only reduce the growth of their host, but also influence allometry, often favouring allocation of biomass to below-ground tissue (see e.g. Press 1995). The parasites also influence carbon gain by the host, not only as a consequence of the altered allometry, which will shift the balance between photosynthetic and respiratory tissue and influence light interception through changes in architecture, but also by directly influencing photosynthetic metabolism (see e.g. Press 1995). For example, for *S. hermonthica*-infected sorghum and maize plants growing in western Kenya, light saturated rates of photosynthesis 46 and 31% lower than those of control uninfected plants were recorded, respectively (Gurney et al. 1995; see also Press et al. 1987).

Far less information is available on the way in which parasites in natural and semi-natural vegetation influence their host. With regard to species present in the subarctic, we examined the influence of *R. minor* on the perennial grass *Poa alpina* ssp. *vivipara* (Seel and Press unpubl.). *Poa alpina* plants infected in their first year of growth accumulated less biomass than uninfected control plants but there was little change in the partitioning of dry matter into stems, roots and leaves. Large changes in biomass partitioning did, however, become apparent in the second year of the study. Plants which had been infected for two subsequent growing seasons with *R. minor* had a shoot:root ratio of 1.98, compared to 3.26 for uninfected control grasses, with differences being largely attributable to less accumulation of above-ground biomass, rather than changes in root biomass.

Parasites also have the capacity to affect the reproductive output of the host plant. The host we studied,

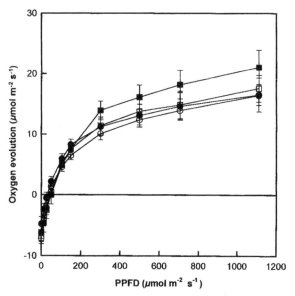

Fig. 4. Relationship between the CO_2-saturated rate of oxygen evolution and photosynthetic photon flux density (PPFD) for leaves of *Poa alpina* plants in either the absence (■ and ○) or presence (● and □) of *Rhinanthus minor*. Means of 20 measurements ± 1 standard error are reported. (Seel and Press unpubl.).

Poa alpina ssp. *vivipara*, produces semi-viviparous plantlets. The buds from which these develop are initiated in the growing season prior to that in which the plantlet becomes visible. Although individual plantlet weight did not differ between infected and uninfected grasses, there was a large difference in the percentage of parents that produced plantlets in the year following infection. All the uninfected plants produced plantlets, compared to only 35% of infected plants. Because the production of the plantlets depends on bud initiation in the previous season, removing the parasite for the duration of the second growing season did not result in plantlet production that year. Instead the plant relieved of parasites diverted their resources into leaf growth (Seel and Press unpubl.).

In contrast with *Striga*, *R. minor* did not appear to exert any major influence on host photosynthetic activity (Fig. 4). Neither did the parasite influence host dark respiration, photosynthetic light compensation point, nor apparent quantum efficiency of photosynthesis.

Conclusion

Hemiparasites are an important component of the vegetation in the Swedish subarctic, with both annual and perennial species present. The variation in host range is great, with some species being restricted to a single partner, and others are capable of infecting a large number of different host species. For the latter group, parasite performance is highly dependent on the nature of the host, with nitrogen-fixing legumes resulting in the most vigorous parasites. Part of the reason for this appears to be that the parasites utilise organic nitrogen supplied by the host to synthesis photosynthetic pigments, and presumably also proteins, in order to enhance autotrophic fixation of carbon. However the plants are also likely to be in receipt of heterotrophic carbon, and the ability of different hosts to supply other organic solutes may be an important determinant of growth. In common with other hemiparasites, the subarctic species that we studied have high rates of transpiration, although we found no evidence to suggest that host nitrogen status could modulate parasite stomatal conductance. Although no evidence has been found to suggest that the parasites can influence host photosynthetic rates, as does the parasitic weed *Striga*, the parasites of the subarctic are capable of influencing host growth and biomass allocation. These two phenomena, along with the striking effect of parasitism on the reproductive output of a host, are undoubtedly of importance to the ecology of the subarctic ecosystem. It is clear from the data presented here and from other studies (e.g. Gibson and Watkinson 1992) that hemiparasites have the capacity to affect community composition by affecting the ability of host plants to compete with neighbours. The parasite affects host competitive ability by altering allometry, so influencing resource capture, and by influencing reproductive success. In terms of success of the parasites, their mode of nutrition is advantageous in nutrient poor habitats such as the subarctic since they do not have to expend valuable resources on the construction of a root system, they can simply tap into one which has already been built. However, a full cost-benefit analysis of this mode of nutrition has yet to be performed, and factors such as the need to maintain metabolically active haustoria must also be taken into account.

Acknowledgements – We gratefully acknowledge the Natural Environmental Research Council for financial support, M. Sonesson and his staff for allowing us to use the facilities at the Abisko Scientific Research Station, and V. Cochrane for technical assistance.

References

Cechin, I. and Press, M. C. 1993. Nitrogen relations of the sorghum-*Striga hermonthica* host-parasite association: growth and photosynthesis. – Plant Cell Environ. 16: 237–247.

Ehleringer, J. R., Schulze, E.-D., Ziegler, H., Lange, O. L., Farquhar, G. D. and Cowan, I. R. 1985. Xylem-tapping mistletoes: water or nutrient parasites? – Science 227: 1479–1481.

Evans, J. R. 1989. Photosynthesis and nitrogen relationships in leaves of C$_3$ plants. – Oecologia 78: 9–19.

Gebauer, G., Rehder, H. and Wollenweber, B. 1988. Nitrate, nitrate reduction and organic nitrogen in plants from different ecological and taxonomic groups of central Europe. – Oecologia 75: 371–385.

Gibson, C. C. and Watkinson, A. R. 1989. The host range and selectivity of a parasitic plant: *Rhinanthus minor* L. – Oecologia 78: 401–406.

– and Watkinson, A. R. 1991. Host selectivity and the mediation of competition by the root hemiparasite *Rhinanthus minor*. – Oecolgia 86: 81–87.

– and Watkinson, A. R. 1992. The role of the hemiparasitic annual *Rhinanthus minor* in determining grassland community structure. – Oecologia 89: 62–68.

Govier, R. N., Nelson, M. D. and Pate, J. S. 1967. Hemiparasitic nutrition in angiosperms I. The transfer of organic compounds from host to *Odontites verna* (Bell.) Dum. (Scrophulariaceae). – New Phytol. 66: 285–297.

Gurney, A. L., Press, M. C. and Ransom, J. K. 1995. The parasitic angiosperm *Striga hermonthica* can reduce photosynthesis of its sorghum and maize hosts in the field. – J. Exp. Bot. 46: 1817–1823.

Jonasson, S. and Michelsen, A. 1996. Nutrient cycling in subarctic and arctic ecosystems, with special reference to the Abisko and Torneträsk region. – Ecol. Bull. 45: 45–52.

Karlsson, P. S., Svensson, B. M. and Carlsson, B. Å. 1996. The significance of carnivory for three *Pinguicula* species in the subarctic environment. – Ecol. Bull. 45: 115–120.

Kuijt, J. 1969. The biology of parasitic flowering plants. – Univ. of California Press, USA.

Lewejohann, K. and Lorenzen, H. 1983. Annotated check-list of vascular plants in the Abisko-area of Lake Torneträsk, Sweden. – Ber. Deutsch. Bot. Ges. 96: 591–634.

Marshall, J. D., Dawson, T. E. and Ehleringer, J. R. 1994. Integrated nitrogen, carbon, and water relations of a zylem-tapping mistletoe following nitrogen fertilization of the host. – Oecologia 100: 430–438.

Panvivi, A. D. and Eickmeier, W. G. 1993. Nutrient and water relations of the mistletoe *Phoradendron leucarpum* (Viscaceae): How tightly are they integrated? – Am. J. Bot. 80: 872–878.

Pate, J. S., Davidson, N. J., Kuo, J. and Milburn, J. A. 1990. Water relations of the root hemiparasite *Olax phyllanthi* (Labill) R. Br. (Olacaceae) and its multiple hosts. – Oecologia 84: 186–193.

–, True, K. C. and Rasins, E. 1991. Xylem transport and storage of amino acids by S. W. Australian mistletoes and their hosts. – J. Exp. Bot. 42: 441–451.

Press, M. C. 1995. How do the parasitic weeds *Striga* and *Orobanche* influence host carbon relations? – Aspects Appl. Biol. 42: 63–70.

– and Whittaker, J. B. 1993. Exploitation of the xylem stream by parasitic organisms. – Phil. Trans. R. Soc. Lond. B. Biol. Sci. 341, 101–11.

– and Graves, J. D. (eds) 1995. Parasitic plants. – Chapman and Hall, London.

–, Tuohy, J. M. and Stewart, G. R. 1987. Gas exchange characteristics of the sorghum-*Striga* host-parasite association. – Plant Physiol. 84: 814–819.

–, Graves, J. D. and Stewart, G. R. 1988. Transpiration and carbon acquisition in root hemiparasitic angiosperms. – J. Exp. Bot. 39: 1009–1014.

–, Parsons, A. N., Mackay, A. W., Vincent, C. A., Cochrane, V. and Seel, W. E. 1993. Gas exchange characteristics and nitrogen relations of two Mediterranean root hemiparasites: *Bartsia trixago* and *Parentucellia viscosa*. – Oecologia 95: 145–151.

Rogers, W. E. and Nelson, R. R. 1962. Penetration and nutrition of *Striga asiatica*. – Phytopathology 52: 1064–1070.

Schulze, E.-D., Turner, N. C. and Glatzel, G. 1984. Carbon, water and nutrient relations of two mistletoes and their hosts: a hypothesis. – Plant Cell Environ. 7: 293–299.

Seel, W. E. and Press, M. C. 1993. Influence of the host on three sub-Arctic annual facultative root hemiparasites. I. Growth, mineral accumulation and above ground dry-matter partitioning. – New Phytol. 125: 131–138.

– and Press, M. C. 1994. Influence of the host on three sub-Arctic annual facultative root hemiparasites. II. Gas exchange characteristics and resource use-efficiency. – New Phytol. 127: 37–44.

– , Cooper, R. E. and Press, M. C. 1993a. Growth, gas exchange and water use efficiency in the facultative hemiparasite *Rhinanthus minor* associated with hosts differing in foliar nitrogen concentration. – Physiol. Plant. 89: 64–70.

– , Parsons, A. N. and Press, M. C. 1993b. Do inorganic solutes limit growth of the facultative hemiparasite *Rhinanthus minor* L. in the absence of a host? – New Phytol. 124: 283–289.

Tennakoon, K. U. and Pate, J. S. 1996. Heterotrophic gain of carbon from hosts by the xylem-tapping root hemiparasite *Olax phyllanthi* (Olacaceae). – Oecologia 105: 369–376.

Yeo, P. F. 1964. The growth of *Euphrasia* in cultivation. – Watsonia 6: 1–24.

Ecological Bulletins 45: 159–169. Copenhagen 1996

Summer air temperatures and tree line dynamics at Abisko

Björn Holmgren and Martin Tjus

Holmgren, B. and Tjus, M. 1996. Summer air temperatures and tree line dynamics at Abisko. – Ecol. Bull. 45: 159–169.

Recorded and calculated summer air temperatures at Abisko (68°21′N, 18°49′E, 388 m a.s.l.) from 1868 to 1994 are discussed in relation to changes of altitudinal tree line and distribution patterns of mountain birch near the beach of Lake Torneträsk. The broad features of the temperature variations are: there are only minor changes from the 1870's to c. 1900 when a rise of c. 1.5°C took place ending in c. 1940. Since then there has been a decline of c. 0.5°C in the long-term trend. The peak air temperatures measured in the 1930's appear to be associated with a 20–50 altitudinal rise of the species limit of mountain birch on the eastern slope of Mt Njulla whereas the potential altitudinal rise, calculated from the temperature increase of 1.0°C and typical mean summer lapse rates, is >200 m. Based on the continuous pine dendrochronologies from the Torneträsk region, it is suggested that the vertical changes of the tree line on the eastern slope of Mt Njulla are small since 500 AD. The reason is that 50 yr mean anomalies of summer temperatures may be too small to induce any significant altitudinal movements. The importance of edaphic and micrometeorological factors for the response to climate of the subalpine heaths below the altitudinal tree line are discussed.

B. Holmgren and M. Tjus, Abisko Scientific Research Station, S-981 07 Abisko, Sweden.

The meteorology of the Torneträsk area is characterised by very variable weather conditions and also, depending upon season, by strong average gradients of many climate parameters. The topographical setting with high fjelds on both sides of the break-through valley across the mountain range in east-west direction, the closeness to the Atlantic coast and also to the large water body of Lake Torneträsk, all contribute to the weather and climate features of the area. Some references to descriptions and reports on this subject are given in the Appendix.

The late Gustaf Sandberg, director of the Abisko station 1949–1973, initiated two major long-term meteorological research projects which, with respect to the demand on field work and the length of the field investigations, probably are outstanding so far: 1) During 1955–1973, measurements of the local air temperature variations were made at six sites in the lower Abisko valley. There were also measurements of precipitation, ground water variations, snow courses and ground temperatures related to the subalpine heaths. There was also a site at the tree line on the eastern slope of Mt Njulla (Andersson et al. 1996). 2) Measure-ments of climate and microclimate gradients in a transect on the southfacing slope of Lullehatjårro in the north-western part of the land north of Torneträsk starting at 345 m a.s.l. close to the beach and ending at 700 m a.s.l. There were in all eight stations in the transect and the project ran during 1966–1970. With the techniques available, mechanical recorders and paper records and manually operated potentiometers for point readings of various parameters, the need for manpower in the evaluations was enormous and the resources for data evaluation were quite insufficient.

The motivation for the first study emanated from 1) observations of the "inverted tree line" near the beach of Torneträsk, 2) the vegetation changes that took place around the altitudinal tree line in the 1930's and later, and finally 3) the puzzle of the subarctic heaths that Gustaf Sandberg was fascinated by, and that he considered to be the most valuable resources of the Abisko National Park.

The second project, that in some respects were more demanding than the first one due among other things to hazardous voyages across Torneträsk late in autumn, stemmed from the altitudinal biotope stratification on Mt Lullehatjårro north of Djupviken in the northwest part of Torneträsk. Here many "south slope" species are found and Gustaf Sandberg took a special interest in studying micrometeorological requirements in situ for these species to survive near the tree line in the subartic climate. His outlook was to understand the development of the many facets in the Torneträsk landscape and how climate change affected the various parts of the natural ecosystem.

The objectives of this study are to present recorded summer temperature variations during 1913–1994 and to extend this data series back in time to 1868 by correlating the Abisko series with those from Karesuando and Tromsö. Some implications of the climate variations on the altitudinal limit of mountain birch are discussed. We also make some remarks on expected past changes of the tree line in relation to pine dendrochronological proxy data of summer air temperatures from the Torneträsk area. The results presented here are to a considerable extent based on some of the records and observations collected in Gustaf Sandberg's projects.

The Abisko observatory

There are no Swedish climate stations in the interior fjeld region that have homogeneous series dating back as far as 1900. In 1859, a net-work of 25 stations was established in Sweden of which a few are situated in the eastern low-fjeld areas (Alexandersson and Eriksson 1987). In the Lake Torne region the first climate records started at the time of the construction of the railroad in about 1900. At Abisko a meteorological observatory was established in 1912. For the Abisko observatory (Rolf 1920–1930) there are detailed (hourly) records of air temperature, humidity, pressure, wind speed and direction, sunshine hours, cloud types and amounts, precipitation types and amounts, and also various weather phenomena published for 1913–1929. Other observations of interest in this context are wind measurements at c. 10 m above the ground, ground temperature measurements (starting in 1925) and snow depths (starting in 1913). On 1 January 1930 the meteorological observatory was moved to Riksgränsen where similar detailed observations as in Abisko were carried out for 7 yr.

At Abisko, the meteorological observations continued but far fewer parameters were measured than before 1930. The quality checks of the records and the observations were poorer. Also, cloud covers and phenomena like halos, thunder, rain bows, auroras etc. were not recorded any longer. As noted, Gustaf Sandberg took a special interest in the collection of meteorological and hydrological parameters. When he became director in 1949 the climate observations were ex-

panded and firm routines were set for checks and calibrations. Several new data series, e.g. of ground temperatures and snow depths were started at new sites near the observatory in order to investigate variations related to the small scale topography, soil and vegetation types. However, there were not means enough for evaluating the graphical records utilised for many of the measurements. A systematic evaluation of the observatory data including transfer of the data to magnetic tapes started c. 1980 when a research position in meteorology was established at the Abisko station. This was a few years after the inauguration of the new director, Mats Sonesson, who gave priority to the evaluation of the old data records.

The site at Abisko (Fig. 1), situated on a small windexposed moraine ridge a few meters above the surroundings, has not been moved since the start in 1913. A few low buildings have been constructed 15–50 m from the thermometer screen in the 1950's. There are no significant nearby vegetation changes. It is therefore judged that the air temperature and humidity observation series are homogeneous with respect to the site characteristics whereas this may not be the case for e.g. the ground temperatures.

The thermometer screen, used during the whole observation period and still in good shape, was manufactured by the iron mining company Luossavaara-Kiirunavaara AB in 1910, probably at the instigation of Hjalmar Lundbom the great benefactor of Kiruna. The laths, allowing air ventilation of the screen, are made of 2 mm thick iron plates which is probably a quite unique feature. The screen has no bottom but the instruments are placed on a shelf at one side. Details of the construction are described in Rolf (1920–1930). Also, since the screen, the instrument types and the calibration technique, although subject to variations in the number of daily checks, are the same throughout the recording period the series must be considered as homogeneous also in this respect.

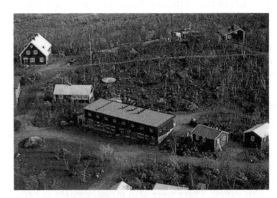

Fig. 1. View of the Abisko research station. In the upper right corner, to the left of a 2-family house for the members of the staff, one may see the small hut of the Meteorological Observatory and also instrument stands. September 1983. Photo: B. Holmgren.

Precipitation amounts have been measured through-out the period at the same point. There may have been some minor changes of the wind exposure due to the growth of nearby mountain birches but for at least the last 30 yr, these have been regularly trimmed. It is not expected that vegetation effects have had any significant influence on the recorded precipitation. However, ex-change of the standard types of precipitation gauges in the past most likely has had a minor influence on the recorded data. Variations of the catch efficiency of the gauges should hardly be any problem since similar types of wind shields have been used throughout the measuring series.

Since 1974 there are parallel measurements at Abisko using two precipitation gauges of the same standard type, one at the original wind-exposed site and the other at a sheltered site c. 100 m to the west and c. 5 m lower. At the sheltered site the catch is 15–20% higher during the winter and 3–5% higher in summer. For the whole year the total precipitation is c. 10% higher at the lower site compared to that on the ridge (Eriksson 1983).

Evaluation of records

The analysis is mainly based on weekly records of thermohygrographs (type Lambrecht) evaluated every three hours from 1913 to the present time. The records after 1930 have been evaluated by staff members at the Abisko Research Station and at the Meteorological Dept, Univ. of Uppsala, over the past 15 yr. Between 1913 and 1930 there were three daily checks of the temperature and humidity diagrams using the dry and wet bulb thermometers of an Assmann psychrometer. After 1930, with very few exceptions, there was at least one daily check until c. 1950. After this period and to the present time the daily routines have involved either 2 or 3 checks.

Missing or obviously erroneous temperature (and humidity) 3-h values have been replaced by interpola-tions when only a few observations are lost taking typical diurnal temperature variations on a monthly basis into account. For the periods with missing data for more than one day the climate records from Riksgränsen (35 km to the west) have been consulted. Corrections based on differences between the daily Assmann readings at Abisko and simultaneous observations at Riksgränsen have been applied when possible. For a few days, the Riksgränsen data have been used together with corrections based on differences in monthly averages for Abisko and Riksgränsen, respectively.

Some as yet not published data on precipitation, amount of sunshine hours, water vapour pressure are referred to in the discussion, but without providing the proper background of evaluation and the correspond-ing curves, which will be presented elsewhere. All data referred to have been checked and analysed. Extensive reanalyses and revaluation of previously published sun-shine data have been carried out in order to ho-mogenise the different evaluation techniques applied throughout the years. The final data are in agreement with the present method of evaluation practised at the Swedish Meteorological and Hydrological Inst. and in conformity with WMO recommendations.

Air temperature series

The time variations of all meteorological parameters discussed here are seasonal values where the 4 seasons are defined as: Spring = March, April, May. Summer = June, July, August. Autumn = September, October, November. Winter = December, January, February.

In order to visualise the climate trends better we have consistently applied a filter with Gaussian weighting coefficients using the same lowpass filtering method and standard deviation values (STD = 3 and 9) according to Alexandersson and Eriksson (1989). The trends corre-sponding to STD = 3 are in the following called short-term, and the trends corresponding to STD = 9 are called long-term. STD = 3 and STD = 9 nearly corre-spond to 10 and 30 yr if normal running averages would be applied. The advantage of using a Gaussian filter is that phase amplifications (cf. Liljequist 1950) and also "artificial" cyclicities are much reduced in comparison with using ordinary running averages. For instance, for a curve showing running 5-yr averages one may expect that more or less regular cycles of the order 7–12 yr may appear because of the method of presentation.

Correlations and estimates of the air temperatures at Abisko 1868–1912

The calculations of air temperature at Abisko before 1913 are based on two stations with much longer records: one in the coastal oceanic climate region west of the main mountain range, Tromsö, and the other in the more continental climate region east of the range, Karesuando. Since Abisko is situated approximately in the middle of the range, although some 200 km further south, the application of these stations for estimating the long-term trends at Abisko are a natural choice given the geography and the records available. The temperature observations in Tromsö and Karesuando started in 1868 and 1879, respectively. The Karesuando series has been tested for homogeneity (Alexandersson and Eriksson 1987). The weather station at Tromsö was moved within the town a few times before 1920. Cor-rections are applied for elevation changes (Aune 1989).

The Abisko estimates for 1868–1912 are based on linear regressions (no significant improvements were obtained by polynomial regression fits) of the recorded seasonal and annual temperatures 1913–1994 at Tromsö and Karesuando as independent, and the temperatures at Abisko as dependent variable. It was found that the highest correlations and the least scatter around the line of regression occurred for Karesuando for the winter and spring seasons, and for Tromsö for the summer and autumn seasons. Except for the winter season the average estimate of temperature, using Tromsö and Karesuando as independent variables respectively, was found to give the closest fit. However, we have used the average estimate also for winter in the final representation of the Abisko temperatures since the degradation of the correlation coefficient and the scatter around the regression line is small. We feel that the winter correlation with Tromsö is high enough ($R^2 = 0.90$) in order to be retained in the calculations although no improvement was obtained in the correlation with the average of the estimated Abisko values in relation to Karesuando and Tromsö for 1913–1994. The R^2-coefficients for the yearly and seasonal values range from 0.80 to 0.95.

The results were tested for stability by doing the same regressions for only half the actual observation period, 1913–1954. Very small differences between the results for the two periods were obtained. Based upon the differences between the calculated and measured temperatures 1913–1994, conservative estimates show that expected maximum deviations for the winter season are $\pm 2°C$, for spring, summer and autumn $\pm 1.5°C$, and for the annual average $\pm 0.7°C$. The expected STD's, including the scatter caused by the uncertainties of the regression coefficients are for the winter 0.7, spring 0.4, summer 0.4, autumn 0.5 and for the annual temperature 0.3°C. A similar analysis for 5-yr averages of annual and seasonal temperatures results in STD's of the differences between the measured and estimated temperatures for Abisko which are approximately half the values of the 1-yr averages.

The comparisons naturally included a test of the reliability of the evaluations during the period 1912–1994 for the Tromsö, Karesuando and Abisko. We found suspiciously high Abisko temperatures for the spring and summer seasons of 1950. A confirmation of this was obtained by a comparison of the climate stations of Kattevuoma, Kiruna and Riksgränsen. Since the suspect Abisko records are obtained during seasons of high incoming solar radiation, the deviations may be due to a radiation error, granted some temporary change of the calibration technique using the Assmann psychrometer or some error with the mercury thermometers (possibly a division of the mercury capillary). However, since we have not been able to locate the reason for the deviation in the diagrams or the notebooks, we retain the Abisko values for 1950, know-

ing that, taking all evidence together, it seems likely that the recorded Abisko temperature values during spring and summer 1950 are approximately one degree too high for some unknown reason.

Air temperature trends

The characteristics of the annual and seasonal temperature curves were found to be similar to those published by Alexandersson and Eriksson (1989) for northern Sweden and by Aune (1989) for Tromsö although in the latter case the curves are not presented using the Gaussian low pass filtering. This shows, together with the high correlations between Abisko, Tromsö and Karesuando, that the general trends of the temperature climate at various stations within a distance of a few hundred kilometres may be quite similar although the local climate settings can be quite different as in the present case. Here only summer temperature variations are shown (Fig. 2).

The broad long-term features of these variations show only minor changes from the 1870's to c. 1910 when a rise of c. 1.5°C took place ending in c. 1940. Since then there has been a decline of 0.5°C in the long term trend. Thus, the average summer temperature at the end of the present century is still about one degree higher than at the end of the 19th century.

Abisko is a rather special site with regard to its topographic setting in the middle of the high mountain range. The adjacent high fjelds, together with the extensive low area formed jointly by the broad Abisko valley and the wide Abisko Basin of Lake Torneträsk, induce marked foehn effects (Eriksson 1987), when the winds aloft are strong. This may most obviously be

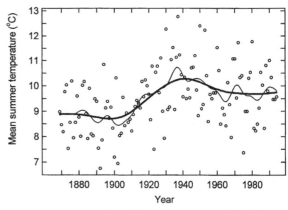

Fig. 2. Calculated (1969–1912) and recorded (1913–1994) air temperatures at Abisko (388 m a.s.l.). The smoothed curves correspond to c. 10 (STD = 3) and 30 yr (STD = 9) means respectively. The calculated temperatures relate to the average value obtained from linear regressions with Tromsö and Karesuando, respectively.

seen from the precipitation amounts and the appearances of the cloud covers, but not necessarily so much in the average cloud amounts. Although the average annual precipitation is in the range 300–350 mm (depending on whether the normal wind-exposed site or the sheltered one is referred to), water stress for, e.g. the mountain birch, is rare at semi-mesic sites (Ovhed and Holmgren 1995). The observatory hill is situated some 200 m from the southern shore of Lake Torne, the cooling effect of which is noticeable by the local climate gradients in the Abisko valley (Sandberg 1965, Josefsson 1990). Local lake and land breezes are common as shown by regular diurnal shifts of the wind direction during fine weather periods in summer.

Some trends during the period 1868–1994 (as appearing from the seasonal and annual curves which, except for summer temperature, are not shown here) are: 1) Similar to the trend in summer temperatures, the mean annual temperature increases sharply by >1°C from 1910 to 1930. There is some covariation between summer temperature and amount of sunshine hours during 1913–1994 for the short-term but not for the long-term trends. 2) A steadily increasing spring temperature of c. 1.5°C from 1870 to the present time. During 1913–1994, when daily amounts of sunshine hours are available, this rise is not accompanied by any regular trend in the seasonal amounts of sunshine hours. On the assumption that the variations of sunshine duration are mainly induced by variations in cloudiness, one may thus deduce that the trend of increasing spring temperatures is not accompanied by any corresponding change of cloudiness. 3) The long-term trends of autumn and winter temperatures are smaller than for the spring and summer seasons.

Lengths of the growing period and snowfree season

The length of the growing period is defined as the time interval during which the air temperatures surpass a given mean daily limit. Here lengths are calculated for 7 limits between 0 and +6°C. The criteria (cf. text of Fig. 3) used for defining the start and the end are similar to those suggested by Odin et al. (1983) and Vedin (1990). Length changes show a wavy pattern for the short-term trends. There is also a noticeable long-term increasing trend varying from a few days to 3 weeks with the greatest increases for the lower limits. The long-term increases may be somewhat exaggerated because of the relatively low short-term values at the start and relatively high short-term values at the end of the observation period. Considering that the spring temperatures have increased steadily by 1.5°C since 1868 and that the autumn temperatures are about one

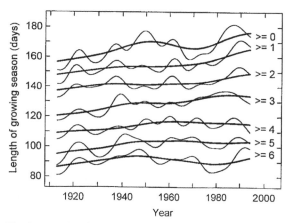

Fig. 3. Lengths of the vegetation season according to 7 temperature limits. The start of the vegetation period requires that 5 consecutive days surpass the indicated limit, and the end of the vegetation period requires that 5 consecutive days are lower than the limit. The curves are smoothed according to the same Gaussian filters as in Fig. 2.

half degree higher on average after 1940 compared with the period 1868–1920, the long-term increase of growing period may be expected to have been even more marked from 1868 to the present time. The effects on the plants of the changes of the duration of the growing seasons for the various temperature limits can be expected to be species-related and highly complex. When the objective is to study the interrelations between the duration of the growing period and the regional zonation of the major vegetation zones, it seems that the limit +6°C or perhaps even higher temperatures are to be preferred (Sjörs 1965).

A preliminary analysis of the snow course data sampled from 1913 (the analysis is somewhat hampered by missing data in some years) shows an increase of the length of the snowfree season in the latter half of the 1930's and in the 1940's and 1950's. During those periods the snowfree season was c. 10 days longer on the average compared to 1913–1925 and after 1970. The duration of snow cover varies somewhat differently than the length of the growth season due to the variations of the accumulated mass of the snow pack and possibly also due to occasional snowfalls which may not be very dependent on the average daily air temperatures.

It may be noted that the lengths of the growing period have been calculated from the temperature series at the Abisko Station and that the lengths of the snow seasons relate to a snow course of ten poles in the birch forest just east of the observatory hill. The absolute values of the changes of the vegetation season and the snow season at the altitudinal tree line may differ substantially from the absolute values on the valley floor. The trends might be expected to be similar, though.

Fig. 4. On the eastern slopes of Mt Njulla (right) and Mt Slåttatjåkka (left) there is a traceable tree line of mountain birch at c. 655 m a.s.l. Mid-September 1974. Photo: N. Å. Andersson.

Observations on recent birch tree line changes

The first observations at Abisko on vegetation responses to the increasing temperatures after c. 1910 were made by Sandberg (1940, 1965). Sandberg noted thickening and expansion of shrubs in the birch forests and at and above the tree line (Fig. 4). The first observation of altitudinal expansion at the tree line was thus not the growth of birch saplings at higher elevation. In the valley, however, saplings of *Pinus sylvestris* were "shooting up almost everywhere" (Sandberg 1965).

The altitudinal expansion of the species tree line was confirmed a few decades later when Sonesson and Hoogesteger (1983) found that there were no birches older than c. 40 yr in the elevation range 20–50 m below the present species line. At 630 m a.s.l. on Mt Njulla, c. 25 m below Sandberg's site of temperature measurements at the tree line (Josefsson 1990), birches of an estimated age class from 1820–1870 or older were prominent.

Using temperature measurements (Josefsson 1990) in various months between 1958 and 1973 at the tree line (655 m a.s.l.) on the eastern slope of the nearby Mt Njulla and the corresponding temperatures at the Abisko Station (388 m a.s.l.) to calculate the average vertical summer temperature gradients (lapse rates), one may estimate that a 1°C change of air temperature corresponds to a vertical isotherm displacement of > 200 m. The observed altitudinal expansion of 20–50 m is thus small compared to the potential rise of > 200 m indicated by the temperature rise since 1870's. This indicates, together with the age class distributions noted above, that the tree line was not noticeably depressed during the cold period at the end of the 19th century.

North of Lake Torneträsk, on the southfacing slope of Lullehatjårro, there are some areas with similar vegetation zonation as on Njulla. Also on Lullehatjårro there are recent expansions of young mountain birch, of similar heights and stem widths as on Njulla, c. to a

maximum of 50 m above a tree line of older mature trees. The young birches on Lullehatjårro also appear within thick mats of willow and dwarf birch.

In Jämtland c. 600 km south of the Torneträsk area, regional studies (> 200 localities) have documented rises of up to 50 m in the altitudinal tree lines for mountain birch *Betula pubescens* ssp. *tortuosa*, spruce *Picea abies* and pine *Pinus sylvestris* during this century (Kullman 1979, 1990). In altogether 125 localities spread over a 8000 km² large area, tree line dynamics has been observed since the mid 1970's. During that time c. 30% of the young birches, mainly in the areas of former tree expansion, have died or transformed to stunted shrubs (Kullman pers. comm.). The tree declines appear to be due mainly to specific weather patterns in winter, causing either too much or too little snow which in the latter case lowered the ground temperature to cause root damage and spring desiccation. Lower summer temperatures during a sequence of years in the 1980's also have contributed to the kills (Kullman 1991).

Similar diebacks on a massive scale as in Jämtland are not observed in the Torneträsk region although some damage to mountain birch near the tree line were noted in 1962 (Tenow 1996). In 1985 and especially in 1991, rather extensive frost damage to mountain birch and also pine occurred within a narrow altitudinal belt between c. 410–430 m a.s.l. (Fig. 5). Some birches (Figs 5 and 6) and pines were injured but very few birches appear to have died so far from the damage. The reasons for the damage in the Abisko valley are not known for certain at the present time but the weather patterns during the preceding spring, winter and autumn are most likely involved.

The climate during the 1930's

On the east slope of Mt Njulla, the subsequent rise of the elevation of the average tree line thus seems to have

Fig. 5. Looking downwards towards Abiskojokka from the eastern slope of Mt Njulla, one recognises a distinct grey zone of frostdamaged birches within the altitudinal interval 410–430 m a.s.l. July 1991. Photo: B. Holmgren.

Fig. 6. A mountain birch tree damaged by frost. The buds of the normal short and long shoots died while they were still in the winter stage. Later in summer only witch broom *Tafrina betulina* infected shoots developed. July 1991. Photo: B. Holmgren.

Fig. 7. Cluster of leaves on a shoot infected by *Tafrina betulina*. Photo: B. Holmgren.

been founded during a peak in the air temperatures in the late 1930's (Fig. 2). Simultaneously, during the 1913–1994 period, there were maxima in the water vapour pressure and in the amount of sunshine hours. The average excess number of sunshine hours in the 1930's corresponds to c. 3 extra cloudfree days, instead of 3 overcast days, in comparison with the average sunshine hours for the summers in 1913–1994. Since there is a positive correlation between sunshine hours and summer temperatures at Abisko, a greater amount of sunshine hours does not necessarily mean improved conditions for the biomass production of birches in comparison with a situation with the same air temperature but with advection of warm air in cloudy weather (cf. Young and Smith 1983). The amount of summer precipitation during the 1930's was about normal and the water vapour pressure deficit somewhat lower than normal for the 1913–1994 period.

On the assumption that the climate changes (air temperature, humidity, sunshine hours, length of the growing and snowfree seasons) show similar trends at the tree line as at the Abisko Station on the valley floor, there are several derived influences that might have contributed to the observed establishment of birch seedlings above the line of mature trees. Ground temperatures might have peaked due both to higher air temperatures and increased sunshine duration. It may be noted that an increase of the incoming solar radiation could be very important for rising air temperatures at the ground surface also below thickets of e.g. shrubs because of the high penetration factor for the near infrared part of the solar spectrum as discussed by Ovhed and Holmgren (1995). A remarkably marked effect of the soil temperature on the growth rates of seedlings grown in small containers was found by Karlsson and Nordell (1996).

Since the production and the germinability of seeds decrease with altitude (Sveinbjörnsson et al. 1996), and probably are related to air temperatures during the growth periods of the preceding and the current years, respectively, it seems that the survival of seedlings above the former tree line might have been facilitated by the peaks in summer temperature and sunshine hours during the 1930's. The sequences of warmest, above average, summers were 1933 and 1934 and in particular 1936, 1937 and 1938. In 1937 there were record maxima of air temperature, $+12.8°C$, compared with the average, $+9.9°C$ and of sunshine for 1913–1994, 822 h compared with the average, 590 h, while the summer precipitation, 103 mm was somewhat below average, 121 mm. The water deficit at the end of the 1937 summer, as calculated for the semi-mesic site near the Abisko station (Ovhed and Holmgren 1996) indicate a moderate water stress (-43 mm). There were no or only small water deficits in 1933, 1934, 1936 and 1938. The 1938 summer was warm and rainy, air temperature was $+11.2°C$ precipitation 176 mm. If one was obliged to pick out a 2-yr period of optimal conditions 1913–1994 for seed production and seedling survival it would be 1937 (seed production) and 1938 (seedling survival).

Eckstein et al. (1991) finds a close association between the mean summer temperature from May to August and the tree-ring width of birches growing in the Abisko valley. Sveinbjörnsson et al. (1996) discuss the physiological aspects of the growth conditions of mountain birch near the tree line. Holmgren et al. (1996) suggest that the positioning of the altitudinal tree line of mountain birch ought to be related to 3 main factors: 1) the seasonal evapotranspiration, 2) the seasonal photosynthate balance and 3) the soil temperatures. The first two factors in particular and the third to a limited extent should be dependent on air temperatures during the birch's growing season.

Tree line of mountain birch and chronologies of pine tree-rings

Proxy summer (April–August) temperatures obtained from ring-width and density chronologies of Scots pine *Pinus sylvestris* in the Torneträsk region, but transferred to summer temperatures of northern Fennoscandia are given by Briffa et al. (1992, Fig. 8) for the period from 500 to 1980. Starting from c. 1750, there was a decreasing trend of the summer temperatures that went on, although with big fluctuations, to c. 1900. The latter half of the 19th century was a cold period in comparison with the reference temperature for 1951–1970. This reference is similar to the mean of the whole period 500–1980 AD. For some periods between 500 and 1750 the proxy chronologies suggest somewhat greater maxima and also minima than for the warm period in the 1930's and the cold period at the end of the 19th century, respectively. Taking 50-yr means (Briffa et al. 1992, Table 3) the maximum positive anomaly is +0.68°C and the maximum negative anomaly −0.67°C between 500 and 1980 AD. Judging from these, after all relatively small anomalies, it does not seem to be out of the question that the tree line altitudinal changes on Mt Njulla may have been rather insignificant since 500 AD to the present time.

The minimum age of the oldest birches in Sonesson and Hoogesteger (1983) appears to be 100–150 yr and Eckstein et al. (1991) report ages up to 200 yr for birches in the Abisko valley. Established birches at the tree line, particularly at sites with favourable edaphic and micrometeorological conditions, may probably survive and "bridge" long periods of cool summers. The position of the present altitudinal tree line may well be in approximate "balance" with the mean summer temperatures since c. 500 AD. An altitudinal depression of the tree line due to intensive reindeer herding in the past few centuries (Emanuelsson 1987) does not appear likely at this particular locality.

Treter (1984) found that isotherms of summer air temperatures (June–September) averaged over periods of several decades, and also July temperatures, may be taken as approximate indices for tree line limits although with frequent regional and also local deviations due to various ecological factors. Using the temperature data collected at the tree line on Mt Njulla in 1955–1973 under the leadership of Gustaf Sandberg, the average July air temperature, as reduced to the 1951–1980 normal period, happens to become +10.0°C (Josefsson 1990). Reduced to 1951–1970, the corresponding July temperature is +9.9°C.

Snow cover, soils and birch forest changes

All observations of recent altitudinal tree line changes in the Torneträsk area, which have been published so far, are either from the Abisko valley, from the Jieprenjåkka area north of Torneträsk (Sonesson and Hoogesteger 1983) or in the western part of the Torneträsk area (Emanuelsson 1987, Rapp 1996). On the lower, gentle, windexposed slopes of Mt Nissuntjårro, Mt Tjuonatjåkka and Mt Vaimuoaivi on the eastern side of the Abisko valley, the tree line is split up less continuous than on Mt Njulla. In many places the highest elevation of the trees clearly relate to the small scale topography of the terrain. The trees appear where snow accumulation may take place in depressions or under escarpments. However, there are also signs that in places curtains of birches may affect the snow accumulation and improve the survival conditions where the snow cover otherwise would have been insufficient (cf. below).

The tree line studies on Mt Njulla are in an area with a good but generally not excessively deep snow cover. Drifting snow over the fjeld slopes may be carried by strong winds with a westerly component to accumulate in the lee of Mt Njulla and Mt Slåttatjåkka. However, a small-scale redistribution of the snow is not indicated at the tree line on Njulla in contrast to the conditions of discontinuous, wind-exposed tree line on the opposite eastern side of the Abisko valley. On the other hand, there are several snow avalanche tracks which distort the tree line on Mt Njulla.

Thickening birch forests in the lower part of the Abisko valley in the period from c. 1910 to the present time are demonstrated by Emanuelsson (1987) using some of the large collection of photos taken by Borg Mesch. As pointed out by Emanuelsson these pictures

Fig. 8. The delta of Abiskojokka from Mt Njulla. In the upper right corner one may see part of a glaciofluvial terrace where the soils consist of sands and gravels. On the terrace there is a uniform, mostly sparse, canopy of mountain birch, 3–4 m high. There are no signs of ground movements because of frost heaving within this area. To the left of the terrace, between two arms of Abiskojokka, there is a small island covered mainly by a dense stand of tall grey alder. The birch forest, along the beach to the left of Abiskojokka and also higher up on both sides of the road, is broken up by subalpine heaths. The soils are silty tills and soil heaving and soil creep due to frost action are commonly observed. August 1986. Photo: B. Holmgren.

are still to be utilised fully in environmental research. A listing of archives where photographs, maps and other documents, which may be useful for future studies of vegetation changes since the turn of the century, is given by Theander (1993).

A striking contrast in birch forest response to climate change, mainly due to edaphic, conditions may be pointed at in the area near the beach of Torneträsk below the STF (Swedish Tourist Organisation) Station. East of the present mouth of Abiskojokka there is a flat delta terrace formed during a period of higher lake level during the deglaciation, probably because of a local lake dammed by glacier ice (Holdar 1957). The soils of this terrace consist of glacio-fluvial material in which the fractions of clay and silt are very small. As indicated from the photographs of Borg Mesch taken at the beginning of this century, the vegetation on the delta terrace was low and bushy. However, the details of the vegetation characteristics on the terrace need verification by closer analysis of the photographs. Now there is a canopy of mountain birch with an estimated average tree heights of 3–4 m on the terrace without any natural treeless areas and without any signs of ground frost activity. It is not known whether the thickening and height increase of the birch canopy is a response to the improved summer climate or if the human land use of this area has changed.

On both sides of the terrace, within 50–100 m from the beach, there are scattered birch stands interspersed with tree-less subartic heaths with numerous signs of active frost movements and also with patches of permafrost. The soils consist of till with high fractions of clay and silt. The particle size distributions confirm that the soils are very frost susceptible (Josefsson 1990). Furthermore, it was found that soil moisture and ground water levels, surface slope angles, the microclimate, especially strong winds, snow drifting and redistribution of snow in winter with less snow in the treeless areas, are important for maintaining the subarctic heaths (Fig. 8).

The exposures for winds from Torneträsk (bringing relatively cool air in spring and summer and relatively warm air in autumn and early winter) are very similar for the two contrasting biotopes: the subartic birch forest with good snow protection of the ground in winter and the treeless tundra with patches of permafrost, patterned ground and a thin, often discontinuous snow cover. The core of the heaths appear to be quite stable features and may have been present since before the Little Ice Age (Josefsson 1990). There are hardly any signs that the open heaths will disappear in the present climate conditions, indicating that the expansion of birch forest limits may be effectively suppressed by the edaphic conditions in combination with a poor snow cover.

Concluding remarks

Summarising the discussion above, there seems to be surprisingly few and rather small recent changes of the altitudinal tree line that might be related to climate change with any degree of certainty in spite of the considerable summer temperature increase of c. +1.5°C from the end of the 19th century to the 1930's. The interpretation of any tree line variations in areas with low snow precipitation and much wind drifting is difficult since sometimes the tree lines appear to become stationary in relation to topographical features and sometimes the edaphic factors apparently have a major effect on the birch forest response to climate change. It thus seems a good strategy to select a tree line study area like the eastern slope of Mt Njulla where the snow cover gives a good protection in winter, where effects of local snow drifting appear minor and where the inorganic soils originally have similar properties above and below the tree line. A nagging question concerns how representative the results, from one locality of this particular type, are on a wider scale considering the complex topography and varying climate, edaphic and geological factors in the region?

From the discussion of the climate changes and the impacts on the birch forest that have occurred during the present century it appears indispensable to keep a number of sites in various environments under regular surveillance in order to document for the future the ongoing environmental changes. One should perhaps question whether, in the Torneträsk region, there is at present a suitable balance in the research between studies of specific biological processes in enclosures, on the one hand, and the documentation of the ongoing environmental changes on the local and regional scales in response to the ever changing climate, on the other?

The present exercise may not have helped in solving any of the intricate problems in the climate-mountain birch relations. For consolation one may point out that the July air temperature at the tree line on the eastern slope of Mt Njulla coincides with +10°C for 1931–1980, and further, by invoking the proxy tree ring data, that the long-term average temperature at the tree line for the period 500–2000 may also be close to +10°C.

Acknowledgements – We thank N. Å. Andersson and A. Temesváry for their contributions to the present work by organizing, scrutinizing and evaluating the data materials for many years. S. Karlsson, O. Nordell, L. Kullman, T. Callaghan, H. Alexandersson, M. Weih, A. Rapp and O. Tenow gave valuable advice and suggested many improvements of the manuscript.

References

Adedokun, J. and Holmgren, B. 1993. Acoustic sounder Doppler measurements of the wind fields associated with a mountain stratus transformed into a valley fog: a case study. – Atmosph. Environ. 27A: 1091–1098.

Andersson, N.-Å., Callaghan, T. V. and Karlsson, P. S. 1996. The Abisko Scientific Station. – Ecol. Bull. 45: 11–14.

Alexandersson, H. and Eriksson, B. 1989. Climate fluctuations in Sweden 1860–1987. – Swedish Meteorological and Hydrological Inst. RMK 58, Norrköping, Sweden.

– , Karlström, C., Larsson-McCann, S. 1991. Temperature and precipitation in Sweden, 1961–90. Reference normals. – Swedish Meteorological and Hydrological Inst. Meteorologi, nr 81. Norrköping, Sweden.

Aune, B. 1989. Lufttemperatur og nerbör i Norge. – Norwegian Meteorological Inst., Report 26/89.

Briffa, K. R., Jones, P. D., Bartholin, T. S., Eckstein, D., Sweingruber, F. H., Karlén, W. and Zetterberg, P. 1992. Fennoscandian summers from AD 500: temperature changes on short and long time scales. – Climate Dynamics 7: 111–119.

Eckstein, D., Hoogesteger, J. and Holmes, R. 1991. Insect-related differences in growth of birch and pine at northern tree line in Swedish Lapland. – Holarct. Ecol. 14: 18–23.

Ekman, S. 1957. Die Gewässer des Abisko-Gebietes und ihre Bedingungen. – Kungl. Vetenskapsakademiens Handlingar, 4 Ser. Band 6: 1–172.

Emanuelsson, U. 1987. Human influence on vegetation in the Torneträsk area during the last three centuries. – Ecol. Bull. (Copenhagen) 38: 95–111.

Eriksson, B. 1981. The potential evapotranspiration in Sweden. – Swedish Meteorological and Hydrological Inst. RMK 28, Norrköping, Sweden.

– 1982. Data concerning the air temperature climate of Sweden. Normal values for the period 1951–80. – Swedish Meteorological and Hydrological Inst. RMK 39, Norrköping, Sweden.

– 1983. Data concerning the precipitation climate of Sweden for the period 1951–80. – Swedish Meteorological and Hydrological Inst. Report 1983: 28. Norrköping, Sweden.

– 1986. The precipitation and humidity climate of Sweden during the vegetation period. – Swedish Meteorological and Hydrological Inst. RMK 46, Norrköping, Sweden.

– 1987. The precipitation climate of the Swedish fells during the 20th century- some remarkable features. – UNGI Report Nr 65. Department of Physical Geography, Uppsala Univ., Sweden.

– 1990. Snödjupsförhållandena i Sverige. Säsongerna 1950/ 51–1979/80. – Swedish Meteorological and Hydrological Inst. RMK 59, Norrköping, Sweden.

Holdar, C.-G. 1957. Deglaciationsförloppet i Torneträskområdet efter senaste nedisningsperioden, med vissa tillbakablickar och regionala jämförelser. – GFF 79: 293–528.

Holmgren, B. and Tenow, O. 1987. Local extreme minima of winter air temperature in high-latitude mountainous terrain. – Dept of Physical Geography, Uppsala Univ., Sweden, Report 6: 25–41.

– , Ovhed, M. and Karlsson, P. S. 1996. Measuring and modelling stomatal and aerodynamic conductances of mountain birch: Implications for treeline dynamics. – Arct. Alp. Res. 28, in press.

Josefsson, M. 1990. The Geoecology of subalpine heaths in the Abisko Valley, northern Sweden. – Dept of Physical Geography, Uppsala Univ. Report 78.

Karlsson, P. S. and Nordell, O. 1996. Effects of soil temperature on the nitrogen economy and growth of mountain birch seedlings near its presumed low temperature distribution limit. – Ecoscience, in press.

Kullman, L. 1979. change and stability in the altitude of the birch tree-limit in the southern Swedish Scandes 1915–1975. – Acta Phytogeogr. Suec. 65: 1–121.

– 1990. Dynamics of altitudinal tree-limits in Sweden: a review. – Norsk Geogr. Tidskr. 44: 103–116.

– 1991. Cataclysmic response to recent cooling of a natural boreal pine (Pinus sylvestris L.) forest in northern Sweden. – New Phytol. 117: 351–360.

Liljequist, G. H. 1950. On fluctuations of the summer mean temperature in Sweden. – Geogr. Annl. Medd. Ser. B. 7.

Odin, H., Eriksson, B. and Perttu, K. 1983. Temperature climate maps for Swedish forestry. – Report For. Ecol. For. Soils. Swedish Univ. Agricult. Sci. Uppsala.

Ovhed, M. and Holmgren, B. 1995. Spectral quality and absorption of solar radiation in a mountain birch forest, Abisko, Sweden. – Arct. Alp. Res. 27: 381–389.

– and Holmgren, B. 1996. Modelling and measuring evapotranspiration in a mountain birch forest. – Ecol. Bull. 45: 31–44.

Parlow, E. and Scherer, D. 1991. Studies of the radiation budget in polar areas using satellite data and GIS-techniques. – Int. Geosci. Remote Sensing Soc. Symp., Helsinki, June 1991.

Raab, B. and Vedin, H. (eds) 1996. National Atlas of Sweden. Climate, lakes and rivers. – Bra Böcker, Höganäs, Sweden.

Rapp, A. 1960. Recent development of mountain slopes in Kärkevagge and surroundings, northern Scandinavia. – Geogr. Ann. Vol. XLII: 2–3: 71–123.

– 1996. Photo documentation of landscape change in northern Swedish mountains. – Ecol. Bull. 45: 170–179.

Rolf, B. (ed.) 1920–1930. Observations Métérologiques A Abisko 1913–1929. Special ed. – Uppsala and Stockholm.

Rydén, B. E. 1980. Climatic representativeness of a project period. Epilogue of a tundra study. – Ecol. Bull. (Stockholm) 30: 55–62.

– , Fors, L. and Kostov, L. 1980. Physical properties of the tundra-soil-water system at Stordalen, Abisko. – Ecol. Bull. (Stockholm) 30: 27–54.

Sandberg, G. 1940. Den pågående klimatförbättringen. – Svenska Vall- och Mosskulturföreningens Kvartalsskrift 2: 163–178.

– 1965. Abisko National Park. – In: Oldertz, C. and Rosén, B. (eds), National Parks of Sweden, pp. 1–44.

Sjörs, H. 1965. Features of land and climate. – Acta Phytogeogr. Suec. 50: 1–12.

Smedman, A.-S. and Bergström, H. 1995. An experimental study of stably stratified flow in the lee of high mountains. – Monthly Weather Review 123: 8.

Sonesson, M. and Lundberg, H. 1974. Late quaternary forest development of the Torneträsk area, north Sweden. 1. Structure of modern forest ecosystems. – Oikos 25: 121–133.

– and Hoogesteger, J. 1983. Recent tree-line dynamics (Betula pubescens Ehrh. ssp. tortuosa (Ledeb.) Nyman) in northern Sweden. – Nordicana 47.

Sveinbjörnsson, B., Kauhanen, H. and Nordell, O. 1996. Treeline ecology of mountain birch in the Torneträsk area. – Ecol. Bull. 45: 65–70.

Tenow, O. 1975. Topographical dependence of an outbreak of Oporinia autumnata Bkh. (Lep. Geometridae) in a mountain birch forest in northern Sweden. – Zoon 3: 85–110.

– 1996. Hazards to a mountain birch forest – Abisko in perspective. – Ecol. Bull. 45: 104–114.

Theander, A. 1993. I rallarnas spår (tracking the navvies). – Ofoten Museum, Narvik.

Treter, U. 1984. Die Baumgrenzen Skandinaviens. Ökologische und dendroklimatologische Untersuchungen. – Franz Steiner Verlag, Wiesbaden.

Vedin, H. 1990. Frequency of rare weather events during periods of extreme climate. – Geogr. Ann. 72 A: 151–155.

Young, D. R. and Smith, W. K. 1983. Effects of cloudcover on photosynthesis and transpiration in the subalpine understorey species Arnica latifolia. – Ecology 64: 681–687.

Appendix

The most authoritative information on the average climate parameters in the Torneträsk area including their regional gradients can be obtained from the report series of the Climate

Section of Swedish Meteorological and Hydrological Inst. (SMHI). Some of these are: Reference normals 1961–1990 for temperature and precipitation (Alexandersson et al. 1991), temperatures for 1951–1980 (Eriksson 1982), precipitation for 1951–1980 (Eriksson 1983), snow depths for 1950/51–1979/80 (Eriksson 1990), the potential evapotranspiration (Eriksson 1981), precipitation and humidity climate during the vegetation period (Eriksson 1986). Most of these references include maps showing the regional gradients of the climate parameters. The new edition of the Swedish National Atlas (Raab and Vedin 1996), is also very useful as a general source although the scale of the maps in some cases is somewhat too large for picking out the details of climate or hydrological gradients in the Torneträsk region.

Descriptions and analyses of various climate features in the Torneträsk area are given in a number of reports where meteorological data are used often for supporting biological or geoscientific studies. Here only a few examples of publications with meteorological information are briefly mentioned. An early account of the main climatic, hydrological and geological features in the Torneträsk area is given by Ekman (1957) in his comprehensive study of the lakes in the Torneträsk area. Rapp (1960) describes climate features of the western part of the Torneträsk area with special regard to geomorphologic processes. Sonesson and Lundberg (1974) provide meteorological background data for studies of the structure of birch forests. Climate features of the Stordalen Mire c. 10 km east of Abisko are given by Rydén (1980) and Rydén et al. (1980). Based partly on unpublished measurements made under the guidance of Gustaf Sandberg, Josefsson (1990) gives an exposé of the local climate of the Abisko valley as a background for her own microclimate studies of the subarctic heaths. Tenow (1975) describes the topographical dependence of local cold air lakes in the Abisko valley and their relations to an outbreak of *Epirrita autumnata* in 1955. Holmgren and Tenow (1987) discuss how local extreme minima of air temperature develop in a mountainous terrain. The application of satellite data and GIS-techniques for studying the radiation budget in polar regions is demonstrated by Parlow and Scherer (1991). Adedokun and Holmgren (1993) investigate local wind pattern and fogs in the Abisko valley using acoustic sounding techniques. Smedman and Bergström (1995) report on the mechanisms behind the development of extreme wind speeds (gap winds) by analysing profiles of wind, temperature and turbulence structure over the lake ice of Torneträsk.

Ecological Bulletins 45: 170–179. Copenhagen 1996

Photo documentation of landscape change in northern Swedish mountains

Anders Rapp

Rapp, A. 1996. Photo documentation of landscape change in northern Swedish mountains. – Ecol. Bull. 45: 170–179.

Selected cases of environmental signals of landscape change are discussed, mainly from the mountain tree line and higher levels in N Sweden. The first two cases are based on comparisons between old photographs from 1905–1907 by O. Sjögren, and repeated photography. The third case shows the short-term glacier-front advance of a small ice cliff in the 1990's. The fourth case is signs and impact of melting permafrost in the Alps, which is of considerable interest for comparative studies in northern Scandinavia as part of a continued research programme of "global change". The extent and processes of permafrost and cold-based glaciers is a neglected field of environmental studies in northern Europe.

A. Rapp, Dept of Physical Geography, Lund Univ., Sölvegatan 13, S-223 62 Lund, Sweden.

Introduction

Impact of climatic fluctuations

The research station at Abisko in N Sweden is owned by the Royal Academy of Sciences and is located in a key area of arctic mountains and subarctic woodlands. Long series of climatic records and ecological proxy data and observations are available e.g. air temperature data of the period 1913–1994. It shows warming in the 1920's and 30's, cooling from c. 1940 to the early 1980's, and again warming in the period 1987 until now. The magnitudes of the temperature fluctuations are more marked than further south in Sweden and Europe as can be expected.

Comparative geomorphology benefits from international cooperation of research at field stations in different climatic zones. The mountains of N Scandinavia and Spitsbergen are key areas in the study of arctic and alpine geomorphology in the North Atlantic climatic area. In our efforts to stimulate international comparative geomorphological research at Abisko we regard that mountain area as part of a north-south transect through Europe and also as a part of a west-east transect across the Atlantic from American to Eurasian mountains.

The time perspective and the historical approach is also very important in comparative geomorphology. In the 1950's, when I began my field studies of mountain landforms and slope processes at Abisko, one of the main scientific questions under discussion was the so-called glacial refugia controversy. It focussed attention on the history of glaciation, deglaciation and the existence of glacial nunataks or other refugia where plants or animals could have survived glaciations.

Many geoscientists believed that long-transported glacial erratics found high up on the assumed nunataks and refugia were proofs of an earlier ice cover in one or several glacial maxima and thus extinction of all life by glacial cover – the so-called "tabula rasa theory". G. Hoppe expressed a modified view: "One of the strongest arguments for the refugia hypothesis is the existence of plants and animals with a bicentric distribution. There may however be other ways of explaining such distributions. Attention is called to the fact that the supposed refugia could have been deglaciated at a much earlier stage than the rest of the country and thus offered both flora and fauna a longer time to get established" (Hoppe 1959). Dahl (1963) verified and developed this hypothesis in north Scandinavia (cf. also Ives 1974).

As early as in the 1960's it was observed and reported by Falconer (1966) that undisturbed vegetation of mosses, lichens and herbs together with patterned ground had been preserved under thin glacier ice caps

170

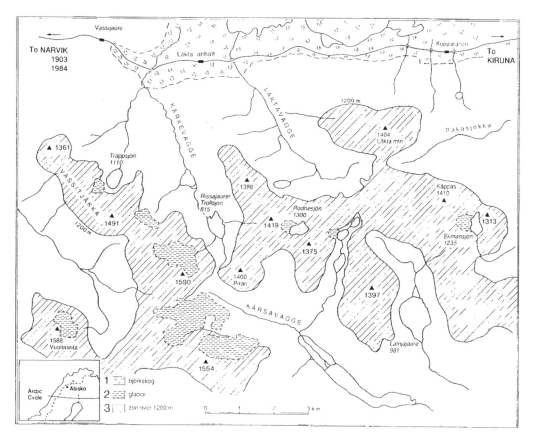

Fig. 1. Map of the western Abisko mountains south of the railway Abisko-Narvik. Mountain birch forest below c. 500–600 m. Mountain ridges above c. 1200 m have probably discontinuous permafrost in areas with thin covers of snow or cold-based glaciers, (Rapp 1992). Railway traffic to Narvik began in 1903 and the highway was opened in 1984. Key: 1 = Birch forest. 2 = Glaciers. 3 = Zone above 1200 m with widespread permafrost.

in the Canadian High Arctic. A thin ice body in northern Baffin Island melted in the margins and revealed undisturbed mosses. The vegetation had a 14C age of 330 ± 75 yr. Similar finds were reported and analysed from other northern glacier areas (cf. Havström et al. 1995). Together with other observations and glaciological evidence the hypothesis of non-scouring, cold-based glacier ice gradually became widely accepted (Kleman and Borgström 1990, Rapp 1992, Dyke 1993, Kleman 1994).

Figure 1 shows the mountain area west of Abisko and the presumed zone of predominant mountain permafrost above 1200 m, where the mean annual air temperature is c. -4 to $-5°C$, as extrapolated from Abisko temperatures.

A drastic illustration of the importance of cold-based ice at high levels in the Alps, was the finding of a completely preserved Stone Age man, a frozen body with clothes and equipment, resting on permafrost and protected during 5 300 yr from decay and animal attacks (e.g. Haeberli 1992, Patzelt 1993). This find gives a lot of new information to the discussion of a

new type of mountain refugia, where organic material and perhaps seeds could be protected and possibly survive in sheltered positions under cold-based ice. Two species of fungi are reported to have survived for 5 300 yr in the frozen hay, used by the Iceman Oetzi as shoe-grass.

Now, in the 1990's, combined with the unsolved problem of biological refugia we have another central topic for discussion on mountain ecology and geomorphology viz. the "global change" issue: "Is there a trend of climatic warming due to the supposed greenhouse effect? If so, how rapid is the change at different latitudes, and what are the impacts, or visible signals of climatic change in the mountain landscape"?

Figure 2 is a curve of annual air temperature at Abisko for the period 1913–1994 measured at the Research Station in the subalpine birch forest belt at 388 m. The curve shows three phases: a) Rapid rise from $-1.5°C$ to just below zero in the period 1913 to 1939. b) Sinking temperatures in steps to c. $-1.0°C$ from 1940–1986. c) Rapid rise again from 1987–1994, mainly due to mild winters, bringing the curve up to

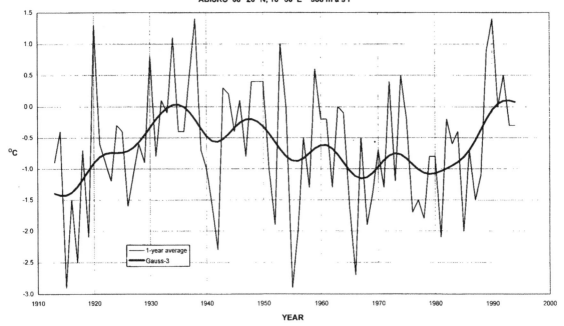

Fig. 2. Mean annual air temperature Abisko 1913–1994, 388 m. Gaussian low-pass filtered curve corresponds to a running average of 10 years (Holmgren and Tjus pers. comm.). Rapid temperature rises are evident in the 1920's and 1930's, followed by periodic decrease in the1940's to 1980's. Then followed another rapid rise from 1987–94, mainly due to mild winters.

slightly above zero. In the 1950's studies and monitoring of slope processes were performed by Rapp (1960) in Kärkevagge valley west of Abisko (Fig. 1). A new monitoring project of actual geomorphic process studies related to climate change in Kärkevagge was started in the 1990's (Schlyter et al. 1993). Important and informative studies of present-day slope processes and related sedimentation in Kärkevagge and other valleys of N Sweden were reported in theses by Nyberg (1985) and Jonasson (1991), dealing particularly with debris flows and slush avalanches. Their studies indicate that slope processes, mainly debris flows, have occurred through the whole Holocene period and with increased frequency during periods of climatic deterioration.

Recovery of birch forest at Kärkevagge shown by photo comparison 1905–1994

Comparative photography is a useful method for analysing landscape change over time, due to natural or human influence. Figures 3a and b are two photographs taken from exactly the same point on a bedrock hill near the mouth of the hanging valley, Kärkevagge. For location, see Fig. 1. The old photograph was taken by O. Sjögren, Stockholm, probably in August 1905, after the opening of the railway to Narvik in 1903. Photo 3a shows that nearly all birch trees were cut down by

railway-workers in the years of railway construction before 1903. Photo 3b was taken by A. Rapp on 3 September 1994. It shows that birches now grow in a mosaic pattern in sheltered sites at the tree line zone up to the same level at c. 600 m as is indicated by dead trees in 1905. The recovery of birch growth and climb of the treeline started in the mild 1930's (G. Sandberg pers. comm.), has progressed since that time and raised the tree line by 70–100 m.

This photo comparison shows that the tree line is recovering after earlier deforestation by man and was probably also "helped" by a warmer climate since the 1920's. The details of birch-forest recovery can be analysed and monitored in the area based on comparisons with this and other old photographs by Sjögren and by dendrochronology as mentioned by Emanuelsson (1987 p. 106). He makes the following comments on the issue: "Several workers have observed that the tree line has risen in altitude significantly in the Scandinavian mountains during the 20th century. This climate change has been proposed as the most important explanation of the increasing altitude of the timberline (Sonesson 1980, Sonesson and Hoogesteger 1983)... Kullman (1984) has recently presented some evidence that the mountain birch requires a number of climatically good years to establish from seeds in environments at the treeline in the southern Swedish Scandes ... Photographs taken by Sjögren ... in the

172

Fig. 3. Ascent of tree line and birch forest recovery illustrated by repeated photographs 1905 (O. Sjögren) and 1994 (A. Rapp). Location, north-facing mouth of Kärkevagge valley, (cf. map. Fig. 1). Photo 3a shows that nearly all birch trees were cut or killed by railway-builders in the years before the opening of the railway traffic in 1903. Photo 3b was taken from the same point on 3 September 1994. Birches have recovered and grow in groups in locally wind-sheltered sites up to the same levels at 600 m as in 1905. Tree-line recovery started in the mild 1930's (Sandberg pers. comm.) and has progressed since.

3a

3b

northern part of Kärkevagge show that the forest had been cut ... but that some trees had survived. Some of them are still alive and grow at a much higher altitude (75 m) than the young, expanding birch forest. This indicates that the birch forest may be able to expand in altitude only during a few good years, but that it does not retreat during periods of bad years, if these periods are not longer than the life span of the trees." The same author also discusses the impact of reindeer grazing during periods of large numbers of reindeer and high browsing pressure on young shoots of trees and bushes. The old traditional practice of reindeer husbandry that existed in the area up to 1920, can have substantially affected the birch forest at the timberline, especially during the last half of the 19th century." (Emanuelsson 1987 p. 107).

Melting of a cirque glacier and growth of a slush avalanche fan in Kärkevagge

Several glaciers in the western Abisko mountains have been reported as receding considerably during this century. Some of them are shown in Fig. 1, e.g. Kårsa glacier, at the upper end of Kårsa valley, the small Ekman glacier with its front forming an ice cliff in Lake

4a

4b

Fig. 4a, b. Repeated photographs showing the melting of the cirque glacier Kärkereppe. (Photo by Sjögren 12 July 1907, above, and by Rapp 6 August 1992, below). All glaciers in the area have melted or retreated considerably since the 1920's. Top plateau of Mt Vassitjåkka at c. 1500 m with blockfields, wide snow fields and widespread permafrost.

Ekman (1235 m) and the small cirque glacier Kärk-erieppe at the west side of Kärkevagge. Figures 4a and b are two photographs of that cirque taken before and after the convex glacier front disappeared by melting. The old photograph was taken by O. Sjögren 12 July 1907, the second by A. Rapp on 6 August 1992. The size and distribution of summer snow patches on the slopes is similar in both pictures. The 1992 photograph was taken about three weeks later than that of 1907. The convex tongue of the cirque glacier, with three dark bands of surface moraine, has disappeared, and bare ground is visible in the 1992 picture as part of the former bed of the cirque glacier.

Figure 4c is an aerial photo of the cirque and slush avalanche fan taken on 7 June 1995, after a powerful slush avalanche (slushflow) on 3 June. The event was video-monitored and time-lapse photographed by Gude

and Scherer (1995). The site was selected for monitoring of snowmelt and possible slushflows in the melting season in May–June 1995. Figure 4c shows the trigger zone with retrogressive cracks in the deep snow at point (1). The upper part of the slushflow track (2–3) is narrow and marked by dirty snow. There the flood waves of water and slush mainly passed below snow cover in the stream channel. At point (3) the flood waves of water and slush erupted through the snow roof in a 15 m high water blowout cascade. It widened the slush avalanche to a 70 m wide lobe of moving, wet snow which eroded boulders up to 2 m size, mixed with earth and deposited a 70 m wide zone of mixed debris on the apex of the fan. The deposit of rock debris on the fan apex has an estimated volume of c. 200 m^3 ($70 \times 30 \times 0.1$ m thick). Slush avalanches were described as important geomorphic and environmental processes in arctic mountains e.g. by Rapp (1960) and by Nyberg (1985). Bull et al. (1995) have shown by lichenometric studies that the actual fan in Kärkevagge had at least c. 15 large slush avalanches per century in the period 1790–1950. The frequency appears to have increased since c. 1950. It is obvious that large, erosive slush avalanches have continued to modify the debris deposition fan also after the cirque glacier front had disappeared from the trigger zone at the mouth of the cirque.

Retreat and advance of the small, cold-based Ekman glacier

The summers of 1992 and 1993 were cool and had unusually large snow fields in this area. This situation is shown by another pair of comparative photographs, Fig. 5a and b of Lake Ekman and Ekman glacier. The two photographs were taken on 15 September 1985 by N. Å. Andersson and on 4 August, 1993 by A. Rapp.

Fig. 4c. Photo of the Kärkereppe cirque and corresponding slush avalanche fan on 7 June 1995 (Gude and Scherer 1995). Cf. Fig. 4a and b. Legend: 1 = Trigger area in snowpack of slushflow of 3 June 1995. 2 = Narrow track of slushflow on snow. 3 = Point of water blowout eruption in 15 m high and 20 m wide water cascade. 4 = 70 m wide slush avalanche track. 5 = Runout zone of slush avalanche erosion and transport. Boulders, fines and snow were deposited in a mixed cover on apex of fan. 6 = Meltout ridge of old colluvial fan with earlier deposits of avalanche debris. 7, 8 = Braided alluvial channels on both sides of colluvial fan.

Both are oblique aerial photographs, taken during helicopter reconnaissances. 1985 and 1986 were years of strong snow-melting in summer and minimum of snow cover at the front cliff of the glacier.

The Ekman glacier is now probably cold-based, frozen to the bed. It was investigated in April 1993 by drilling and by radar sounding of ice and snow thickness: P. Holmlund, A. Bodin and C. Richardsson performed the echo-sounding and V. Pohjola, assisted by A. Lindskog and A. Rapp the thermal drilling through 7 m of snow, 5 m of firn and 3 m of glacier ice, which contained 4 dirt layers of eolian silt and vegetation fragments. The hole was drilled in the upper part of the glacier. The temperature in the ice was measured by thermistors 7 August 1993. The ice temperature was $-0.8°C$ at 12 m depth and $-1.3°C$ at 15 m indicating cold-based ice, frozen at the bottom of 20 m to a a maximum depth of 30 m of ice and firn, according to the radar soundings. Completely clear water in the lake, even during summer time with rather quick melting indicates cold-based ice. Layers of vegetation fragments in the ice occurred at depths of 12.07 m, 13.45 − 0.58 m, 13.74 − 0.85 and at 14.37 m. They consisted mainly of mosses, a few seeds of vascular plants and two *Coleoptera* fragments (Lindskog 1993).

The uppermost two dirt layers in the frontal ice cliff were sampled in early September 1992. According to radioactivity tests by the T. Swedberg Lab., Uppsala, these layers originated from storms in the summers of 1956 and 1958.

Due to very large amounts of wind-driven snow and thick lee-side snow drifts in the years 1992 and 1993 the glacier front advanced more than c. 20 m per year, as is indicated by Fig. 5b. This is a very interesting observation showing how cold-based glaciers can advance into a lake basin without glacier flow and basal erosion or push, just by a growing, lee-side snow accumulation. Furthermore, it shows an immediate response by advance of small glaciers and snowfields after cool years with high snow precipitation. Small valley glaciers in the Alps are presumed to have a lag time of c. 30 yr before they react at the front with an advance after a climatic fluctuation with higher snow accumulation. The main glaciers at Tarfala, Lapland, are attributed a time lag of c. 60 yr (Karlén pers. comm.). The Greenland inland ice is said to have an average time lag of c. 200 yr (Weidick pers. comm.).

The frontal recession of the Ekman glacier at Lake Ekman can be documented by existing aerial photographs. Lake Ekman was all covered by ice and snow on 15 September 1943, the date of the first available aerial photographs of the area. It began to emerge as a narrow strip of open water, visible on air photographs of 31 July 1959 widening by ice-cliff recession into a 600 m long and 400 m wide lake in the late 1980's (e.g. 1985 and 1986). Snow and firn cover increased during the period 1991–1994.

5a

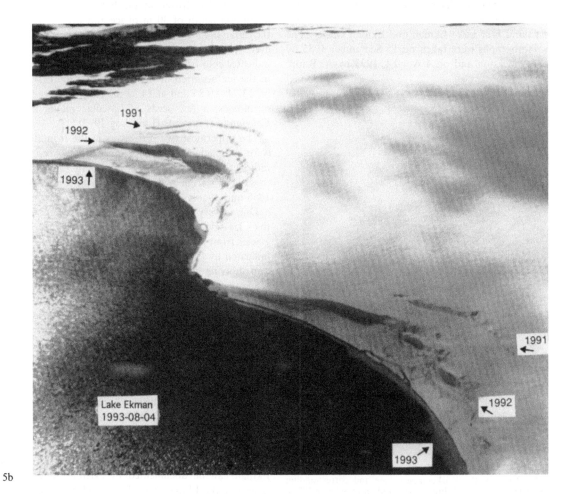

5b

Fig. 5a, b.The frontal ice-cliff of the small Ekman glacier, Låkta mts, Abisko and Lake Ekman, elevation 1235 m a.s.l. The lake began to emerge in the 1950's, due to glacier recession. It was at maximum width in September 1985 (photo 5a, by Andersson) and advanced due to high snow accumulation by lee-side snow drifts in the period 1991–1994, maybe due to the so-called "Pinatubo volcanic dust effect" since 1991.

Fig. 5b. Photo: Rapp 4 August 1993. Two years of much snow accumulation have caused the glacier front to advance into the lake up to 50 m.

176

Fig. 6. Borehole temperatures in the permafrost of the active rock glacier Murtel, Swiss Alps, at various depths. The permafrost table is between 2.6 and 3.6 m depth. Recent warming trends from the exceptionally warm 1980's and early 1990's are most clearly visible at 11.6 and 21.6 m depth. The accuracy of the measurements is ± c. 0.05°C (from Haeberli 1992).

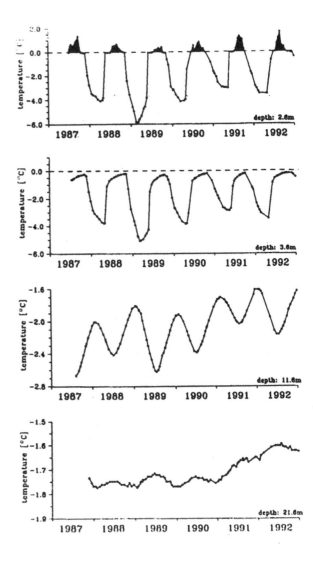

Oetzi, a Stone Age man found in the Alps after melting of ice and permafrost

The melting of the marginal part of a plateau glacier in the Alps exposed the frozen body of a stone-age man (the Oetzi iceman). He was found on 19 September 1991 in a pass near the Italian–Austrian border at an altitude of 3 200 m. The body, clothes and equipment had been extremely well preserved by ice over a period of 5 300 yr according to several 14C datings (Patzelt 1992). The site is located 500 m above the regional lower limit of mountain permafrost in the area, at c. 2 700 m.

In a paper Haeberli (1992), makes the following remarks on climatic fluctuations in the Alps: "With due consideration to the feedback mechanisms affecting glacier mass balance and of the delayed response related to permafrost warming and degradation, secular Alpine glacier and permafrost changes seem to be comparable with estimated rates of anthropogenic green-house forcing. The recent emergence of a stone-age man from cold ice/permafrost on a high altitude ridge of the Oetztal Alps confirms results of earlier moraine investigations: the extent of Alpine ice is probably more reduced to-day than ever during the past 5 000 yr and now seems to pass beyond the "warm" limit of the range known for natural holocene fluctuations".

According to Haeberli three well-preserved wooden bows and other archaeological objects had been discovered as early as 1934 and 1944. Recent 14C-AMS dating of the three bows found at 2 700 m in Lötschenpass, the Swiss Alps, gave dendro-chronologically corrected ages of c. 4 000 yr.

Figure 6 shows four temperature curves measured in permafrost in the Swiss Alps from 1987–1992 in a rock glacier at depths from 2.6 m to 21.6 m below ground surface. Recent warming trends from the exceptionally warm 1980's and early 1990's are most clearly visible at 11.6 and 21.6 m (Haeberli 1992).

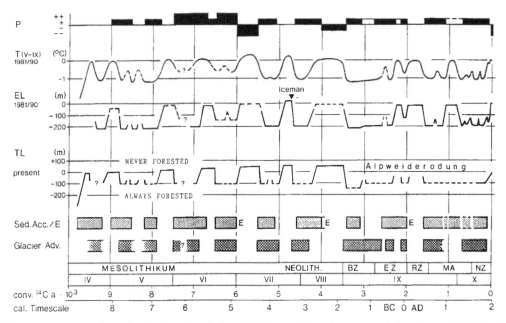

Fig. 7. Air temperature curve from the Austrian Alps during the Holocene. Note the warm period ending c. 5 300 BP with special indication of the find of the Iceman Oetzi, a frozen and totally conserved man from the Stone Age. P = Annual precipitation. T = summer mean temperature. EL = glacier equilibrium line altitude. TL = tree-line. From Patzelt 1993. Note the marked fluctuations between warm and cold periods in the Neolithicum, with the find of the frozen Iceman Oetzi marked in the diagram.

Figure 7 shows reconstructed air temperatures in summer, glacier equilibrium line, mountain tree line and other environmental factors during the last 10 000 yr, according to Patzelt (1993). Note the position of the Oetzi Iceman at the end of a "warm" period of rather short duration, followed by several cold periods and rapid warming in our time.

The combined evidence of the information provided by Fig. 6 and Fig. 7 is that glaciers and permafrost at high levels in the Alps are now at their minimum extent as regards the last 5 000 yr, owing to recent warming in our century. The melting of permafrost in connection with intensive rainstorms has increased the frequency of landslides and debris flows in the Swiss Alps, as another kind of environmental signal of climatic warming.

Figure 8, a photograph of recent large-scale debris flows in the mountains of Lapland is added here as a reminder of open questions for future research in northern Sweden. Are the recent Skanatjåkka landslides a sign of thawing permafrost and increase in intensity of rainstorms or are they expressions of "normal" extreme events? If the triggering mechanism is connected with thawing of permafrost, these high mountain processes may represent a different climatic signal compared with some of the previous finds (Nyberg 1985, Rapp and Nyberg 1988, Jonasson 1991). They indicated increased activity of slope processes during periods of cold and wet conditions.

We hope to be able to analyse the environmental signal of this group of debris slides in more detail. By comparative photo documentation we have dated the

occurrence of landsliding at Skanatjåkka, eastern Sarek mountains, to some time before 1986, since they appear on photos taken by O. Svenningsen in that year. They are later than 1980, as can be checked on air photographs.

Discussion and concluding remarks

Comparative photography is a very useful method for documentation and analysis of landscape change in

Fig. 8. Large debris slides and flows from high-level block-fields, probably triggered by heavy summer rainfall in 1986 or earlier and perhaps also due to melting permafrost at high levels. Mt Skanatjåkkå (crest at 1 767 m, valley bottom at 800 m, Kukkes valley) at the eastern border of Sarek National Park, N Sweden. Photo: Rapp 26 August 1994.

mountain areas or other remote landscapes where the roads and weather stations are far apart. Old photographs, either ground or aerial photographs are of large value for documentation of changes in our century, either due to natural processes or due to impact by man and his grazing animals. In this paper some examples of landscape changes are discussed based on photo comparisons and discussions of possible impact of man-made factors or climatic fluctuations or extremes.

In all mountain areas with cold-based glaciers and permafrost, geoscientists and ecologists should be encouraged to pay attention to possible new finds of glacially non-scoured landforms, soils and frozen vegetation, which may be exposed if climate warming continues. Signals or signs of long-term warming are tree lines moving upslope, retreating glacier fronts and shrinking perennial snow fields. Small glaciers can have a short response time to climatic fluctuations, large glaciers show longer time-lags of several decades. Small glaciers can react with frontal expansion due to high snowfall and much snow-drifting even in the early 1990's with general annual warming in north Scandinavia, as shown by our case study of the small Ekman glacier. In the Alps a general warming in the 1980's and 1990's has resulted in marked melting of glaciers and permafrost and increased frequency of intensive rainstorms which have triggered many slides and debris flows from high altitudes. It is possible that conditions are similar in the N Scandinavian mountains, but increased awareness of researchers and better documentation is recommended to evaluate the existing signals of secular trends (tree lines, glacier recession, permafrost temperature) and of short-term episodic events (extreme winds, rainstorms causing debris flows in mountains, extreme snow-melt and slushflows).

Acknowledgements – Field research in the Abisko mountains has been performed by the author since the 1950's, always with much support and help from the director and staff of the Abisko Research Station. Since 1990 annual international geosymposia have been coordinated by the author at Abisko, with financial support from the Crafoord Foundation, Lund, by Bröderna Wallenberg Foundation, Royal Swedish Academy of Sciences, Stockholm and with participation of international and Swedish colleagues. L. Lidström, helicopter pilot is warmly thanked for his instructive and skilful flights with our research teams and symposia groups to many inspiring mountain sites in northern Lapland. The late O. Sjögren, Stockholm, allowed me to use documentary photographs from his collection. B. Holmgren, C. Jonasson, and P. Schlyter made critical and helpful comments on the manuscript. M. Greenwood-Petersson checked the English language. My warm thanks to many persons and institutions for help and support.

References

Bull, W. B., Schlyter, P. and Brogaard, S. 1995. Lichenometric analyses of a slush-avalanche fan, Kärkevagge, Sweden. – Geogr. Annaler 77A: 231–240.

Dahl, R. 1963. Shifting ice culmination, alternating ice covering and ambulant refuge organisms. – Geogr. Annaler 45A: 122–138.

Dyke, A. S. 1993. Landscapes of cold-centered Late Wisconsinan ice caps, Arctic Canada. – Progr. Physical Geog. 17: 223–247.

Emanuelsson, U. 1987. Human influence on vegetation in the Torneträsk area during the last three centuries. – Ecol. Bull. 38: 95–111.

Falconer, G. 1966. Preservation of vegetation and patterned ground under a thin ice body in Northern Baffin Island, N. W. T. – Geograph. Bull. (Ottawa) 8: 194–200.

Gude, M. and Scherer, D. 1995. Snowmelt and slush torrents. Preliminary report from a field campaign in Kärkevagge, Swedish Lapland. – Geogr. Annaler 77A: 199–206.

Haeberli, W. 1992. Accelerated glacier and permafrost changes in the Alps. – Int. Conf. Mountain Environments in Changing Climates, Davos, Oct. 1992, Report.

Havström, M., Callaghan, T. V., Jonasson, S. and Svoboda, J. 1995. Little Ice Age temperature estimated by growth and flowering differences between subfossil and extant shoots of *Cassiope tetragona*, an arctic heather. – Funct. Ecol. 9: 650–654.

Hoppe, G. 1959. Några kritiska kommentarer till diskussionen om isfria refugier. – Svensk Naturvetenskap, Årsbok 1959 (Stockholm): 123–134, in Swedish.

Ives, J. D. 1974. Biological refugia and the nunatak hypothesis. – In: Ives, J. and Barry, R. (eds), Artic and Alpine Environments. Methuen and Co., London, pp. 605–636.

Jonasson, C. 1991. Holocene slope processes of periglacial mountain areas in Scandinavia and Poland. – Uppsala Univ. Dept Phys. Geogr. Rapport 79.

Kleman, J. 1994. Preservation of landforms under ice sheets and ice caps. – Geomorphology 9: 19–32.

– and Borgström, I. 1990. The boulder fields of Mt. Fulufjället, west-central Sweden – Late Weichselian boulder blankets and interstadial periglacial phenomena. – Geogr. Annaler 72A: 63–78.

Kullman, L. 1984. Transplantation experiments with saplings of *Betula pubescens* spp. *tortuosa* near the tree limit in central Sweden. – Holarct. Ecol. 7: 289–293.

Lindskog, A. 1993. Plant dispersal and succession in glacial areas. – Dept of Syst. Botany, Gothenburg, Sweden: 1–15, Report.

Nyberg, R. 1985. Debris flows and slush avalanches in northern Lappland. Distribution and geomorphological significance. – Lund Univ. Geogr. Inst. Avhandl. 97.

Patzelt, G. 1992. Neues vom Ötztaler Eismann. – Österr. Alpen Verein Mitteil. 2.

– 1993. Holocene mudflow chronology deduced from the development of alluvial cones in Tyrolian mountain valleys. – E.S.F. Workshop, European Paleoclimate and Man. Mainz, Oct. 1993, Report.

Rapp, A. 1960. Recent development of mountain slopes in Kärkevagge and surroundings, northern Scandinavia. – Geogr. Annaler 42A: 65–200.

– 1992. Kärkevagge revisited. Field excursions on geomorphology and environmental history in the Abisko mountains, Sweden.– Swed. Geol. Survey. Ser Ca 81: 269–276.

– and Nyberg, R. 1988. Mass movements, nivation processes and climatic fluctuations in northern Scandinavian mountains. – Norsk Geografr. Tidsskr. 42: 245–253.

Schlyter, P., Jönsson, P., Nyberg, R., Persson, P., Rapp, A., Jonasson, C. and Rehn, J. 1993. Geomorphic process studies related to climate change in Kärkevagge, northern Sweden Status of current research. – Geogr. Annaler 75A: 55–60.

Sonesson, M. 1980. Klimatet och skogsgränsen i Abisko.- Fauna och Flora 75: 1, in Swedish.

– and Hoogesteger, J. 1983. Recent tree-line dynamics (*Betula pubescens* Ehrh. ssp. *tortuosa* (Ledeb.) Nyman) in northern Sweden. – Nordicana 47: 47–54.

Ecological Bulletins 45: 180–191. Copenhagen 1996

Direct and indirect effects of increasing temperatures on subarctic ecosystems

Sven Jonasson, John A. Lee, Terry V. Callaghan, Mats Havström and Andrew N. Parsons

Jonasson, S., Lee, J. A., Callaghan, T. V., Havström, M. and Parsons, A. N. 1996. Direct and indirect effects of increasing temperatures on subarctic ecosystems. – Ecol. Bull. 45: 180–191.

This study summarizes the responses of soil and plant processes to experimental manipulation of air temperatures and nutrients in three subarctic ecosystems at Abisko, N Sweden and compares the treatment effects across a wider latitudinal gradient ranging from the temperate zone in the U.K. to the High Arctic in Svalbard. The temperature enhancement of 2–4°C is within the most commonly predicted range of temperature increase in the Arctic if atmospheric CO_2 doubles, and the fertilizer addition simulated the expected increase of nutrient mineralization rates in warmer arctic soils.

Soil nutrient mineralization was not, however, affected by the temperature increase, probably because the soil warming was slight, and the soils were even cooled in some circumstances. Fertilizer application increased the N and P content in soil microbial biomass indicating a substantial potential for microbial nutrient immobilization.

Both the temperature enhancement and the application of fertilizer caused responses in plant growth and reproductive output. The responses to fertilizer application were often stronger than those to temperature. Responses to temperature were greatest at the sites closest to the northern or upper climatic distributional limits for the species studied. For instance, boreal dwarf shrubs responded strongly at a forest understorey site which was close to their distributional limit, although this site did not experience extreme climatic conditions in comparison to the other sites along the gradient. Low arctic shrubs and dwarf shrubs did not respond at a subarctic tree-line site, whereas all species investigated hitherto responded by increased growth at the climatically most harsh site at a high altitude fellfield. Reproductive output was increased by temperature perturbation. Vegetatively, plants responded to enhanced temperature with greater branching or shoot formation, and although individual leaves and branches also grew more in most cases, their response to temperature manipulation was less marked.

Nutrient concentration declined in species which showed enhanced growth in response to increased temperature, probably because constrained nutrient uptake due to the lack of increased soil nutrient mineralization resulted in low availability of plant nutrients.

The integrated response of plants and microbes is discussed in the context of possible sink to source status for CO_2 in a warmer Arctic.

S. Jonasson, Dept of Plant Ecology, Univ. of Copenhagen, Ø. Farimagsgade 2D, DK-1353 Copenhagen K, Denmark. – J. A. Lee, T. V. Callaghan and A. N. Parsons, Centre for Arctic Biology, Williamson Bldg, Univ. of Manchester, Manchester, U.K. M13 9PL. – M. Havström, Inst. of Botany, Univ. of Göteborg, Carl Skottsbergs Gata 22, S-413 19 Göteborg, Sweden. (Present address of J. A. L. and T. V. C.: Sheffield Centre for Arctic Ecology, Dept of Animal and Plant Sciences, The Univ. of Sheffield, 26 Taptonville Rd, Sheffield, U.K. S10 5BR. Present address of A. N. P.: Natural Resource Ecology Lab., Colorado State Univ., Fort Collins, CO 80523, USA).

The current man-induced increase of atmospheric CO_2 concentration is predicted by general circulation models (GCMs) to have indirect effects of raised global temperature with particularly high temperature increases in northern areas (Mitchell et al. 1990, Maxwell 1992). As a result of this, and the fact that the species in northern, arctic ecosystems are operating close to their lower physiological temperature tolerance limits, arctic areas

Table 1. Dominant species, air and soil temperatures during the growing season at the Abisko sites, the temperate site in the U.K. and the High Arctic sites on Svalbard. Temperature enhancement is that induced by greenhouse treatments (data for both a low and high temperature enhancement are given for two of the Abisko sites) and nutrients refer to the added amounts in a fertilizer treatment. Data are from Werkman (pers. comm.) for Site 1, Wookey et al. (1993) (Sites 2 and 6), Havström et al. (1993) (Sites 3, 4, 5 except temperature data from Site 5 which are from Coulson et al. 1993). Modified from Callaghan and Jonasson (1995b).

| Site and location | Dominant species | Temperature | | Temperature Enhancement | | Nutrients (N, P, K) $g\,m^{-2}\,yr^{-1}$ |
		air	soil	air	soil	
1. Temperate uplands Upper Teesdale, U.K. 54°39′N, 2°13′W	*Pteridium aquilinum*, *Calluna vulgaris*	12.2	10.8	0.8	0.1	5, –, –
2. Subarctic forest understorey, Abisko 68°21′N, 18°49′E	*Vaccinium/Empetrum* under *Betula pubescens* ssp. *tortuosa*	11.8	5.3	2.7	−0.3	10, 10, 12.6
3. Subarctic heath above treeline, Abisko 68°21′N, 18°49′E	*Cassiope tetragona* *Vaccinium*, *Empetrum* *Rhododendron*, *Salix*	11.0	9.1	3.9 2.7	1.0 1.0	10, 2.6, 9
4. Subarctic high altitude fellfield, Abisko, 68°21′N, 18°49′E	*Cassiope tetragona* *Salix polaris, S. herbacea*, *Aulacomium* sp.	6.9	7.9	4.8 2.1	1.9 1.2	10, 2.6, 9
5. High Arctic heath Ny Ålesund, Svalbard 78°56′N, 11°50′E	*Cassiope tetragona* *S. polaris, Racomitrium lanuginosum*	8.0	4.8	2.5	0.3	10, 2.6, 9
6. High Arctic polar semidesert, Ny Ålesund Svalbard, 78°56′N, 11°50′E	*Dryas octopetala, S. polaris, Saxifraga oppositifolia*	5.4	6.1	3.5	0.7	5, 5, 7.5

will probably be the first parts of the world where biological responses to atmospheric heating will be visible. Furthermore, the responses to the changes will probably be stronger than in most other ecosystems because even relatively small changes of temperature are likely to result in strong biological responses.

Biological processes in arctic ecosystems can also feed back to the global climate either by acting as moderators (sinks) or accelerators (sources) of the rate of increase of atmospheric CO_2 levels. Enhanced sink action is likely if bare ground is colonized by plants in a warmer environment or if biomass production increases in presently vegetated areas. A source action is likely if the emission of CO_2 and other trace gasses (e.g. CH_4) from microbial decomposition of presently rather inactive organic matter deposits in tundra regions increases more rapidly than the fixation of CO_2 by plants (Callaghan and Jonasson 1995a). The potential for arctic areas to influence global atmospheric gas composition is evident because they contain c. 14% of the global amount of soil organic carbon (Post et al. 1982).

The concern about possible, future drastic changes in biological processes in the Arctic due to man-induced perturbations of the environment led to the establishment of a series of experiments at the Abisko Scientific Research Station in the late 1980's. Perturbation experiments that simulated predicted direct or indirect changes were established in 1989 and measurements were initiated on an array of biological responses to various environmental changes, including enhanced air temperature (Havström et al. 1993, Jonasson et al.

1993). Since then, additional experiments have been initiated at Abisko (Wookey et al. 1993) and similar experiments have been expanded to other arctic (Havström et al. 1993, Wookey et al. 1993) and non-arctic sites (Werkman and Callaghan 1995, Werkman et al. 1996) so that the whole series covers a local altitudinal gradient at Abisko (Havström et al. 1993, Jonasson et al. 1993) and a north-south transect across arctic (Coulson et al. 1993, Wookey et al. 1994, 1995), subarctic (Havström et al. 1993, Parsons et al. 1994, Strathdee et al. 1995) and temperate ecosystems (Werkman and Callaghan 1995, Werkman et al. 1996). The arctic transect approach of manipulations was also subsequently extended into the ITEX (International Tundra Experiment) programme.

At present, our experiments are established in tundra-like British moorlands at Upper Teesdale, in the subarctic with measurements at Abisko, and in the High Arctic at Ny-Ålesund, Svalbard. At Abisko, three sites cover an altitudinal gradient ranging from the birch forest at c. 400 m a.s.l., across the lower part of the treeless tundra heaths with a site located just above the treeline at c. 450 m a.s.l. and ending at a fellfield c. 1150 m a.s.l. (Table 1).

This gradient approach recognizes that species and communities in different locations may respond differently to the same environmental constraints and that the nature of the environmental constraints are likely to vary with geographical locations. The approach also leads to a greater representativeness of the results for the arctic region as a whole.

Responses to the manipulations have been measured in a number of plant and microbial processes including plant growth and reproduction, plant chemistry, plant nutrient uptake, microbial decomposition and mineralization. In the following we summarise the main biological responses to the manipulations emphasising direct and indirect effects of the temperature enhancements observed at Abisko. We compare our observations with data from the other sites along the latitudinal gradient and from relevant experiments elsewhere in the Arctic and discuss the possible impacts of the responses on the arctic biota and on the global atmosphere.

The perturbation experiments

At each site, air temperature was increased to simulate the levels and vertical gradients predicted by the general circulation models. In addition, nutrients were added alone or in combination with the air heating to simulate increased nutrient mineralization rates postulated as an effect of the warming (Melillo et al. 1990) and to identify the degree of nutrient limitations of the ecosystems at the various sites. At some sites, other environmental perturbation treatments such as shading and water addition were also performed, but these perturbations will not be discussed here.

Direct temperature manipulations

Since 1989 (the heath and fellfield sites) and 1991 (the forest understorey site), temperature has been increased within dome-shaped polythene passive greenhouses described in detail by Havström et al. (1993). They are erected at the start of the growing season in early June at the birch forest, in mid June at the tree-line site and in late June at the fellfield and removed at the end of the growing season in late August or early September. The size of the greenhouses varies from $1.2 \times 1.2 \times 0.5$ (height) to $1.5 \times 1.5 \times 0.7$ m among the sites. Vents are cut out of the tops to prevent extreme temperature, to reduce humidity differentials and allow precipitation to enter the plots. At the heath and fellfield site, an additional treatment with smaller temperature increases is achieved by creating a gap between the base of the greenhouse and the ground.

This method of raising temperature is used because it provides a correct gradient of temperature along a vertical profile from air to soil and because the technique is simple, robust and flexible enough to use in remote locations. The constructions are small enough to allow sufficient replication of treatments. In addition, by considering a latitudinal gradient, the greenhouses not only simulate the correct vertical gradient of air warming, but also simulate a greater increase in the number of degree days above zero at the coldest sites,

which is also predicted by climate models (Maxwell 1992).

Manipulations simulating indirect effects of temperature

The simulation of enhanced microbial mineralization in warmer soils, is achieved by application of fertilizer (N, P, K) once (heath and fellfield sites) or six times (forest site) per year. The application rates are 10, 10 and 12.6 g m^{-2} yr^{-1} of N, P and K at the forest site and 10, 2.6 and 9 g m^{-2} yr^{-1} at the heath and fellfield (Table 1; for details, see Havström et al. 1993 and Parsons et al. 1994). In addition, Mg was added at a rate of 0.8 g m^{-2} yr^{-1} at the two latter sites in the first three years.

Environmental responses

At Abisko, the air temperature is raised between c. 2.5 and 4.5°C by the greenhouses with the lowest temperature enhancement at the forest understorey site and in the low temperature enhancement at the heath and fellfield sites (Table 1). The soil temperature is raised considerably less than the air temperature, c. 0.5–2.0°C, and can even be reduced as at the forest site (Table 1).

The plastic greenhouses were optically neutral but reduced the photosynthetically active radiation by c. 9% (Havström et al. 1993). The loss of precipitation in the greenhouses at the forest site was first compensated for by watering with amounts equivalent to the precipitation, but later precipitation was allowed to enter by increasing the size of the vent at the top of the greenhouses. Also, the greenhouses at the heath and the fellfield are placed on sloping ground so that soil water can flow through them. Repeated measurements in several years have shown only a slight reduction of a few percent soil water in the greenhouses, which are within the range that can be expected by higher evaporation and transpiration in a warmer environment (Chapin et al. 1995). Furthermore, measurements of $\delta^{13}C$ in the plants after six years of treatments did not show any differences in soil water use that could be ascribed to water stress in the covered plots (Michelsen et al. 1996).

The responses of soil processes to increased temperature

The effects of temperature on mineralization of nitrogen (N) was investigated at all three Abisko sites and phosphorus (P) mineralization was quantified at the heath and fellfield sites. The measurements took place during the third growing season after the initiation of the perturbations, by using the buried bag method (Eno 1960). The increased temperature did not lead to any markedly enhanced net soil nutrient mineralization or changes in the pool sizes of plant available nutrients (Jonasson et al. 1993, Robinson et al. 1995). This was the expected response considering the low warming of the soil in the greenhouses. Transplantation of fellfield soils to the heath site with c. 4°C higher ambient soil temperature led, however, to a flush of mineralization of N but not of P. Hence, the increase of inorganic N suggests raised activity of the decomposing microorganisms at the same time as the unchanged mineralization rate of P indicates that this element had been immobilized, probably by incorporation in microbial biomass (Jonasson et al. in press).

These results give two indications of the likely responses of soils to enhanced atmospheric temperature. Firstly, as the soil is apparently buffered against temperature changes, microbial processes may not be strongly affected by increased air temperatures within the interval predicted by the GCMs, unless the microorganisms are sensitive to small temperature changes (Jonasson in press). Data from the transplantation experiment of Jonasson et al. (1993) suggest that changes in mineralization rates can take place in soils heated to c. 4°C above ambient. This temperature enhancement is approximately the lower level of temperature increase at which net nutrient mineralization increased notably in Alaskan tundra soils (Nadelhoffer et al. 1991). Consequently, any increase in soil nutrient availability may be dependent on how resistant the soil is to increased heating in the long term. This probably varies from site to site with variations in soil type, soil moisture content, depth to the permafrost, and will also depend on how the radiation to the soil surface changes as a feedback from altered reflection by the vegetation and accumulated litter.

Alternatively, even if the soil warms and the microbial activity increases, any changes in gross mineralization do not necessarily lead to a corresponding increase of net mineralization and an immediate increase in amounts of plant available nutrients. The nutrients may instead be immobilized by soil microorganisms (Jonasson et al. in press). If so, the main effect of increased decomposer activity will not be visible in plant production until the carbon to nutrient ratio has declined to non-limiting levels for the soil microorganisms or during periodic nutrient flushes after die-back of the microbial populations (Malmer 1962, Barèl and Barsdate 1978, Jonasson in press, Jonasson et al. in press).

Analyses of the nutrient content in soils and microbial biomass at the heath and fellfield revealed a low soil pool of inorganic N and P and high content of these elements fixed in the microbial biomass (Jonasson et al. in press). After five years' temperature enhancement, the content of microbial N and P remained unchanged, whereas the inorganic fraction had increased (Jonasson et al. unpubl.). In contrast, both the microbial and the inorganic N and P contents had, however, increased significantly in plots with added fertilizer, showing that the microorganisms had responded to the increased availability of nutrients and that part of the nutrients added had been sequestered by the microbial populations.

The microbial and soil inorganic nutrient content varied slightly between measurements made at spring thaw, mid growing season and autumn freezing, and without any trend that indicated any pronounced seasonal peak or decline. Hence, if the microorganisms are efficient in absorbing inorganic nutrients, the microbial populations could, potentially, act as competitors for plant nutrients (Harte and Kinzig 1993) and there was no indication of periodic microbial dieback that could increase the availability of nutrients to the plants seasonally.

The responses of plants to temperature and nutrients

Growth responses to temperature enhancement and fertilizer addition

At the forest understorey site, the plant species demonstrated rapid and large responses to temperature. After two seasons of treatment, the dominant dwarf shrubs (*Empetrum hermaphroditum*, *Vaccinium uliginosum*, *V. myrtillus* and *V. vitis-idaea*) had greater leaf and stem mass, longer stems and, for two of the species measured, a greater leaf area. The response to nutrient addition was even greater, particularly when combined with temperature enhancement. The combined treatments had a synergistic effect on stem length, stem mass, leaf mass and leaf area in one or more of the four species (Parsons et al. 1994). The most abundant graminoid at the site, *Calamagrostis lapponica*, also showed strong responses. After two seasons of perturbation, there was an increased above ground biomass and increased shoot height of this species in response to both temperature and nutrient addition. These treatments also acted synergistically on shoot height and above ground biomass (Parsons et al. 1995).

At the heath, the shrub *Betula nana*, and the dwarf shrubs *Arctostaphylos alpina*, *Empetrum hermaphroditum*, *Rhododendron lapponicum* and *Vaccinium uliginosum* did not have greater leaf mass or area per branch after six years of temperature enhancement. However the number of active meristems, the leaf area index and the leaf mass per unit ground area was significantly greater in *A. alpina* and *V. uliginosum*, indicating that these two species responded to the temperature enhancement by increasing the branch density (Graglia 1995, Graglia et al. unpubl.).

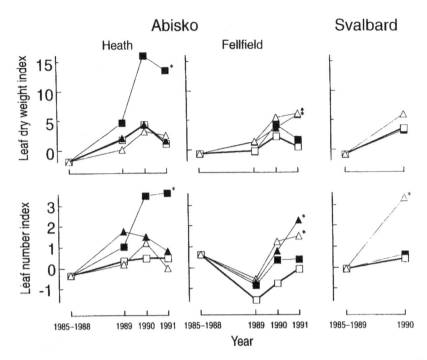

Fig. 1. Comparison of branch growth of *Cassiope tetragona* in unperturbed controls (open squares) low and high temperature enhancement plots (open and filled triangles, respectively) and in plots with added fertilizers (filled squares) at a tree-line heath and a fellfield at Abisko, N Sweden, and at a High Arctic polar semi desert at Svalbard. The data illustrate treatment effects on growth, expressed as indices, after each of three years following the perturbations at the Abisko sites and two years after the perturbations at the Svalbard site. Significant treatment effects (p < 0.05) are marked with an asterisk. The starting points are the average leaf indices during five or six (Svalbard) years before the onset of the treatments. The graph is modified from Havström et al. (1993).

In the plots with added fertilizer, *A. alpina* and *V. uliginosum* had significantly or close to significantly (*V. uliginosum*) greater leaf area and leaf mass per branch. *Arctostaphylos alpina*, *E. hermaphroditum* and *V. uliginosum* had significantly more meristems and leaf mass per unit ground area and greater leaf area index. Hence, fertilizer addition caused both a branching response and a leaf growth response (Graglia 1995, Graglia et al. unpubl.).

At the fellfield, both temperature enhancement and fertilizer addition generally led to greater leaf growth, leaf area and mass per branch and per unit ground area in the deciduous *Salix polaris* × *herbacea* and the evergreen *Vaccinium vitis-idaea*, except for a non-significant response to increased temperature in leaf area index and mass per unit ground area in *S. polaris* × *herbacea*. While fertilizer addition generally stimulated growth more than the temperature enhancement at the heath in the responding species, temperature enhancement often explained a larger proportion of the variance in growth across treatments than fertilizer amendments at the fellfield (Graglia 1995, Graglia et al. unpubl.). This indicates that the response to temperature enhancement

increased relative to that of fertilizer addition along the climatic severity gradient.

Nutrient amendments and temperature enhancement during three years to another dwarf shrub, *Cassiope tetragona*, at both the heath and fellfield sites resulted in a similar site dependent pattern of response (Fig. 1). At the heath, the main branches formed more leaves with greater biomass after nutrient addition but not in response to increases in air temperature. At the fellfield, a similar growth response occurred after temperature enhancement but not after fertilization; this was also the case at a Svalbard site with similar vegetation and climate as the fellfield site at Abisko (Havström et al. 1993).

In 1993, *C. tetragona* had greater above ground biomass in the temperature enhanced and fertilized plots treated since 1989 than in the controls at both the heath and the fellfield (Michelsen et al. 1996). Furthermore, at the fellfield, the proportion of the total biomass allocated to green branches was larger in both fertilized and temperature enhanced plots than in controls with particularly strong effects in the combined treatment (Table 2). In contrast, the allocation pattern

Table 2. Effects of temperature enhancement and fertilizer addition on the proportions (%) of green biomass (leaves plus stems) on leafy branch parts of *Cassiope tetragona* at a tree-line heath and a high altitude fellfield after five years of treatment. The leaves remain green for c. 3–4 yr and persist on the branches for c. 20 yr. Fertilizer addition increased the growth significantly at both the tree-line heath (p = 0.0001) and at the fellfield (p = 0.006) whereas the temperature effect was non-significant at the tree-line heath (p = 0.4) but significant (p = 0.05) at the fellfield (three-factor ANOVA with fertilizer addition, temperature enhancement and block as factors; data are from Michelsen et al. 1996).

Treatment	Green biomass (%)	
	Treeline heath	Fellfield
Control	34.2 ± 2.2	24.9 ± 2.9
Fertilized	38.3 ± 3.7	34.3 ± 2.5
Low temperature	34.2 ± 1.5	33.1 ± 4.4
High temperature	31.4 ± 1.3	33.4 ± 2.9
Fertilized + low temperature	44.0 ± 3.1	42.4 ± 4.3
Fertilized + high temperature	46.1 ± 2.0	42.1 ± 4.7

to the green branches had changed only in response to fertilizer addition but not to temperature enhancement at the heath.

The higher above ground biomass in treated plots in 1993 could be confounded by differences in the plots existing at the start of the treatments in 1989. However, changes among treatments in the proportions of biomass on branch sections with green and older, brown leaves reflect undisputable changes in biomass production during the last 3–4 yr. This is because we assume that the leaves remained green for 3–4 yr (Havström et al. 1993) irrespective of treatment. Hence, an increased ratio indicates an enhanced accumulation of biomass during the period of 1990/1991 to the year of sampling in 1993 as compared to the biomass accumulation during a long period of time before 1990/1991, i.e. mostly during the years before the treatments started in 1989.

The lack of change of the ratio in temperature enhanced plots at the heath but the strong response to temperature enhancement at the fellfield follows a similar pattern for *C. tetragona* as for most other dwarf shrubs for which data are available. That is, temperature enhancement apparently has the strongest overall effects on growth at the climatically most severe sites, even though individual species may also respond at climatically benign sites (Parsons et al. 1994). The response to temperature enhancement appears particularly strong for branch (Parsons et al. 1994, Graglia 1995) or shoot (Tissue and Oechel 1987) formation. This emphasizes recent observations that changes in environmental conditions may have greater effects on the rate of module formation than on the growth of the individual modules (Chapin and Shaver in press). Consequently, any lack of short term effect on the growth of individual modules does not necessarily mean that

there are no longer term effects on the growth of the whole plants.

At the community level, both the temperature enhancement and the fertilizer addition resulted in greater biomass of the whole vegetation of vascular plants with particularly strong response of graminoids after fertilizer addition and after the combined fertilizer addition and temperature enhancement (Parsons et al. 1995, Jonasson et al. unpubl.). In contrast, the mosses declined strongly after fertilizer addition and the combined treatment at the heath but not at the fellfield (Jonasson et al. unpubl.) and moss cover decreased and relative proportions of species changed in response to fertilizer addition at the forest understorey site (Potter et al. 1995).

The changes of fine root biomass of the heath and fellfield mirrored the changes in biomass production above ground with the largest increase found in the combined treatments. Since the fine root biomass was generally greatest in the treatments where the above ground biomass was also greatest, there was no consistent pattern among treatments in the ratio of shoot to below ground fine root biomass (Jonasson et al. unpubl.).

Plant phenology

Air warming in the birch forest understorey vegetation generally, but not always, accelerated rates of development of three dwarf shrubs (Andrews 1994). Each of the dwarf shrubs had a phenology which responded individualistically to air warming. Air warming stimulated earlier bud burst in *Empetrum hermaphroditum* and bud burst was completed one week earlier than in the control treatment (Andrews 1994). Bud burst was also accelerated by air warming in *Vaccinium myrtillus*, but not in *Vaccinium vitis-idaea*. In contrast to nutrient addition, however, air warming did not significantly increase the rate of extension of annual shoots of *E. hermaphroditum*. This might have been due to a reduction in shoot turgor pressure associated with the air warming (Pantis pers. comm.). The rate of leaf development was surprisingly unaffected by air warming in the deciduous *V. myrtillus* but, in contrast, it was significantly stimulated in the evergreen *V. vitis-idaea* (Andrews 1994).

Plant reproduction

Observations on effects of the environmental manipulations on reproduction from the sites at Abisko are available for four species, *Empetrum hermaphroditum*, *Calamagrostis lapponica*, *Carex bigelowii* and *Cassiope tetragona*.

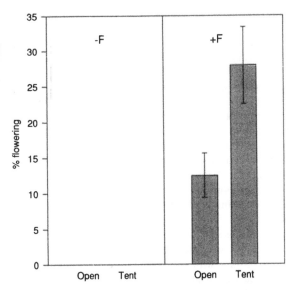

Fig. 2. Effects on flowering frequency (means ±SE) of *Calamagrostis lapponica* three years after initiation of temperature enhancement and fertilizer addition at a subarctic forest understorey site. Data are from Parsons et al. (1995).

After one season of treatments at the forest understorey site, the reproductive output of the dominant dwarf shrub species, *E. hermaphroditum*, was assessed by collecting fruits from all plots at the end of the season and measuring the fresh and dry weight. There was a significant effect of nutrient addition on both fresh and dry weight, with heavier fruit produced in the plots with an enhanced nutrient supply, but there was

no effect of temperature on fruit weight (Wookey et al. 1993).

In the second and third season, the reproductive output of the grass *Calamagrostis lapponica* was assessed by determining the proportion of plants that produced flowers. The flowering apparently was strongly limited by nutrients. After two seasons of nutrient application, there was a flowering rate of 1.2% in plots without nutrient addition compared with 21.5% flowering in plots where nutrients were added. In the third season (Fig. 2) the effect was similar: 0% flowering in unamended plots and 20.1% flowering in nutrient amended plots (Parsons et al. 1995).

The flowering frequency of *Cassiope tetragona* at the heath and fellfield sites did not differ significantly (Havström et al. 1995a). Flowering increased in response to fertilizer application by c. 30% but with no significant response to temperature enhancement (Fig. 3). However, the phenology of flowering as well as fruit maturation and seed production are most likely enhanced by increased temperature (Havström unpubl.). Interestingly, flowering of *Cassiope tetragona* at these sites seems more affected by changes in light than nutrient and temperature enhancement, as shading caused a 80–85% decline of flowering frequency. All effects on flowering frequency were detected in the third year of manipulation.

Long term monitoring of flowering (% of tillers with flowers) of the sedge *Carex bigelowii* at the fellfield showed an apparent cyclicity of production with a 3–4 yr period (Carlsson and Callaghan 1994). Intensity of flowering was significantly correlated with the July temperature of the previous year, i.e. the year of flower

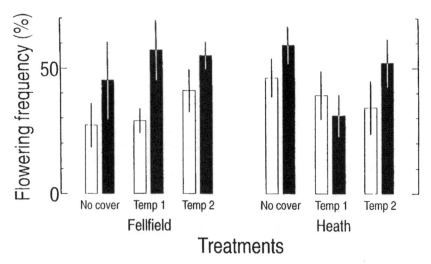

Fig. 3. Flowering frequency (means ±SE) of *Cassiope tetragona* at a tree-line heath and a high altitude fellfield at Abisko three years after fertilizer addition (filled bars), low and high temperature enhancement. No cover means plots without temperature enhancing plastic cover. Data are relative numbers of 5 × 5 m squares containing flowering shoots. Flowering was significantly affected by fertilizer addition (p = 0.015) but not by temperature enhancement or site in a three-factor ANOVA. The graph is modified from Havström et al. (1995a).

initiation. The increase in flowering in the +4°C temperature enhancement treatment plots of Havström et al. (1993) was 22.4%, which is very close to the predicted increase of 25% from the long term monitoring. This provides a validation of the experimental methodology.

Treatment effects on plant nutrition

Plant nutrient concentration has been analyzed in plant tissues from the heath and the fellfield site, whereas no data are available yet from the forest site. In *Cassiope tetragona* and *Empetrum hermaphroditum*, tissue nutrient concentration declined significantly in warmed plots and increased in fertilized plots (Michelsen et al. 1996). Hence, air warming led to a dilution of the nutrient concentration similar to an earlier reported decline of nutrient concentrations in species of the same vegetation type during naturally occurring warm years (Jonasson et al. 1986). This was an expected response, considering the lack of treatment effects on microbial nutrient mineralization (see above). It appears, therefore, that the plants used nutrient reserves present in their biomass to support most of their extra growth in warmed plots, at least initially during the first years of the treatments. Consequently, it is not certain that any increased biomass production can be sustained during a longer period of time unless the net mineralization rate also increases. As a result, the nutritional quality of the forage plants of herbivores and of the litter may decline in a warmer Arctic and feed back on both the grazer and decomposer food chains (Berendse and Jonasson 1992).

Temperature effects on insects

Air warming experiments carried out on a *Dryas octopetala* feeding aphid, *Acyrthosiphon brevicorne*, at two contrasting subarctic sites (upland and lowland) in the Abisko area, showed that, as in the plants (see above), responses in aphid density and overwintering egg production were greatest at the upland site with the most severe climate (Strathdee et al. 1995). The temperature manipulation resulted in an air warming of 1.8°C equivalent to 184 day degrees at the upland site, and 2.3°C equivalent to 250 day degrees at the lowland site. This warming produced far greater responses in the invertebrate populations than even greater experimental warming produced in slow growing arctic clonal plants (Havström et al. 1993, Wookey et al. 1993).

Other data on natural rather than experimentally induced changes in temperature also suggest strong temperature control on insect populations. For instance, Tenow and Holmgren (1987) found a significant negative correlation between the incidence of low win-

ter temperature and insect damage during outbreaks of *Epirrita autumnata*, a moth caterpillar which defoliates sub-arctic birch trees. Overwintering eggs were killed in cold depressions and in cold winters when temperature dropped below −36°C. This suggests that increased temperature could lead to increased survival of eggs and greater destruction of birch forest.

Inter-site comparison of responses

Direct and indirect effects of temperature on plant performance

The data collected from the experiments at Abisko give, firstly, strong indications that both the direct effect of increased temperature and indirect effects of increased nutrient availability have impacts on ecosystem processes. In general, enhanced nutrient levels have given more pronounced responses than the temperature enhancements at the climatically most benign sites, except that the dwarf shrubs responded strongly also to temperature at the forest site, and some species increased the leaf area index and leaf biomass at the treeline heath. In contrast, both increased temperature and fertilizer addition affected growth (all dwarf shrub species investigated) and generally also reproduction (*Carex bigelowii*) at the fellfield.

Comparisons of treatment effects of the perturbations across the altitudinal gradient at Abisko and across the sites located along the latitudinal gradient (Table 1) reveal a similar response pattern. It appears that the relative climatic severity of the sites may influence the responses strongly. The response to temperature enhancement was greater in plants towards their northern or altitudinal distribution limits (Havström et al. 1993, Wookey et al. 1993). For instance, the predominantly boreal *Vaccinium* species with a northern distributional limit within the low Arctic responded strongly to the temperature enhancement in the subarctic forest understorey (Parsons et al. 1994), and *Vaccinium uliginosum* also had greater growth in temperature enhanced plots at the treeline heath (Graglia 1995). *Rhododendron lapponicum* and *Betula nana*, which have their main distribution on low arctic heaths, did not, however, respond at the tree-line site. *Vaccinium vitis-idaea*, which has its distribution centre in the boreal forest and *Salix polaris × herbacea* increased growth strongly at the climatically harsh fellfield, and *Carex bigelowii*, which is common in the low Arctic, increased flowering. The same pattern was observed for the predominantly high arctic *C. tetragona*, which was relatively insensitive to temperature enhancement at the heath site, close to its southern distributional limit (Havström et al. 1993), but responded with increased growth at the climatically more severe fellfield and in the High Arctic at Svalbard. In fact, this species

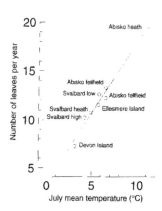

Fig. 4. The numbers of leaves produced per year in shoots of *Cassiope tetragona* as a function of the average July temperature reported for various arctic sites. The regression line represents the linear relationship: Leaf No. $= 3.43 + 1.46 \times$ July temperature; $r^2 = 0.98$, and emphasizes the importance of temperature as a growth limiting factor in *C. tetragona*. Modified from Havström et al. (1995b).

has shown a nearly perfect correlation between leaf growth and July temperature at sites across the entire Arctic (Havström et al. 1995b, Fig. 4).

Similarly, at the most benign site on the gradient (the temperate site at Upper Teesdale), temperature enhancement stimulated most responses in bracken which occurred near its upper altitudinal limit, rather than in the heather which can grow at higher elevations (Werkman and Callaghan 1995, Werkman et al. 1996). Increased temperature stimulated earlier emergence of bracken fronds in spring and delayed senescence in autumn; i.e. the phenological response was similar to the most common response in the subarctic understorey. Also, above ground biomass and productivity of bracken were stimulated by increased air temperature through its effects of increasing both density and size of fronds whereas N addition had little impact on bracken's performance. In heather only the density of annual shoot responded to the environmental perturbations: density was increased by the temperature treatments.

The transect data indicate particularly strong temperature responses in reproductive output at the climatically severe sites (Callaghan and Jonasson 1995b). In a polar semi-desert on Svalbard, *Dryas octopetala* did not respond vegetatively to fertilizer application the first year but showed a strong response to air warming, particularly in reproductive parameters (Wookey et al. 1993). Similarly, the combination of fertilizer and air warming enhanced vegetative growth of the forb *Polygonum viviparum* but only air warming enhanced the reproductive output (Wookey et al. 1994), a pattern that corresponds to previously observed responses along natural temperature gradients (Callaghan and Collins 1981). In contrast, fertilizer addition led to greater flowering and reproductive output of graminoids (Jonasson 1992, Parsons et al. 1995) and

dwarf shrubs (Wookey et al. 1993, Havström et al. 1995a) at the southern subarctic sites.

The differentiated vegetative responses along the arctic severity gradients are not surprising, because it is known that plant growth even in species adapted to cold environments is set back by low temperatures (Körner and Larcher 1988). Similarly, the success of sexual reproduction is closely related to temperature, both directly, as it influences propagule (seed, bulbil etc.) maturation, and indirectly because it influences pollinator activity. Hence, air warming should have the proportionally greatest effect in climatically severe environments.

Plant community responses

The treatment effects of both temperature enhancement and nutrient addition on plant growth were similar across all sites: a greater growth of all responding species and life forms of the vascular plants, i.e. there was no compensatory decline of any species as other species increased. These results also agree with earlier reported increases of biomass among most vascular species in this region after fertilizer addition (Jonasson 1992).

The response pattern is different, however, from treatment effects of similar manipulations of Alaskan tussock tundra. Chapin and Shaver (1985) and Chapin et al. (1995) reported strong perturbation effects of either increased or decreased production of individual species and life forms and with species-specific responses to each manipulation type. Furthermore, community biomass production remained constant in the tussock tundra as increased biomass production of some species was compensated by about equally large decreases of other species. Hence, there is no common response of the plant communities across the Arctic.

The time-scale of responses

Arctic plants are typically slow growing, perennial species which are subjected to proportionally great stochastic environmental fluctuations between years. It could be expected, therefore, that they exhibit a considerable plasticity acting as a buffer against periodically adverse conditions (Jonasson in press) and that long term responses to changes in the environment may be different from short term responses. The treatment effects of the perturbations at Abisko have been measured both as short term responses on time-scales of between one to three years and as effects on longer terms of up to 6 yr. Responses within the shorter time-scale give valuable information on immediate reactions of the plants to events during the life-time of individual modules, e.g. shoots or branches, whereas the longer time-scale of 5–6 yr gives additional information on the whole plant level and initial responses at the community level (Oechel and Billings 1992, Chapin et al. 1995).

Responses of communities (e.g. in species abundance or species composition) and ecosystems (e.g. in interactions between species within or between trophic levels) to environmental changes typically occur after considerably longer time spans than responses in individual plants or plant modules. Anticipated changes could be both responses in plants and plant populations and feedbacks between different components of ecosystems or even feedbacks between the biota, the soil and the atmosphere.

The interactions and long term effects of temperature increase on the CO_2 exchange between the subarctic biota and the atmosphere have been considered in the experiments at Abisko by the analyses of the integrated responses in both the primary producers and the decomposers. These two components represent the main terrestrial sinks and sources, respectively, of atmospheric carbon. The sink strength of the ecosystems will increase if the production of biomass increases, or if increased plant litter production leads to increased accumulation of organic matter. The source action of the decomposers will increase if their respiratory activity increases. Hence, the net sink to source status of the ecosystem depends on the balance between CO_2 emission from the decomposers and the net CO_2 uptake by the plants together with the accumulation pattern of soil organic matter (Callaghan and Jonasson 1995a).

Our results hitherto suggest a likely increase in biomass production both as an effect of increased temperature and particularly as a result of increased nutrient availability, i.e. an increased sink strength in the vegetation. We do not know, however, if this is a short term response that is dependent on stores in the plants already available at the initiation of the experiment, or if the increased rate of biomass production will continue. Since the growth response in at least sub- and low arctic ecosystems are almost always regulated by the availability of nutrients, particularly N and P (Shaver and Chapin 1980), the sustainability depends on the extent to which nutrients are supplied to the plants by soil microbial activity. The lack of response in nutrient mineralization due to low degree of soil heating in our temperature manipulated treatments indicates that the flux of nutrients from decomposers to plants may be relatively unaffected by air warming. The nutrient dilution found in the plant tissues within our warmed plots could well be a reflection of enhanced plant biomass production due to the air temperature enhancement at the same time as the nutrient flux to the plants was proportionally less affected.

If, however, soil temperature increases to the extent of increasing mineralization (Nadelhoffer et al. 1991), the response will depend on to what extent the extra nutrients released will be traded off to the plants or immobilized in the microbial biomass. If a large part of the nutrients are kept in a closed cycle by soil microorganisms, their activity will increase above the present level and more CO_2 will be released to the atmosphere at the same time as the plant biomass production will level off due to nutritional constraints. Hence, this suggests a soil-plant interaction leading to an increased emission of CO_2 to the atmosphere at the same time as the CO_2 consumption by plant tissue production changes less, i.e., a net increase of the CO_2 flux from the ecosystem and a positive feedback on the atmospheric warming. This scenario seems realistic, because it is well known that only small proportions of nutrients added to arctic ecosystems are recovered in the vegetation, probably because of strong immobilization in the soil microbial biomass (e.g. Marion et al. 1982, Shaver et al. 1986). It is also supported by the increased microbial biomass N and P we found in the fertilized plots. It should be stressed, however, that these and other long-term responses, e.g. feedbacks due to changed mycorrhizal action or long term changes in plant quality for herbivores and decomposers are poorly investigated and need specialized process oriented studies and a maintenance of the experiments over a longer period of time.

Acknowledgements – Many colleagues have allowed access to material in press and in preparation. We are particularly grateful to E. Graglia, A. Michelsen, C. H. Robinson and A. T. Strathdee. For hospitality and logistic support, we thank M. Sonesson and N. Å. Andersson at the Abisko Scientific Research Station. S.J. and M.H. wish to thank the Swedish Natural Science Research Council, the Danish Natural Science Research Council and the Swedish Environmental Protection Board for financial support (Grant Nos.: B-BU 4903-300, 11-0421-1 and 127402, respectively) and J.A.L, T.V.C. and A.N.P. were supported by grant GST/02/531 from the U.K. Natural Environment Research Council Arctic Terrestrial Ecology Special Topic Programme.

References

Andrews, M. J. 1994. The effects of simulated climate change perturbations on arctic dwarf shrub phenology. – M. Sc. Diss., Univ. of Manchester, U.K.

Barèl, D. and Barsdate, R. J. 1978. Phosphorus dynamics of wet coastal tundra soils near Barrow, Alaska. – In: Brisbin, I. L. and Adriano, D. C. (eds), Environmental chemistry and cycling processes. DOE Symposium series Conf. 760429, Washington D.C.

Berendse, F. and Jonasson, S. 1992. Nutrient use and nutrient cycling in northern ecosystems. – In: Chapin, F. S., III, Jefferies, R. L., Reynolds, J. F., Shaver, G. R. and Svoboda, J. (eds), Arctic ecosystems in a changing climate, an ecophysiological perspective. Academic Press, San Diego, pp. 337–356.

Callaghan, T. V. and Collins N. J. 1981. Life cycles, population dynamics and the growth of tundra plants. – In: Bliss, L. C., Heal, O. W. and Moore, J. J. (eds), Tundra ecosystems: a comparative analysis. Cambridge Univ. Press, pp. 257–284.

– and Jonasson, S. 1995a. Arctic terrestrial ecosystems and environmental change. – Phil. Trans. Roy. Soc. Lond., A. 352: 259–276.

– and Jonasson, S. 1995b. Implications for changes in arctic plant biodiversity from environmental manipulation experiments. – In: Chapin, F. S., III and Körner, C. (eds), Arctic and alpine biodiversity: patterns, causes and ecosystem consequences. Ecol. Stud. 113: 151–166, Springer, Berlin.

Carlsson, B. Å. and Callaghan, T. V. 1994. Impact of climate change factors on the clonal sedge Carex bigelowii: implications for population growth and vegetative spread. – Ecography 17: 312–330.

Chapin, F. S. III and Shaver, G. R. 1985. Individualistic response of tundra plant species to environmental manipulations in the field. – Ecology 66: 564–576.

– and Shaver, G. R. in press. Physiological and growth responses of arctic plants to a field experiment simulating climatic change. – Ecology.

–, Shaver, G. R., Giblin, A. E., Nadelhoffer, K. G. and Laundre, J. A. 1995. Responses of arctic tundra to experimental and observed changes in climate. – Ecology: 76: 694–711.

Coulson, S., Hodkinson, I. D., Strathdee, A., Bale, J. S., Block, W., Worland, M. R. and Webb, N. R. 1993. Simulated climate change: the interaction between vegetation type and microhabitat temperatures at Ny Ålesund, Svalbard. – Polar Biol. 13: 67–70.

Eno, C. F. 1960. Nitrate production in the field by incubating the soil in polyethylene bags. – Soil Sci. Soc. Am. Proc. 24: 277–279.

Graglia, E. 1995. Effects of shading, nutrient application and warming on leaf parameters and shoot densities in two sub-arctic plant communities. – M. Sc. Diss., Bot. Inst., Univ. of Copenhagen.

Harte, J. and Kinzig, A. P. 1993. Mutualism and competition between plants and decomposers: implications for nutrient allocation in ecosystems. – Am. Nat. 141: 829–846.

Havström, M., Callaghan, T. V. and Jonasson, S. 1993. Differential growth responses of Cassiope tetragona, an arctic dwarf shrub, to environmental perturbations among three contrasting high- and subarctic sites. – Oikos 66: 389–402.

–, Callaghan, T. V. and Jonasson, S. 1995a. Effects of simulated climate change on the sexual reproductive effort of Cassiope tetragona. – In: Callaghan, T. V., Oechel, W. C., Gilmanov, T., Molau, U., Maxwell, B., Tyson, M., Sveinbjörnsson, B. and Holten, J. I. (eds), Global change and arctic terrestrial ecosystems. Ecosystem report 10, EUR 15519 EN. European Comm., Luxenburg, pp. 109–114.

–, Callaghan, T. V., Jonasson, S. and Svoboda, J. 1995b. Little ice age temperature estimated by growth and flowering differences between subfossil and extant shoots of Cassiope tetragona, an arctic heather. – Funct. Ecol. 9: 650–654.

Jonasson, S. 1992. Growth responses to fertilization and species removal in tundra related to community structure and clonality. – Oikos 63: 420–429.

– in press. Buffering of arctic plant responses in a changing climate. – In: Oechel, W. C., Callaghan, T. V., Gilmanov, T., Holten, J. I., Maxwell, B., Molau, U. and Sveinbjörnsson, B. (eds), Global change and arctic terrestrial ecosystems. Proc. Int. Conf., 21–26 Aug. 1993, Oppdal, Norway, Springer.

–, Bryant, J. P., Chapin, F. S. III and Andersson, M. 1986. Plant phenols and nutrients in relation to variations in climate and rodent grazing. – Am. Nat. 128: 394–408.

–, Havström, M., Jensen, M. and Callaghan, T. V. 1993. In situ mineralization of nitrogen and phosphorus of arctic soils after perturbations simulating climate change. – Oecologia 95: 179–186.

–, Michelsen, A., Schmidt, I. K., Nielsen, E. V. and Callaghan, T. V. in press. Microbial biomass C, N, and P in two arctic soils and responses to addition of NPK fertilizer and sugar: Implications for plant nutrient uptake. – Oecologia.

Körner, C. and Larcher, W. 1988. Plant life in cold climates. – In: Long, S. F. and Woodward, F. I. (eds), Plant and temperature. Symp. Soc. Exp. Biol. Vol. 42. The Company of Biologists Limited, Cambridge, pp. 25–57.

Malmer, N. 1962. Studies of mire vegetation in the Archaean area of southwestern Götaland (south Sweden). II. Distribution and seasonal variation in elementary constituents on some mire sites. – Opera Bot. 7: 1–67.

Marion, G. M., Miller, P. C., Kummerow, J. and Oechel, W. C. 1982. Competition for nitrogen in a tussock tundra ecosystem. – Plant and Soil 66: 317–327.

Maxwell, B. 1992. Arctic climate: Potential for change under global warming. – In: Chapin, F. S. III, Jefferies, R. L., Reynolds, J. F., Shaver, G. R. and Svoboda, J. (eds), Arctic ecosystems in a changing climate. An ecophysiological perspective. Academic Press, San Diego, pp. 11–34.

Mitchell, J. F. B., Manabe, S., Meleshko, V. and Tokioka, T. 1990. Equilibrium climate change – and its implications for the future. – In: Houghton, J. T., Jenkins, G. T. and Ephraums, J. J. (eds), Climate change. The IPCC scientific assessment. Cambridge Univ. Press, pp. 131–172.

Michelsen, A., Jonasson, S., Sleep, D., Havström, M. and Callaghan, T. V. 1996. Shoot biomass, $\delta^{13}C$, nitrogen and chlorophyll responses of two arctic dwarf shrubs to in situ shading, nutrient application and warming simulating climatic change. – Oecologia 105: 1–12.

Melillo, J. M., Callaghan, T. V., Woodward, F. I., Salati, E. and Sinha, S. K. 1990. Effects on ecosystems. – In: Houghton, J. T., Jenkins, G. T. and Ephraums, J. J. (eds), Climate change. The IPCC scientific assessment. Cambridge Univ. Press, Cambridge, pp. 282–310.

Nadelhoffer, K. J., Giblin, A. E., Shaver, G. R. and Laundre, J. L. 1991. Effects on temperature and substrate quality on element mineralization in six arctic soils. – Ecology 72: 242–253.

Oechel, W. C. and Billings, W. D. 1992. Effects of global change on the carbon balance of arctic plants and ecosystems. – In: Chapin, F. S. III, Jefferies, R. L., Reynolds, J. F., Shaver, G. R. and Svoboda, J. (eds), Arctic ecosystems in a changing climate. An ecophysiological perspective. Academic Press, San Diego, pp. 139–168.

Parsons, A. N., Welker, J. M., Wookey, P. A., Press, M. C., Callaghan, T. V. and Lee, J. A. 1994. Growth responses of four sub-arctic dwarf shrubs to simulated environmental change. – J. Ecol. 82: 307–318.

–, Wookey, P. A., Welker, J. M., Press, M. C., Callaghan, T. V. and Lee, J. A. 1995. Growth and reproductive output of Calamagrostis lapponica in response to simulated environmental change in the sub-arctic. – Oikos 72: 61–66.

Post, W. M., Emanuel, W. R., Zinke, P. J. and Stangenberger, A. G. 1982. Soil carbon pools and world life zones. – Nature 298: 156–159.

Potter, J. A., Press, M. C., Callaghan, T. V. and Lee, J. A. 1995. Growth responses of Polytrichum commune (Hedw.) and Hylocomium splendens (Hedw.) Br. Eur. to simulated environmental change in the Subarctic. – New Phytol. 131: 533–541.

Robinson, C. H., Wookey, P. A., Parsons, A. N., Potter, J. A., Callaghan, T. V., Lee, J. A., Press, M. C. and Welker, J. M. 1995. Responses of plant litter decomposition, soil inorganic nutrient concentrations and nitrogen mineralization to simulated environmental change in a high arctic polar semi desert and subarctic dwarf shrub heath. – Oikos 74: 503–512.

Shaver, G. R. and Chapin, F. S. III, 1980. Response to fertilization by various plant growth forms in an Alaskan tundra: nutrient accumulation and growth. – Ecology 61: 662–675.

–, Chapin, F. S. III and Gartner, B. L. 1986. Factors limiting seasonal growth and peak biomass accumulation in *Eriophorum vaginatum* in an Alaskan tussock tundra. – J. Ecol. 74: 257–278.

Strathdee, A. T., Bale, J. S., Strathdee, F. C., Block, W. C., Coulson, S. J., Webb, N. R. and Hodkinson, I. D. 1995. Climate severity and the response to temperature elevation of Arctic aphids. – Global Change Biol. 1: 23–28.

Tenow, O. and Holmgren, B. 1987. Low winter temperatures and an outbreak of *Epirrita autumnata* along a valley of Finnmarksvidda, the "cold pole" of northern Fennoscandia. – In: Axelsson, H. and Holmgren, B. (eds), Climatological extremes in the mountain, physical background, geomorphological and ecological consequences. UNGI Rapport 15: 203–216, Dept of Physical Geography, Univ. of Uppsala, Sweden.

Tissue, D. T. and Oechel, W. C. 1987. Response of *Eriophorum vaginatum* to elevated CO_2 and temperature in the Alaskan tussock tundra. – Ecology 68: 401–410.

Werkman, B. R. and Callaghan, T. V. 1995. Bracken growth in a changing environment. – In: Smith, R. T. and Taylor, J. A. (eds), Bracken: an environmental issue. Proc. Third Int. Bracken Conf. 18–21 July 1994, pp. 90–93.

– , Callaghan, T. V. and Welker, J. M. 1996. Responses of bracken to increased temperature and nitrogen availability – Global Change Biol. 2: 59–66.

Wookey, P. A., Parsons, A. N., Welker, J. M., Potter, J. A., Callaghan, T. V., Lee, J. A. and Press, M. C. 1993. Comparative responses of phenology and reproductive development to simulated environmental change in sub-arctic and high arctic plants. – Oikos 67: 490–502.

– , Welker, J. M., Parsons, A. N., Press, M. C., Callaghan, T. V. and Lee, J. A. 1994. Differential growth, allocation and photosynthetic responses of *Polygonum viviparum* L. to simulated environmental change at a high arctic polar semidesert. – Oikos 70: 131–139.

– , Robinson, C. H., Parsons, A. N., Welker, J. M., Press, M. C., Callaghan, T. V. and Lee, J. A. 1995. Environmental constraints on the growth, photosynthesis and reproductive development of *Dryas octopetala* at a high arctic polar semi-desert, Svalbard. – Oecologia 102: 478–489.

Ecological Bulletins 45: 192–203. Copenhagen 1996

Effects of enhanced ultraviolet-B radiation on terrestrial subarctic ecosystems and implications for interactions with increased atmospheric CO_2

C. Gehrke, U. Johanson, D. Gwynn-Jones, L. O. Björn, T. V. Callaghan and J. A. Lee

Gehrke, C., Johanson, U., Gwynn-Jones, D., Björn, L. O., Callaghan, T. V. and Lee, J. A. 1996. Effects of enhanced ultraviolet-B radiation on terrestrial subarctic ecosystems and implications for interactions with increased atmospheric CO_2. – Ecol. Bull. 45: 192–203.

Two predominating types of ecosystems in the Subarctic were exposed to simulated environmental perturbations. A heathland ecosystem was exposed to enhanced UV-B (corresponding to 15% ozone depletion) combined with either increased CO_2 (600 ppm) or additional watering. An ombrotrophic peatland ecosystem was exposed to only enhanced UV-B. Responses both at a plant species level, including different growth forms and life strategies, and at a trophic level (decomposition of organic matter) were studied.

There were differences both in the magnitude and direction of plant responses to enhanced UV-B. The four dwarf shrub species in the heathland developed shorter stems, though not at a significant level in the two deciduous species. The leaves of the evergreen, thick-leaved *V. vitis-idaea* grew thicker under enhanced UV-B, while leaves of the two deciduous species *V. myrtillus* and *V. uliginosum* grew thinner. The heathland moss *H. splendens* showed reduced growth after two and three years under enhanced UV-B but when water was applied simultaneously growth was stimulated by enhanced UV-B. The peat moss *S. fuscum* had 20% less height increment during the first growing season under enhanced UV-B.

Mosses tended to respond quicker to a change in UV-B regime than long-lived dwarf shrubs did. They responded in growth and phenological development already after a few weeks of treatment.

Enhanced UV-B in the heathland affected decomposition of organic matter. It had direct negative effects on decomposer community function and structure and indirect negative effects on turnover of *V. uliginosum* leaf litter by changing the tissue quality of the litter. This was confirmed by studies in the field with another deciduous dwarf shrub (*V. myrtillus*).

Increased growth due to enhanced CO_2 was recorded in *V. myrtillus* during the first growing season. No change in growth was apparent in any of the dwarf shrubs on a longer-term perspective but the number of flowers and berries were increased in *V. myrtillus*. There was, however, reduced growth in *V. myrtillus* but enhanced growth in the moss *H. splendens* when UV-B and CO_2 were enhanced simultaneously.

C. Gehrke, Dept of Ecology, Plant Ecology, Lund Univ., Ecology Building, S-223 62 Lund, Sweden and Abisko Scientific Research Station, S-981 07 Abisko, Sweden. – U. Johanson and L. O. Björn, Dept of Plant Physiology, Lund Univ., Box 117, S-221 00 Lund, Sweden. – D. Gwynn-Jones and T. V. Callaghan, Sheffield Centre for Arctic Ecology, Univ. of Sheffield, Tapton Gardens, 26 Taptonville Rd, Sheffield, U.K. S10 5BR. – J. A. Lee, Dept of Animal and Plant Sciences, Univ. of Sheffield, Sheffield, U.K. S10 2TN.

The influence of man on the global climate is of extremely serious and quickly growing international concern. Ultraviolet-B radiation (280–320 nm) which can be harmful to human health and organisms in ecosystems is increasing as a consequence of the depletion of the stratospheric ozone layer (Frederick et al. 1989, Blumthaler and Ambach 1990). Ground level emission of carbon dioxide (CO_2) has both direct effects on plants (Farrar and Williams 1991, Baker and Allen 1994) as well as contributing to global warming (Walsh 1993), which will feed back on terrestrial primary production (Melillo et al. 1993). As increasing atmospheric CO_2 is suggested to be accompanied by surface warming it also gives rise to higher evaporation rates and a change in precipitation. In fact, during the twentieth century an increase in precipitation has already been recorded for the northern extratropics (Walsh 1993).

Ever since the first warnings about a decrease in the stratospheric ozone layer (Farman et al. 1985), the globally most pronounced depletion has been recorded over the Antarctic (Hofmann and Deshler 1991, Waters et al. 1993). Even above the northern hemisphere a steady thinning of the ozone layer has been recorded during the last years (Proffitt et al. 1990, Hofmann and Deshler 1991, Brune et al. 1991, Gleason et al. 1993) with an 1% annual increase in UV-B radiation (Blumthaler and Ambach 1990, Stolarski et al. 1992). High latitude ecosystems are adapted to a low UV-B regime and thus expected to be exceptionally susceptible to UV-B enhancement (Caldwell et al. 1980, Robberecht et al. 1980).

The level of atmospheric CO_2 has increased from preindustrial 280 ppm to c. 350 ppm. It is predicted that it will reach 460–560 ppm by the year 2100 (Watson et al. 1990). Over the northern hemisphere atmospheric CO_2 increases with latitude so that high latitude ecosystems are exposed to a slightly higher CO_2 regime (Goreau 1990). However, an enhancement may not directly affect plants at high latitudes (Tissue and Oechel 1987) as productivity is nutrient limited due to the often infertile soils. On the other hand, a CO_2 enhancement may have an indirect fertilizing effect as CO_2-induced global warming may increase nutrient availability and thus stimulate productivity.

During the last decades effects of enhanced UV-B or CO_2 applied separately on plants have been studied intensively. We know about stimulative effects of CO_2 upon physiology and growth in a short-term perspective (Strain 1992, Oechel et al. 1994) and about both negative and positive responses to enhanced UV-B radiation, differing between plant families, but even among species within the same family and between varieties. Enhanced UV-B may strike at targets which are not affected by enhanced CO_2 or vice versa. In fact, UV-B radiation is likely to be an oxidative stress factor on a cellular level (Larson 1988, Panagopoulus et al. 1990) while CO_2 is not effective on that level. On the other hand, UV-B radiation and CO_2 can also strike at the same targets in plants (photosynthesis, growth, phenology), thus leading to additive or synergistic interactive effects.

For a few years plant responses to enhanced UV-B in an increased CO_2 environment have been studied. It has been found that a beneficial effect of enhanced CO_2 upon growth or photosynthesis can be reduced or even eliminated if enhanced UV-B radiation is applied simultaneously (Teramura et al. 1990, Hoddinott and Yakimchuk 1991, Stewart and Hoddinott 1993, Yakimchuk and Hoddinott 1994). Teramura et al. (1990) found that a CO_2-induced increase in growth of soybean was maintained and even further increased when CO_2 was applied in combination with enhanced UV-B radiation. However, UV-B-induced modifications of CO_2 responses appear to be rather interactive than additive (Teramura et al. 1990) and species or cultivar specific.

All studies on interactive effects published so far have been carried out under artificial background light conditions (greenhouse or phytotron experiments). This is of particular concern as it was shown by Takayanagi et al. (1994) that plants grown in their natural UV-B environment are less sensitive to UV-B induced damage than would be predicted from studies based on plants grown under unnaturally weak UV-B. Also the ratio between photorepairing wavelengths (UV-A, 320–400 nm, and blue light) and damaging wavelengths (UV-B) is known to influence the magnitude and even direction of UV-B effects (Teramura et al. 1980, Cen and Bornman 1990). This emphasizes the need for experiments under natural light conditions, i.e. outdoor experiments.

So far studies on the combined effects of UV-B radiation and elevated CO_2 have been concentrated on the plant species level (Rozema et al. 1990, Ziska and Teramura 1992, van de Staaij et al. 1993, Stewart and Hoddinott 1993, Sullivan and Terramura 1994). None to date has looked at how these variables influence natural ecosystems. Effects on the species level may have impacts on the interaction between plant species (e.g. via changes in timing of phenological development, reproduction, competition), leading to changes in the plant canopy structure. This may influence interactions between trophic levels (decomposition, herbivory etc.) in a long-term perspective (Fig. 1). In order to estimate UV-B and CO_2 impacts on whole ecosystem structure and function, experiments on natural vegetation systems are necessary.

Another motivation for field experiments is the fact that other environmental factors like water, nutrient or temperature stress can vary between years and may mask or reinforce single or interactive UV-B/CO_2 responses in plants as was shown for just UV-B responses (Murali and Teramura 1986, Pang and Hays 1991). Although CO_2 responses in plants tend to show up already in short-term experiments, it was shown that the magnitude and even direction of such responses can change in a long-term perspective (Tissue and Oechel 1987). If UV-B responses are cumulative it can take years until the first responses to UV-B radiation become visible (Sullivan and Teramura 1992, Johanson et al. 1995b). This is especially important in ecosystems characterized by low productivity and mainly clonal propagation like subarctic and arctic ecosystems.

Even though c. 300 plant species or cultivars have been tested for their UV-B sensitivity the list of life forms investigated for either single or interactive UV-B/CO_2 responses mainly consists of economically important crops, herbs and coniferous trees. Studies of effects on long-lived clonal perennials and cryptogams which dominate high latitude ecosystems are missing. The first practical approaches to investigate responses on trophic levels (nutrient cycling, herbivory etc.) to seperately applied UV-B and CO_2 were started recently, showing effects upon the secondary metabolism in plants and thereby litter quality (McCloud and Berenbaum 1993). In some cases herbivory or litter decomposition were subsequently affected (Coûteaux et al. 1991, Lincoln 1993, Lambers 1993) as secondary compounds function in a range of trophic levels, including allelopathy and deterrence of herbivores. Possible interactive effects of UV-B and CO_2 on trophic levels in an ecosystem are unknown. How changes at different trophic levels may influence each other is a matter of pure speculation (SCOPE 1993) which emphasizes the need for long-term field experiments.

The Arctic and Subarctic are dominated by natural terrestrial ecosystems (7 million km^2) which play an important role in global energy balance and sequestering atmospheric carbon. Thus any changes in climate factors can have a vast effect on ecosystem function and structure at high latitudes as damage cannot be prevented by breeding and cultivating insensitive species. Also feedback effects on the global climate are likely to occur.

Arctic and subarctic ecosystems are supposed to be extraordinarily vulnerable to disturbances as they oper-

ate at a temperature and nutrient availability limit as discussed elsewhere in this volume (Jonasson et al. 1996) and because of their low biodiversity, short food chains and slow adaptation capacity due to slow growth, longevity and infrequent sexual reproduction. Thus global climate changes might most easily be recognized in high latitude ecosystems using plant responses as an indicator.

So far only one investigation on CO_2 effects on a high latitude ecosystem has been done (Tissue and Oechel 1987). Here we present the first field-study approaches where enhanced UV-B radiation is either applied solely or combined with increased atmospheric CO_2 at subarctic ecosystems. We investigated two different ecosystem types (a heathland and an ombrotrophic peat bog) at two trophic levels (primary production and decomposition) for their response in ecologically important parameters (plant performance, phenology, species distribution and composition).

Field experiments at Abisko

In three experiments different life forms (mosses and vascular plants) with a range of life strategies (deciduous, evergreen, poikilohydric, homoihydric) are compared in their sensitivity to enhanced UV-B radiation or/and enhanced CO_2. After the first responses of growth, plant performance, phenological development, secondary metabolism and litter turnover were identified (see below) it is now neccessary to study their impacts on each other in a long-term perspective.

Material and methods

Experiment 1: Effects of UV-B on a heathland ecosystem

The field site and experimental design (Fig. 2)
In 1991 the first UV-B field experiment on naturally growing plants worldwide was established in a subarctic heath close to Abisko Scientific Research Station, 68.35°N, 18.82°E. Four plots in a natural heath ecosystem were irradiated with additional UV-B simulating 15% ozone depletion under clear sky conditions. They were compared to four plots which were just exposed to ambient UV-B (Johanson et al. 1995a).

The plots were situated in a birch forest community (*Betula pubescens* Ehrh. ssp. *tortuosa* Ledeb. Nyman) corresponding to the "*Empetrum-Vaccinium myrtillus* variant" described by Sonesson and Lundberg (1974). The heathland contains many life forms from lichens (*Peltigera apthosa* L. Willd., *Nephroma articum* L. Torss.) and mosses (*Hylocomium splendens* Hedw. B.S.G., *Polytrichum commune* L. ex Hedw.) to decidu-

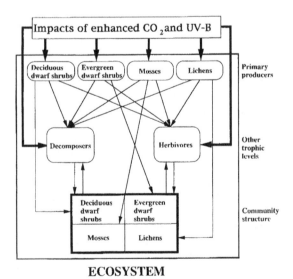

Fig. 1. Possible impacts of UV-B and CO_2 enhancement at an ecosystem level. Both direct (→) and indirect (→) effects are shown with particular emphasis on interactions between trophic levels.

Table 1. Generalized overview of single and interactive effects of enhanced UV-B and CO_2 on growth and decomposition in subarctic ecosystems (L = length increment of shoot, T = thickness of leaves, − = effect showing a decrease, + = effect showing an increase, 0 = no effect, nm = not measured)

Life form or process	Species	Response to UV-B	Response to CO_2	Response to UV-B×CO_2
Evergreen dwarf shrub	*Vaccinium vitis-idaea*	L−, T+	nm	nm
Evergreen dwarf shrub	*Empetrum hermaphroditum*	L−	0	0
Deciduous dwarf shrub	*Vaccinium myrtillus*	L0, T−	L+	L−
Deciduous dwarf shrub	*Vaccinium uliginosum*	L0, T−	nm	nm
Moss	*Hylocomium splendens*	L−	L+	L+
Moss	*Sphagnum fuscum*	L−	nm	nm
Litter turnover	*Vaccinium myrtillus*	−	nm	nm

ous (*Vaccinium myrtillus* L., *V. uliginosum* L.) and evergreen dwarf shrubs (*V. vitis-idaea* L., *Empetrum hermaphroditum* Hagerup).

UV-B treatment

Above each plot a metal frame was erected (2.5×1.3 m, 1.5 m high) holding six parallelly mounted UV-B fluorescent tubes (Q-PANEL UVB 313, Cleveland, OH, USA). In the UV-B treatment ecologically irrelevant shortwave UV radiation (UV-C, < 280 nm) emitted by the tubes was cut off by cellulose diacetate filters (0.13 mm, Courtaulds, Derby, U.K.) placed on UV-transmitting plexiglas (Röhm GmbH, Darmstadt, Germany) under each tube. In the control wavelengths shorter than 320 nm (UV-B and UV-C) were excluded by means of 4 mm thick window glass filters. The cellulose diacetate was pre-solarized until it reached a UV-B transmittance that was stable for 50 h irradiation from the tubes. The tubes were pre-burned for 100 h until their output was stable. During use in the field their output was checked regularly with a spectroradiometer (Optronics 742, Orlando, FL, USA).

The extra UV-B exposure required to simulate 15% ozone depletion at Abisko was calculated for clear sky conditions, no aerosols and 50% air humidity with a computer model (Björn and Murphy 1985, Björn and Teramura 1993) using a modification of Caldwell's generalized plant action spectrum (Caldwell 1971), normalized at 300 nm. The ambient UV-B_{BE} (max. 4.6 kJ m^{-2} d^{-1}) and the UV-B_{BE} simulating 15% ozone depletion (max. 5.8 kJ m^{-2} d^{-1}) calculated by the computer model were confirmed by measuring the solar UV-B and the UV-B from the tubes on clear days each season. Taking cloud cover for the year 1994 into account, the extra UV-B radiation given corresponded to 19% ozone depletion as described elsewhere in this volume (Björn and Holmgren 1996).

The irradiation started around snowmelt, usually in late April, and continued until mid September each year. Every second week the irradiation time was adjusted to follow the seasonal course of natural UV-B radiation. The daily UV-B radiation from the tubes was centered around noon and controlled by timers. Every second lamp was switched on and off at a time to allow

a stepwise increase and decrease of the daily UV-exposure. For more detailed information see Johanson et al. (1995a).

Experiment 2: Effects of UV-B, CO_2 and H_2O on a heathland ecosystem

The field site and experimental design (Fig. 2)

In 1993 a new project was initiated looking at the interactive effects of enhanced UV-B with other variables of climate change. The heathland was exposed to enhanced UV-B simulating 15% ozone depletion under clear sky conditions (corresponding to 19% under average cloudy conditions at Abisko during 1994) whilst simultaneously being exposed to increased atmospheric carbon dioxide (600 ppm) or increased precipitation (420 mm per annum compared to the average 320 mm per annum). Each treatment combination (±UV/± CO_2 and ±UV/±water) was applied to four randomly chosen plots (2.5×1.3 m).

The UV-B treatment was done as previously described for experiment.

CO_2 treatment

One open top chamber (0.85 m in diameter, 0.5 m in height, 0.73 m^2 area and 0.36 m^3 volume) was placed under each of the sixteen frames. The chambers were constructed from UV-transmitting plexiglass (see above) to minimise absorption of natural radiation. Atmospheric air was blown into each chamber from mixing boxes housing fans. The air entered the chamber through a plenum with numerous holes ensuring even distribution of CO_2 with c. 5–7 air changes min^{-1}. In CO_2 enriched chambers (600 ppm) pure CO_2 was trickled into the air supply via flow controllers. The CO_2 concentration within the chambers was monitored continuously at the top of the vegetation canopy using an IRGA linked to a data logger. The elevated level was 600 ppm ± 7% and the ambient level was 360 ppm ± 2%. Sensors connected to the datalogger were placed within the chambers to measure photosynthetically active radiation, air temperature (thermocouples) and relative humidity. The mean daily air temperature within the chambers was only 0.93°C and 1.08°C higher than outside during the seasons 1993 and 1994, respectively.

The experimental site was established during August 1992 and exposure to the respective treatments initiated during snow melt, May 1993. The treatments were maintained during the growing season and stopped in early September. The same sequence of events was adopted during the 1994 growing season.

Water treatment
Each season the mean annual precipitation of 320 mm was enhanced with a total of 100 mm, applied as tap water in 0.3 l doses m^{-2} three times each week during the growing season.

Experiment 3: Effects of UV-B on a peatland ecosystem

The field site and experimental design (Fig. 3)
Since 1994 five plots in an ombrotrophic peat bog ecosystem, dominated by *Sphagnum fuscum* (Schimp.) Klinggr. and including scattered herbs (*Rubus chamaemorus* L., *Drosera rotundifolia* L.) and dwarf shrubs (*E. hermaphroditum* Hagerup, *Vaccinium microcarpum* (Rupr.) Schmalh.), were exposed to extra UV-B radiation simulating 15% ozone depletion under clear sky conditions. Another five plots received only ambient UV-B.

Fig. 2. The experimental design used in the heath ecosystem: UV-B tubes for simulating 15% ozone depletion are suspended above each plot (experiment 1). One half of each plot is used for additional water treatment and the other half contains an open top chamber for enhanced CO$_2$ treatment (experiment 2). Photo: C. Gehrke.

UV-B treatment
A frame was erected at 25 cm height above each plot holding four UV-B tubes (Philips TL/4W) arranged as the sides of a square. The evenly irradiated plot area used for data sampling was 10 × 10 cm. In the UV-B treatment ecologically irrelevant short-wave UV radiation (UV-C) emitted by the tubes was excluded by cellulose diacetate filters (pre-solarized and changed after 30 irradiation hours in the field) wrapped around each tube. In the controls, wavelengths shorter than 318 nm were cut off by a Mylar film wrapped around each tube. The extra UV-B dose required to simulate 15% ozone depletion under clear sky conditions at Abisko during the growing season was calculated by a computer model following the procedure and settings as described for experiment 1. Taking cloud cover for the year 1994 into account, the extra UV-B radiation given corresponded to 19% ozone depletion (Björn and Holmgren 1996). Every second week the irradiation time was adjusted to follow the seasonal course of natural UV-B radiation. The timer-controlled daily UV-B irradiation from the tubes was centered around noon.

Results and dicussion

Effects of enhanced UV-B on mosses

Cryptogam species form an increasing part of the vegetation with increasing latitude. At subarctic latitudes, especially bryophytes can dominate the understorey of heathlands, thus creating a specific temperature and humidity microclimate for the dwarf shrub canopy and the underlying root and soil system. Further north they insulate permafrost ground from heat generated by solar radiation. They can also predominate whole ecosystems like swamps, bogs or fens. Boreal regions contain the largest areas of peatland worldwide (346 × 10^6 ha, Gore 1983), accumulating carbon as organic matter which amounts to one third of the total world pool of soil carbon (Gorham 1991). Peatlands sequester the major greenhouse gas, atmospheric CO$_2$, in photosynthetically produced organic matter, stored in deep peat layers. They have a huge water holding capacity and thereby stabilize the local water level and serve as water supply for surrounding ecosystems. Thus any change in shoot density, morphology, productivity and peat decomposition induced by changing environmental factors may have a vast effect on the water budget of the surrounding biosphere, on carbon sequestration from the atmosphere and on CO$_2$ and methane release to the atmosphere with feedback effects on a larger geographical scale.

Mosses are expected to be extraordinarily susceptible to UV-B radiation because firstly they have rather thin leaves without protection of a cuticle and secondly they show a rather homogeneous leaf anatomy which does not allow a plastic response of one type of cell layer to protect the underlying cells from environmental perturbation. As many mosses are not able to control their water content the periods of highest UV-B radiation overlap or even coincide with drought, i.e. physiological inactivity, leaving sensitive targets such as cell organells, nuclei or DNA exposed to UV-B radiation without enabling instant repair mechanisms like photoreactivation.

To the best of our knowledge there is just one investigation about UV-B effects on mosses published. Markham et al. (1990) showed a correlation between the amount of UV absorbing flavonoids in herbarium specimens of the moss *Bryum argenteum* collected from Antarctica over the period 1957–1989 and the levels of stratospheric ozone for the same period.

In experiment 2 the moss *H. splendens*, which dominates in the bottom layer of the investigated dwarf shrub heath, was exposed to enhanced UV-B radiation corresponding to 15% ozone depletion combined with either natural precipitation or additional watering. Due to shortage of material the only destructive measurements done were pigment analyses (chlorophylls, carotenoids and UV-screening pigments). After two years of exposure no change was detectable.

Non-destructive observations showed that enhanced UV-B alone had no effect upon length increment after the first year of exposure, but after the second and third growing season length increment was reduced by 25% and 18%, respectively. When enhanced UV-B was applied in combination with additional water stimulative effects of UV-B upon annual length increment were obvious after all three growing seasons (15%, 31% and 27%, respectively) (Gehrke unpubl.). Such stimulative effects are also evident from a parallel greenhouse experiment where the same species was exposed to four different UV-B levels under optimal water conditions (Gehrke et al. unpubl.).

Stimulating responses to UV-B radiation on a few vascular plant species are mentioned in the literature (Tosserams and Rozema 1995) but hardly discussed. They may be induced by experimental designs where an unnatural level and quality of background light is used but this does not apply to this field study. Stimulative UV-B effects might be short-term responses, forcing stressed plants to exhaust themselves in growth or reproduction in order to survive genetically. If positive effects show up even in a long-term perspective, the mechanisms behind a truly stimulative UV-B effect need to be investigated.

In experiment 3 an ombrotrophic peat bog mainly consisting of a dense mat of *S. fuscum* was exposed to enhanced UV-B radiation corresponding to 15% ozone depletion under clear sky conditions. Before the experiment shoot density and biomass production were investigated, parameters that also will be studied at the end

of the experiment in order to estimate any changes in productivity due to enhanced UV-B radiation. Already during the first growing season enhanced UV-B radiation reduced the height increment of the *S. fuscum* layer, measured with the cranked-wire-method (Clymo 1970), by 20% (Gehrke unpubl.). A reduction in height increment may result from a change in biomass partitioning between stem and capitulum in individual shoots. It may affect shoot density and thus the amount of atmospheric carbon dioxide sequestered as phytomass. A thinning of the green phytomass layer may lead to more insolation from solar irradiation penetrating into the dead peat layer underneath and the resulting increase in temperature might stimulate peat decompostion and thus the release of carbon dioxide or methane, depending on the position of the water table.

Effects of enhanced UV-B on dwarf shrubs

The dwarf shrub canopy of the heath ecosystem consists mainly of four species, *V. myrtillus* and *V. uliginosum* (deciduous), *V. vitis-idaea* and *E. hermaphroditum* (evergreen). These species cover >60% of the area investigated. All four species are long-lived clonal plants with only rare occurrence of seedling establishment by sexual reproduction. Seedlings are found very infrequently and they occur usually only on disturbed grounds. Any major effect of enhanced UV-B should therefore be on growth patterns and changes in morphology, rather than on sexual reproduction.

There might be differences in susceptibility between the different life strategies due to differences in leaf performance. The deciduous leaves are thinner and have a less waxy cuticle compared to the evergreen leaves which are leathery and shiny. This may makes deciduous leaves more susceptible to enhanced UV-B due to higher epidermal transmittance and depth of penetration of UV-B. More radiation can therefore penetrate into sensitive parts of the leaf such as chloroplasts and DNA. In fact, the mean epidermal transmittance and depth of penetration of UV-B were 28% and 75 µm for deciduous species and only 4% and 32 µm for evergreen species (Day 1993). This might indicate that evergreen species have a stronger defence against emergence of UV-B lesions in the leaves. One of the advantages of keeping the leaves is that evergreen species can start photosynthesizing earlier in the spring, while the deciduous plants have to rely on stored assimilates in the initial stages of shoot sprouting (Karlsson 1985a, b). *Vaccinium vitis-idaea* starts capturing CO_2 directly after snow-melt (Karlsson 1989). Even *V. myrtillus* can achieve some photosynthesis early in the season due to the green stem, but not to the same extent as *V. vitis-idaea*. Evergreen species therefore have an advantage early in the season over the deciduous species which cannot start photosyntesizing to any greater extent until the new leaves have broken their buds and unfolded. The same is true at the end of the season when the deciduous species shed their leaves but the evergreens can continue for a few more weeks until the temperature drops below a critical level. On the other hand the evergreens might suffer from accumulated damage while they keep their leaves from one season to another. They are also exposed during a longer period each season to enhanced UV-B. Even if the damage is negligable after one season it might turn out to be important after several seasons of enhanced UV-B.

Since the start of the irradiation in experiment 1, a range of parameters have been measured on the dwarf shrubs, including phenology, photosynthesis, relative growth, leaf thickness, chemistry of leaves and composition of species.

Until 1994 no change in phenological development has been found. When an acceleration of development was found by earlier investigators, it was in species with completely different life strategies (Staxén and Bornman 1994, Nikolopoulos et al. 1995, Holmes pers. comm.).

The thickness of leaves was measured throughout the seasons of 1992 and 1993 using a caliper. A change in thickness due to UV-B enhancement (corresponding to 15% ozone depletion) was found in all three species investigated. The thickness increased in new leaves of the already thick-leaved evergreen *V. vitis-idaea* (from 353 to 385 µm in 1992 and from 390 to 407 µm in 1993) but decreased in the two thin-leaved deciduous species *V. myrtillus* (from 193 to 185 µm in 1992 and from 238 to 216 µm in 1993) and *V. uliginosum* (from 286 to 268 µm in 1992 and from 316 to 283 µm in 1993), due to enhanced UV-B (Johanson et al. 1995b).

After one year there was no reduction in relative growth (the growth of current year divided by the growth of 1990, the year before the treatment started) for the two deciduous species and *V. vitis-idaea*, but a 14% reduction for *E. hermaphroditum* was found. After two years of treatment, plants under enhanced UV-B had a lower relative growth than plants under ambient UV-B. This was less pronounced in the deciduous species (11% reduction for *V. myrtillus* and 10% reduction for *V. uliginosum*) than in the evergreen species (27% reduction for *V. vitis-idaea* and 33% reduction for *E. hermaphroditum*) which may accumulate UV-B damage over longer periods (Johanson et al. 1995b). For the evergreens the effect increases with prolonged exposure time, implying that some responses might take a few years to develop and to become visible.

These differences in responses to UV-B in different life strategies might lead to a shift in competitive balance and species composition with further effects on herbivory, litter quantity and quality, decomposing microorganisms and litter turnover.

Effects of enhanced UV-B on decomposition

One limiting factor for plant growth, particularly in high latitude ecosystems, is the availability of nutrients in the soil. Nutrient availability depends on the rate of litter turnover which is water and temperature dependent. In arctic and subarctic ecosystems the low nutrient availability, together with low temperature, limit plant productivity compared to productivity at lower latitudes. UV-B may further slow down litter turnover, both by damaging UV-B intercepting microbial organisms active in the decomposition process and by changing the chemistry of the leaves so that they become less attractive to microorganisms involved in decomposition. An increase in UV-B could, on the other hand, speed up photodegradation of compounds such as lignins (Moorhead and Callaghan 1994). The anticipated greenhouse effect would also increase the litter turnover rate by means of increasing the soil temperature, but it can also lead to decreased turnover in drier soils due to higher evaporation rates.

An experiment was conducted where the indirect and direct effects of enhanced UV-B on decomposition were investigated. Indirect effects were studied by collecting *V. uliginosum* leaves, grown under either ambient UV-B or enhanced UV-B (corresponding to 15% ozone depletion) from experiment 1, and allowing them to decompose without UV-B for 62 days in a growth chamber. It was found that the relative mass loss was lower in leaf litter that was grown under enhanced UV-B (27.9% compared with 33.5% for leaf litter grown under ambient UV-B). CO_2-release from the microorganisms was sampled at different intervals and at the first sampling occasion (after 8 days) it was 35% lower from the microorganisms decomposing leaves grown under enhanced UV-B. These findings are probably due to a change in leaf litter chemistry, since it was found that tannins had increased and cellulose had decreased (Gehrke et al. 1995).

Direct effects of UV-B on decomposition were studied by collecting leaf litter from *V. uliginosum* grown under ambient UV-B and allowing them to decompose under no UV and under high UV-B (10 kJ m^{-2} d^{-1} biologically effective UV-B (UV-B$_{BE}$)) for 62 days in a growth chamber. Between day 21 and 62 CO_2-release decreased significantly (13%) from the microorganisms decomposing under high UV-B. The proportions of lignins decreased from 49.5 to 45.5% of dry weight and cellulose decreased from 33.5 to 29.6% of dry weight when decomposed under UV-B (Gehrke et al. 1995). This could be due to photodegradation.

Growth and colonization of three fungal species were studied on leaves decomposed with (10 kJ m^{-2} d^{-1}, UV-B$_{BE}$) and without UV-B. Two of the species, *Truncatella truncata* and *Mucor hiemalis*, showed significantly impaired colonization rate under enhanced UV-B radiation. *Mucor hiemalis* showed only 35% of the colonization rate of the control, while *T. truncata* did not colonize at all under high UV-B. The third species, *Penicillium brevicompactum*, was tolerant to UV-B (Gehrke et al. 1995).

The negative indirect effect of enhanced UV-B on litter turnover was also shown under field conditions. Leaf litter of *V. myrtillus* from the different treatments in experiment 1 (ambient UV-B and UV-B corresponding to 15% ozone depletion) was allowed to decompose in litter cups placed outside any treatment plot. After 10, 12 and 24 months of decomposition the litter turnover was significantly lower (5, 9 and 11%, respectively) in the leaves that were grown under enhanced UV-B compared to the controls (Gehrke et al. 1995).

In summary, UV-B affects plant chemistry as well as microbial activity and fungal colonization. All these three factors together will slow down an already low decomposition rate at high latitudes. On the other hand there is some evidence that enhanced UV-B might increase the decomposition rate by direct photodegradation of certain compounds in those ecosystems where litter intercepts solar radiation. This together with the increased global warming might counteract the negative effects on litter turnover.

Interactive effects of enhanced UV-B and CO$_2$ on cryptogams

The responses of cryptogams to enhanced UV-B and elevated concentrations of CO$_2$ may be different from those in higher plants. Cryptogams occupy a niche in the understorey of higher plants where the levels of PAR and UV-B can be low and concentrations of CO$_2$ high (Sonesson et al. 1992). Most of our studies on the effects of UV-B and CO$_2$ on cryptogams have been limited to laboratory experiments due to the limitation of harvestable material. However long-term field experiments will be initiated during 1995 to investigate the responses of both CO$_2$ and UV-B on bryophyte and lichen species.

In a short-term laboratory experiment on three cryptogam species chlorophyll fluorescence yield (Fv:Fm) was measured following 45 h exposure to enhanced/natural levels of UV-B (enhancement simulating 60% ozone depletion) and elevated/present concentrations of CO$_2$ (1000 and 360 ppm respectively). In all species an increase was observed in fluorescence at enhanced UV-B whilst no effects were observed at elevated CO$_2$ (Sonesson et al. 1995). This implies that there can be a short-term stimulation of the photosynthetic machinery at enhanced UV-B radiation. This phenomenon was also seen in the net photosynthesis of other cryptogam species (Gwynn-Jones and Sonesson unpubl., Gehrke unpubl.). However in an ecological sense it is probably more important to see whether UV-B and CO$_2$ have long-term effects on growth in the field.

The effects of UV-B and CO$_2$ on the moss *H. splendens* have been addressed in the field during 1993 (see experiment 2). Results collected following the first season of exposure to the treatments suggest that there is a UV-B × CO$_2$ interaction whereby a stimulation of length increment in one year old shoots (c + 1) occurs when the moss is exposed to both enhanced CO$_2$ and UV-B.

Preliminary evidence suggests that the concentration of total non structural carbohydrates (TNC) in current and one year old shoots of *H. splendens* increases following one season's exposure to elevated CO$_2$. Such increases are also evident in a short-term experiment where increases of up to 30% in TNC have been reported following exposure for only 96 h at elevated CO$_2$ (Sonesson unpubl.). Increased TNC in the moss *H. splendens* is consistent with that found in higher plants (Farrar and Williams 1991), although more pronounced effects would be anticipated in this species due to its inability to translocate between segments (Callaghan et al. 1978).

Interactive effects of enhanced UV-B and CO$_2$ on dwarf shrubs

Increases in growth (17% increase in stem length and 19% increase in shoot dry weight) were observed in the field in the deciduous dwarf shrub *V. myrtillus* following exposure to elevated CO$_2$ from beginning of June to middle of September in 1993 (see experiment 2), whilst no effects were apparent in the evergreen dwarf shrub *E. hermaphroditum*. Increased growth of *V. myrtillus* was not maintained at elevated CO$_2$ during 1994. Indeed the only response observed was an interaction between CO$_2$ and UV-B which together reduced growth. Similarly there were no effects of CO$_2$ on the growth of *E. hermaphroditum* during the 1994 season.

Measurement of the photosynthetic capacity of *V. myrtillus* during late July 1993 showed a 36% reduction at elevated CO$_2$ suggesting that stimulation of growth must have occurred during the early growing season in 1993. A reduced photosynthetic capacity has been widely reported in the literature and was attributed to limitations in plant sink capacity (Arp 1991). The occurrence of this condition may be anticipated in plants growing in infertile subarctic soils due to nutrient availability limiting sink capacity. Initial increases in photosynthesis at elevated CO$_2$ may result in carbohydrate accumulation (mainly starch) which may result in a down-regulation of photosynthesis.

During 1993 there appeared to be no effects of enhanced UV-B on the amount of UV-B absorbing leaf pigments in neither deciduous nor evergreen dwarf shrubs following enhanced UV-B exposure during 1993. Similarly there were no reductions in growth due to UV-B in both shrub species. The lack of clear cut UV-B growth and chemical responses in the dwarf shrubs compared to the previous experiment (see experiment 1) may be explained by the temperature sensitivity of enzymatic UV-B repair mechanisms. Although the light environment was similar within the chambers the mean daily air temperature was around 1°C higher and could be up to 5°C higher at mid-day. Enzymatic repair mechanisms are dependent on temperature (Takeuchi et al. 1993), a factor which may have more influence in colder subarctic regions.

During both the 1993 and 1994 seasons the number of flowers and berries per unit area produced by three dwarf shrub species was estimated (*V. myrtillus*, *V. vitis-idaea* and *V. uliginosum*). During the first season (1993) no effects were observed as flower primordia were initiated in the previous year i.e. the year before start of treatment. During the second growing season increases in the number of *V. myrtillus* flowers and berries produced were recorded (236 and 250% increase, respectively). No effects of UV-B radiation were apparent on the reproduction of the other dwarf shrub species (*V. uliginosum* and *V. vitis-idaea*), whilst no effects of elevated atmospheric CO_2 or any $CO_2 \times$ UV-B interaction were observed. The stimulation of reproductive growth at enhanced UV-B in *V. myrtillus* may have been due to stimulation of a photoreceptor which may be instrumental in controlling flower bud initiation the year before flowering.

Results collected to date show that both cryptogams and dwarf shrub species are in some ways responsive to elevated CO_2 and enhanced UV-B. It appears that UV-B responses in the field may occur over longer time periods whilst CO_2 responses are more immediate affecting the growth of dwarf shrubs and cryptogams in as little as one season.

Future objectives

Most of our studies to date have looked at the effects of UV-B and CO_2 on the primary producers although we aim to expand our interests by looking at influences on below ground processes. In looking at soil microbial responses and mycorrhizal development we may gain further understanding into the responses of plants to these perturbations. The need to look at the responses of other trophic levels is also a major objective in our research strategy as a whole; these will include studies on decomposition and herbivory within the heathland.

Changes in leaf quality as a result of plant exposure to elevated CO_2 and UV-B may influence the degree of insect herbivory in the subarctic heathland. Hence a project has been initiated to determine the effect on herbivory following exposure of plants to these perturbations. The moth *Epirrita autumnata*, which feeds on both deciduous and evergreen dwarf shrubs, will be

Fig. 3. The design used in the ombrotrophic bog ecosystem, dominated by *Sphagnum fuscum*, (experiment 3): Small UV-B tubes are arranged in a square for simulating 15% ozone depletion. Photo: C. Gehrke.

among the organisms included in the study. Preferential feeding on certain species due to different leaf chemistry may influence both insect populations and plant competition within the heathland. We therefore aim to continue current research in the future with the aim to determine long-term effects of these perturbations on each individual trophic level in an attempt to understand the ecosystem as a whole.

Conclusions

Responses and interactions between different responses to environmental perturbations applied to ecosystems are much more complex than the working hypothesis suggested. As shown in Table 1 there is contradicting evidence for instance in growth responses in dwarf shrubs and mosses due to enhanced UV-B and elevated CO_2. There are differences both in direction of response as well as in magnitude. This indicates that responses to realistic environmental perturbations are subtle and may be masked or reinforced by other environmental changes or short-term stresses such as drought, low summer temperature or herbivory. It emphasizes the need of long-term field experiments where effects of climatic variations between years can be ruled out.

Mosses tended to respond in growth to enhanced UV-B already in the first year of exposure while it took longer time to detect responses in the dwarf shrubs. This might be due to their different morphological features such as very thin leaves in mosses versus thicker leaves in dwarf shrubs. Especially in long-lived tissues like evergreen leaves, conifer needles and stems, UV-B responses caused one year can be preserved and accumulate to changes caused in consecutive years (Sullivan and Teramura 1992).

High latitude ecosystems have low biomass productivity due to low temperatures and slow nutrient cycling, leading to slow growth rates. Thus responses to environmental perturbations take time to develop or become detectable. This was shown for UV-B effects on stem growth of evergreen dwarf shrubs.

Effects of enhanced UV-B on decomposition of organic matter were shown. UV-B has both negative and positive direct effects by decreasing the respiration of microbial decomposers but increasing the degradation of chemical compounds. It has indirect negative effects by changing the quality of the litter and hereby changing decomposer community function and structure. Direct effects of UV-B presume that the litter layer is exposed to solar radiation. This is just the case for ecosystems with a low leaf area index or for patchy habitats enabling sporadic sunflecks. Thus it is most likely that UV effects on litter turnover at a global level will mainly be indirect, meaning that enhanced UV-B will slow down litter turnover and nutrient cycling.

With the increase of CO_2 and other greenhouse gases in the atmosphere, the global mean temperature will increase. This will directly facilitate plant growth and reproductive output as described elsewhere in this volume (Jonasson et al. 1996). It might also speed up the rate of nutrient cycling assuming that a warmer climate will not reduce precipitation or dry out soils due to higher evaporation. A higher decomposition rate might positively feed back on the greenhouse effect by a higher microbial release of CO_2. It might also increase the amount of available nutrients for plants and facilitate plant growth and productivity. Increased plant biomass leads to more photosynthetic fixation of CO_2 and a negative feedback on the greenhouse effect.

Beyond the obvious fertilization effect of elevated atmospheric CO_2 on plants, an increase can lead to a range of feedback effects on climate change. Elevated levels of CO_2 and UV-B can change the plant cover, height and structure of the canopy and thus have impacts on reflectivity and absorption of both UV-B and PAR and on advective and convective transfer of heat. Thus evaporation rates and heat exchange rates between biosphere and atmosphere may change. Higher temperature due to more greenhouse gases can increase evaporation rates and cloud formation, which in turn will decrease the amount of UV-B reaching the biosphere. More greenhouse gases will obstruct infrared radiation reflection up to the stratosphere affecting the temperature at this altitude. A lower stratospheric temperature leads to a more favourable environment for ozone destruction (Austin et al. 1992).

Acknowledgements – M. Sonesson, to whose honour this volume is edited, is a member of our research group. The projects are financed by the Comm. of the European Communities (Contract No. EV5V-CT91-0032), by the Swedish Environmental Protection Board (SNV) and by Astra Draco AB. The frames and electricity supply used in the heathland experiments were constructed by P. Westergren. T. Murath (Timco AB, Lund) built the transformers and rectifier used in the bog experiment. Finally we want to thank the staff at ANS (N. Å. Andersson, K. Ericsson, B. Wanhatalo, A. Eriksson and T. Westin) for logistical and technical support and all the students for helping as field assistants over the years.

References

Arp, W. J. 1991. Effects of source-sink relations on photosynthetic acclimation to elevated CO_2. – Plant Cell Environ. 14: 869–877.

Austin, J., Butchard, N. and Shine, K. P. 1992. Possibility of an Artic ozone hole in a doubled-CO_2 climate. – Nature 360: 221–225.

Baker, J. T. and Allen Jr. L. H. 1994. Assessment of the impact of rising carbon dioxide and other potential climate changes on vegetation. – Environ. Pollut. 83: 223–235.

Björn, L. O. and Murphy, T. M. 1985. Computer calculation of solar ultraviolet radiation at ground level. – Physiol. Vég. 23: 555–561.

– and Teramura, A. H. 1993. Simulation of daylight ultraviolet radiation and effects of ozone depletion. – In: Young, A. R., Björn L. O., Moan, J. and Nultsch, W. (eds), Environmental UV Photobiology. Plenum Press, New York, pp. 41–71.

– and Holmgren, B. 1996. Monitoring and modelling the radiation climate in Abisko. – Ecol. Bull. 45: 204–209.

Blumthaler, M. and Ambach, W. 1990. Indication of increasing solar ultraviolet-B radiation flux in alpine regions. – Science 248: 206–208.

Brune, W. H., Anderson, J. G., Toohey, D. W., Fahey, D. W., Kawa, S. R., Jones, R. L., McKenna, D. S. and Poole, L. R. 1991. The potential for ozone depletion in the arctic polar stratosphere. – Science 252: 1260–1266.

Caldwell, M. M. 1971. Solar UV irradiation and the growth and development of higher plants. – In: Giese, A. C. (ed.), Photophysiology, Vol. 6. Academic Press, New York, pp. 131–177.

–, Robberecht, R. and Billings, W. D. 1980. A steep latitudinal gradient of solar ultraviolet-B radiation in the arctic-alpine life zone. – Ecology 61: 600–611.

Callaghan, T. V., Collins, N. J. and Callaghan, C. H. 1978. Photosynthesis and reproduction of *Hylocomium splendens* and *Polytrichum commune* in Swedish Lapland. – Oikos 31: 73–88.

Cen, Y.-P. and Bornman, J. F. 1990. The response of bean plants to UV-B radiation under different irradiances of background visible light. – J. Exp. Bot. 41: 1489–1495.

Clymo, R. S. 1970. The growth of *Sphagnum*: Methods of measurement. – J. Ecol. 58: 13–49.

Coûteaux, M., Mousseau, M., Célérier, M. and Bottner, P. 1991. Increased atmospheric CO_2 and litter quality: decomposition of sweet chestnut leaf litter with animal food webs of different complexities. – Oikos 61: 54–64.

Day, T. A. 1993. Relating UV-B radiation screening effectiveness of foliage to absorbing-compound concentration and anatomical characteristics in a diverse group. – Oecologia 95: 542–550.

Farman, J. C., Gardiner, B. G. and Shanklin, J. D. 1985. Large losses of total ozone in Antarctica reveal seasonal ClO_x/NO_x interaction. – Nature 315: 207–210.

Farrar, J. F. and Williams, M. L. 1991. The effects of increased atmospheric carbon dioxide and temperature on carbon partitioning, source-sink relations and respiration. – Plant Cell Environ. 14: 819–830.

Frederick, J. E., Snell, H. E. and Haywood, E. K. 1989. Solar ultraviolet radiation at the earth's surface. – Photochem. Photobiol. 50: 443–450.

Gehrke, C., Johanson, U., Callaghan, T. V., Chadwick, D. and Robinson, C. H. 1995. The impact of enhanced ultraviolet-B radiation on litter quality and decomposition processes in *Vaccinium* leaves from the Subarctic. – Oikos 72: 213–222.

Gleason, J. F., Bhartia, P. K., Herman, J. R., McPeters, R., Newman, P., Stolarski, R. S., Flynn, L., Labow, G., Larko, D., Seftor, C., Wellemeyer, C., Komhyr, W. D., Miller, A. J. and Planet, W. 1993. Record low global ozone in 1992. – Science 260: 523–526.

Gore, A. J. P. (ed.) 1983. Ecosystems of the world: Mires: swamp, bog, fen and moor. 4A, General studies. – Elsevier, Amsterdam.

Goreau, T. J. 1990. Balancing atmospheric carbon dioxide. – Ambio 19: 230–236.

Gorham, E. 1991. Northern peatlands: role in the carbon cycle and probable responses to climate warming. – Ecol. Appl. 1: 182–195.

Hoddinott, J. and Yakimchuk, R. 1991. Responses of conifer seedlings to increased CO_2 partial pressures with or without UV-B light. – In: Randall, D. D., Blevins, D. G. and Miles, C. D. (eds), Current topics in plant biochemistry and physiology, Vol. 10. Univ. of Missouri-Columbia.

Hofmann, D. J. and Deshler, T. 1991. Evidence from balloon measurements for chemical depletion of stratospheric ozone in the arctic winter of 1989–90. – Nature 349: 300–305.

Johanson, U., Gehrke, C., Björn, L. O., Callaghan, T. V. and Sonesson, M. 1995a. The effects of enhanced UV-B radiation on a subarctic heath ecosystem. – Ambio 24: 106–111.

–, Gehrke, C., Björn, L. O. and Callaghan, T. V. 1995b. The effects of enhanced UV-B radiation on the growth of dwarf shrubs in a subarctic heathland. – Funct. Ecol. 9: 713–719.

Jonasson, S., Lee, J. A., Callaghan, T. V., Havström, M. and Parsons, A. N. 1996. Direct and indirect effects of increasing temperatures on subarctic ecosystems. – Ecol. Bull. 45: 180–191.

Karlsson, P. S. 1985a. Photosynthetic characteristics and leaf carbon economy of a deciduous and an evergreen dwarf shrub: *Vaccinium uliginosum* L. and *V. vitis-idaea* L. – Holarct. Ecol. 8: 9–17.

– 1985b. Patterns of carbon allocation above ground in a deciduous (*Vaccinium uliginosum*) and an evergreen (*Vaccinium vitis-idaea*) dwarf shrub. – Physiol. Plant. 63: 1–7.

– 1989. In situ photosynthetic performance of four coexisting dwarf shrubs in relation to light in a subarctic woodland. – Funct. Ecol. 3: 481–487.

Lambers, H. 1993. Rising CO_2, secondary plant metabolism, plant-herbivore interactions and litter decomposition. – Vegetatio 104/105: 263–271.

Larson, R. A. 1988. The antioxidants of higher plants. – Phytochemistry 27: 969–978.

Lincoln, D. E. 1993. The influence of plant carbon dioxide and nutrient supply on susceptibility to insect herbivores. – Vegetatio 104/105: 273–280.

Markham, K. R., Franke, A., Given, D. R. and Brownsey, P. 1990. Historical antarctic ozone level trends from herbarium specimen flavonoids. – Bull. Liaison Groupe Polyphenols 15: 230–235.

McCloud, E. S. and Berenbaum, M. R. 1993. Effects of increased UVB radiation on plant-insect interactions: *Plantago lanceolata* and *Junonia coenia*. – Bull. Ecol. Soc. Am. 74 (2 suppl.): 350.

Melillo, J. M., McGuire, A. D., Kicklighter, D. W., Moore III, B., Vorosmarty, C. J. and Schloss, A. L. 1993. Global climate change and terrestrial net primary production. – Nature 363: 234–240.

Moorhead, D. L. and Callaghan, T. V. 1994. Effects of increasing ultraviolet-B radiation on decomposition and soil organic matter dynamics: a synthesis and modelling study. – Biol. Fertil. Soils 18: 19–26.

Murali, N. S. and Teramura, A. H. 1986. Effectiveness of UV-B radiation on the growth and physiology of field-grown soybean modified by water stress. – Photochem. Photobiol. 44: 215–219.

Nikolopoulos, D., Petropoulou, Y., Kyparissis, A. and Manetas, Y. 1995. The effects of enhanced UV-B radiation on the drought semi-deciduous Mediterranean shrub *Phlomis fruticosa* L. under field conditions are season-specific. – Aust. J. Plant Physiol. 22: 737–745.

Oechel, W. C., Cowles, S., Grulke, N., Hastings, S. J., Lawrence, B., Prudhomme, T., Riechers, G., Strain, B., Tissue, D. and Vourlitis, G. 1994. Transient nature of CO_2 fertilization in Arctic tundra. – Nature 371: 500–503.

Panagopoulus, I., Bornman, J. F. and Björn, L. O. 1990. Effects of ultraviolet radiation and visible light on growth, fluorescence induction, ultraweak luminescence and peroxidase activity in sugar beet plants. – J. Photochem. Photobiol. B. Biol. 8: 73–87.

Pang, Q. and Hays, J. B. 1991. UV-B-inducible and temperature sensitive photoreactivation of cyclobutane pyrimidine dimers in *Arabidopsis thaliana*. – Plant Physiol. 95: 536–543.

Proffitt, M. H., Margitan, J. J., Kelly, K. K., Loewenstein, M., Podolske, J. R. and Chan, K. R. 1990. Ozone loss in the Arctic polar vortex inferred from high-altitude aircraft measurements. – Nature 347: 31–36.

Robberecht, R., Caldwell, M. M. and Billings, W. D. 1980. Leaf ultraviolet optical properties along a latitudinal gradient in the arctic-alpine life zone. – Ecology 61: 612–619.

Rozema, J., Lenssen, G. M. and van de Staaij, J. W. M. 1990. The combined effects of increased atmospheric CO_2 and UV-B radiation on some agricultural and salt marsh species. – In: Goudriaan, J., van Keulen, H. and van Laar, H. H. (eds), The greenhouse effect and primary productivity in European agroecosystems. Pudoc, Wageningen, pp. 68–71.

SCOPE (Scientific Committee on Problems of the Environment). 1993. Effects of increased ultraviolet radiation on global ecosystems. – Proc. Tramariglio, Sardinia, October 1992. SCOPE Secretariate, Paris, France.

Sonesson, M. and Lundberg, B. 1974. Late Quaternary forest development in the Torneträsk area, North Sweden. 1. Structure of modern forest ecosystems. – Oikos 65: 121–133.

–, Gehrke, C. and Tjus, M. 1992. CO_2 environment, microclimate and photosynthetic characteristics of the moss *Hylocomium splendens* in a subarctic habitat. – Oecologia 92: 23–29.

–, Callaghan, T. V. and Björn, L. O. 1995. Short-term effects of enhanced UV-B and CO_2 on lichens at different latitudes. – Lichenologist 27: 547–557.

Staxén, I. and Bornman, J. F. 1994. A morphological and cytological study of *Petunia hybrida* exposed to UV-B radiation. – Physiol. Plant. 91: 735–740.

Stewart, J. D. and Hoddinott, J. 1993. Photosynthetic acclimation to elevated atmospheric carbon dioxide and UV irradiation in *Pinus banksiana*. – Physiol. Plant. 88: 493–500.

Stolarski, R., Bojkov, R., Bishop, L., Zerefos, C., Staehelin, J. and Zawodny, J. 1992. Measured trends in stratospheric ozone. – Science 256: 342–349.

Strain, B. R. 1992. Atmospheric carbon dioxide: a plant fertilizer? – New Biologist 4: 87–89.

Sullivan, J. H. and Teramura, A. H. 1992. The effects of ultraviolet-B radiation on loblolly pine 2. Growth of field-grown seedlings. – Trees 6: 115–120.

– and Teramura, A. H. 1994. The effects of UV-B radiation on loblolly pine. 3. Interaction with CO_2 enhancement. – Plant Cell Environ. 17: 311–317.

Takayanagi, S., Trunk, J. G., Sutherland, J. C. and Sutherland, B. M. 1994. Alfalfa seedlings grown outdoors are more resistant to UV-induced DNA damage than plants grown in a UV-free environmental chamber. – Photochem. Photobiol. 60: 363–367.

Takeuchi, Y., Ikeda, S. and Kasahara, H. 1993. Dependence on wavelength and temperature of growth inhibition induced by UV-B irradiation. – Plant Cell Physiol. 34: 913–917.

Teramura, A. H., Biggs, R. H. and Kossuth, S. 1980. Effects of ultraviolet-B irradiances on soybean. II. Interaction between ultraviolet-B and photosynthetically active radiation on net photosynthesis, dark respiration, and transpiration. – Plant Physiol. 65: 483–488.

– , Sullivan J. H. and Ziska, L. H. 1990. Interaction of elevated ultraviolet-B radiation and CO_2 on productivity and photosynthetic characteristics in wheat, rice and soybean. – Plant Physiol. 94: 470–475.

Tissue, D. T. and Oechel, W. C. 1987. Response of *Eriophorum vaginatum* to elevated CO_2 and temperature in the Alaskan tussock tundra. – Ecology 68: 401–410.

Tosserams, M. and Rozema, J. 1995. Effects of ultraviolet-B radiation (UV-B) on growth and physiology of the dune grassland species *Calamagrostis epigeios*. – Environ. Pollut. 89: 209–214.

van de Staaij, J. W. M., Lenssen, G. M., Stroetenga, M. and Rozema, J. 1993. The combined effects of elevated CO_2 levels and UV-B radiation on growth characteristics of *Elymus athericus* (= *E. pycnanathus*). – Vegetatio 104/105: 433–439.

Walsh, J. E. 1993. The elusive arctic warming. – Nature 361: 300–301.

Waters, J. W., Froidevaux, L., Read, W. G., Manney, G. L., Elson, L. S., Flower, D. A., Jarnot, R. F. and Harwood, R. S. 1993. Stratospheric ClO and ozone from the microwave limb sounder on the upper atmosphere research satellite. – Nature 362: 597–602.

Watson, R. T., Rodhe, H., Oeschger, H. and Siegenthaler, U. 1990. Greenhouse gases and aerosols. – In: Houghton, J. T., Jenkins, G. J. and Ephraums, J. J. (eds), Climate change: The IPCC scientific assessment. Cambridge Univ. Press, Cambridge, pp. 1–40.

Yakimchuk, R. and Hoddinott, J. 1994. The influence of ultraviolet-B light and carbon dioxide enrichment on the growth and physiology of seedlings of three conifer species. – Can. J. For. Res. 24: 1–8.

Ziska, L. H. and Teramura, A. H. 1992. CO_2 enhancement of growth and photosynthesis in rice (*Oryza sativa*). Modification by increased ultraviolet-B radiation. – Plant Physiol. 99: 473–481.

Ecological Bulletins 45: 204–209. Copenhagen 1996

Monitoring and modelling of the radiation climate at Abisko

Lars Olof Björn and Björn Holmgren

Björn, L. O. and Holmgren, B. 1996. Monitoring and modelling of the radiation climate at Abisko. – Ecol. Bull. 45: 204–209.

Photosynthetically active radiation (PAR), ultraviolet-B (UV-B) radiation and sunminutes per hour have been monitored at Abisko Scientific Research Station in northern Sweden during the summer of 1994. The PAR and UV-B values have been compared to models, and from the comparison cloud transmission factors could be determined and compared for the two wavebands. For the whole period the clouds, on average, decreased the UV-B to 74.2% of what it would have been without clouds. This means that in simultaneous experiments in which vegetation was irradiated with extra UV-B corresponding to 15% depletion under clear skies, taking cloud cover into account, the radiation corresponds to 19.0% ozone depletion. Ozone column at Abisko during the summer 1994 estimated from measurements at Vindeln to the south of Abisko and Tromsø to the north showed no depletion compared to a model based on values determined 3–4 decades earlier.

L. O. Björn, Abisko Scientific Research Station, S-981 07 Abisko, Sweden and Dept of Plant Physiology, Lund Univ., Box 117, S-221 00 Lund, Sweden. – B. Holmgren, Abisko Scientific Research Station, S-981 07 Abisko, Sweden and Dept of Meteorology, Uppsala Univ., Box 516, S-751 20 Uppsala, Sweden.

In connection with the ongoing studies of ultraviolet-B radiation effects on vegetation in Abisko (Gehrke et al. 1995, Johanson et al. 1995a, b, Sonesson et al. 1995, Björn et al. 1996, Gwynn-Jones et al. 1996), it has become desirable to study the ambient radiation and its variations. From the start of the project occasional spectral measurements of ambient UV-B have been made. The design of the experiments has relied heavily on modelling of the UV-B climate, but in this modelling we have not been able to take cloud effects into account. From the summer of 1994 both PAR (photosynthetically active radiation) and UV-B have been monitored using broad band sensors, and here we report the first results. Also UV-A has been monitored, but we are planning to report UV-A values in a future communication, covering a longer period.

Equipment for radiation measurement

All the instruments from which the values reported here were obtained are installed on the roof of the observatory at Abisko Scientific Research Station (68.35°N,

18.82°E). There is no significant shading by nearby buildings or tall vegetation. All measurements refer to irradiances incident on the upper side of a horizontal surface.

UV-B was measured by a UV-Biometer (Model 501A, Ver. 3). The meter utilizes the same principle as that of the Robertson-Berger instrument (Berger 1976). The solar radiation enters the detector through a quartz hemisphere and an input filter. The filtered radiation excites fluorescence in a thin layer of phosphor. The visible light emitted by the phosphor is detected by a GaAsP diode. The detector (filter, phosphor and diode) is encapsulated in a metal enclosure and temperature is stabilized at 25°C by a Peltier element. Also UV-B was monitored in a similar manner, but the results will be reported in another context.

From the diagrams provided by the manufacturer, the cosine errors of the biometers are negligible for zenith angles up to 60°, and increase to c. 10% at 70°. The maximum deviations allowed between individual sensor calibrations are < 6% for solar zenith angles between 0 and 70° and ozone columns between 280 and 320 DU (Dobson units).

The spectral response of the UV-B meter resembles, but is relatively lower than, the McKinlay and Diffey (1987) erythema spectrum for wavelengths above c. 330 nm. The UV-B meter has negligible sensitivity above 375 nm. Because the daylight UV-A is so much more intense than the daylight UV-B, the contribution of UV-A to the reading, although small, is not completely negligible.

The spectral response (per incident power) of the UV-A meter is rather flat with values >80% of maximum from 350 to 380 nm. At 330 nm the response is c. 40% and at 400 nm c. 3% of the maximum sensitivity. The output of the UV-B sensor is in MED h^{-1} (minimum erythemal dose per hour) and the output of the UV-A sensor in W m^{-2}.

Throughout the measurement period the temperature controllers of both sensors were continuously operating. The outputs of the photodiodes as well as of the inbuilt thermistors were sampled by a Campbell CR21 logger every 10 s and averaged over 30 min intervals. In order to diminish errors due to adverse weather conditions, the biometers were housed in a box (50 × 70 × 50 cm) with two circular holes in the lid for the quartz hemispheres. The receiving surfaces of the sensors were in line with the upper surface of the lid. The air inside the box is heated by electric elements when the air temperature falls below c. 20°C. The bottom of the box had an opening of c. 100 cm^2. Convection currents (through slots c. 3 mm wide between the receiving surfaces of the biometers and the lid of the box) affect the quartz hemispheres. This arrangement, besides being increasing the temperature stability of the sensors (with recorded maximum variations of < 0.1 around 25°C during the summer period), so far appears to effectively prevent the formation of dew and/or rime on the quartz hemispheres. This may otherwise constitute a considerable measuring problem in most climates at high latitudes.

PAR was recorded by a Li-190SA quantum sensor (LiCor) as photon irradiance in μmol s^{-1} m^{-2}. The recording is made via an automatic weather station (of the standard type used by the Swedish Meteororological and Hydrological Inst.). The sampling interval is 15 s and average hourly values are recorded. According to the specifications of LiCor, the absolute calibration is ± 5% relative to the US National Bureau of Standards (NBS). The sensor, according to the manufacturer, is "cosine corrected up to 80° angle of incidence". The temperature dependence is < ±0.15% per °C. During the period studied, the average daytime temperature typically varies ± 5°C from avereage summer temperature, and temperature averaged over hours typically varies ± 15 °C. This temperature interval corresponds to an error in the order of ± 1% for daily values and ± 2.5% for hourly values.

In conclusion, errors due to temperature variations should be negligible in the present analysis. Also, the

analysis of PAR and UV irradiance emphasizes solar zenith angles < 65°. Therefore, the deviations due to cosine errors are likely to have small effects on the results.

Photosynthetically active radiation has been measured every hour. UV-B radiation has been sampled every 10 s and averaged over half-hourly intervals. Since our aim here is to compare variations in PAR and UV-B levels, we have here used only UV-B data corresponding to the hourly measurements of PAR. PAR values were available from the first of March, UV-B values from the middle of June, with some interruptions. The last values used here are from 14 August.

Sun-minutes is the number of minutes that the direct solar beam reaches a detector consisting of a glass sphere focusing the light on a paper. When computing a daily weighted sum, the sun-minutes for a particular hour multiplied by the model-value of UV-B for that hour, and divided by the modeled UV-B sum for the whole day.

Ozone data

Column ozone data were obtained from measurements by the Swedish Meteorological and Hydrological Inst. (SMHI) taken at Vindeln (64.2°N, 19.6°E) and by Nordlysobservatoriet (Univ. of Tromsø, Tromsø, Norway) (69.7°N, 19.0°E). On a few days, when values were missing either from Vindeln or Tromsø, values from Sodankylä Meteorological Observatory (67.4°N, 26.6°E) managed by The Finnish Meteorological Inst. were used. The three stations are part of the Nordic ozone monitoring programme.

We estimate that the DU values interpolated for Abisko are within 5 units from what measurements in Abisko would have given, except on a few days with exceptional horizontal gradients.

Models

For modelling of PAR we have used the spectral daylight model of Bird and Riordan (1986), and computed approximate integrals from 400 to 700 nm. The program was written in QuickBasic and executed on a Macintosh computer. For UV-B we have used the computer program of Björn and Murphy (1985), slighly modified to weight the spectral values according to the function

$$EXP(-6.861 - 3.631x - 13.106x^2 - 35.784x^3$$

$$+ 217.015x^4 + 59.639x^5 - 499x^6)$$

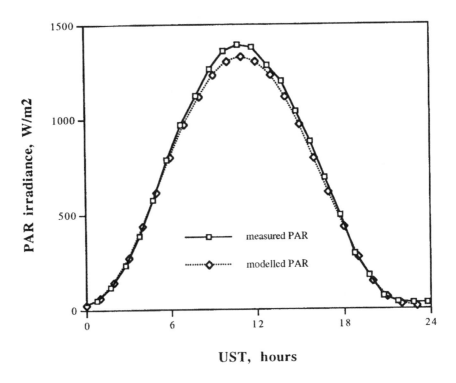

Fig. 1. Comparison of measured and modelled PAR during a clear day (5 July 1994). Modelled PAR is plotted directly, without any normalization to measured PAR.

where $x = $ (wavelength in nm $-350)/100$, using only wavelengths from 285 to 350 nm. This gives no exact spectral match to the broadband sensor used for monitoring, but is estimated to give the same balance between short and long wavelength components, which is the important consideration for the temporal variations. The UV-B model contains a subprogram which estimates ozone column, and this subprogram was also used separately for comparison with the measured ozone column (see below). The ozone subprogram relies on the annual mean ozone column as given by Gebhart et al. (1970) for the period 1957–1966, and annual variations as described by Hilsenrath and Schlesinger (1981) for the period 1970–1977. It is part of the program for UV-B modelling described by Björn and Murphy (1985) and Björn and Teramura (1993).

It is not feasible to theoretically compare the absolute values of the UV-B monitoring and the UV-B modelling. Therefore the model was normalized to the monitored values by a 13 h long comparison on an exceptionally clear day (day 186, i.e. 5 July). During this period the ratio between monitored and model values varied by $\pm 2\%$ from the mean. The model value of the ozone column for this day was 335 DU, while the ozone column measured at Vindeln, latitude c. 4.2

degrees south of Abisko, was 306 DU. This has been taken into account when calculating the equivalent ozone depletition (Fig. 6).

The performance of the models is also illustrated by Figs 1 and 2, where a comparison is made between models and monitoring from midnight to midnight on day 186. The UV-B model was normalized as described above, while the PAR model values have been obtained by computation only.

Results and discussion

One important question is to what extent is it possible to predict UV-B irradiance from PAR. Of course, there is no simple relation between them, since UV-B is affected by solar elevation in a different way than is PAR. This is evident from a comparison of Figs 1 and 2.

It is clear, that models for PAR and UV-B radiation have to be taken into account if one wants to estimate the latter from the former. By combining measured and modelled values, "cloud transmission values" can be computed. We did this in the following way: For data from UST (Universal standard time) 7 to 15 h (local

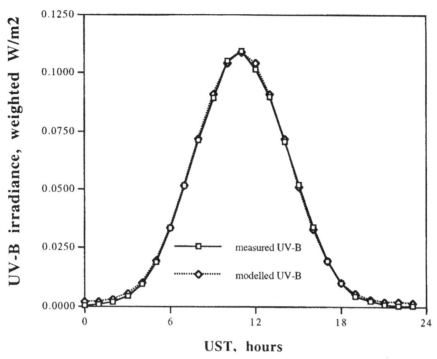

day 186

Fig. 2. Comparison of measured and modelled UV-B radiation on a clear day (5 July). The modelled UV-B was normalized to the measured UV-B over a 13 h period.

solar time c. 9–16 h) measured hour-averaged values of PAR and of UV-B were divided by the corresponding modelled values to obtain values of "cloud transmission".

The correlation between "cloud transmission factors" (i.e. the ratio between measured and modelled values (Fig. 3) was surprisingly low, as clouds are usually considered to be rather neutral in a spectral sense. One possible reason for the low correlation is that UV-B consists of relatively more sky (diffuse) radiation and less direct radiation from the sun than PAR, and the two spectral bands therefore do not encounter the same cloud distribution in a sky with partial cloud cover.

For plants it is thought that the daily UV-B exposure (or, more exactly, the daily UV-B fluence) is more important than the values of irradiance or fluence rate per se. Therefore we also computed the daily UV-B exposure, of daily sun-minutes and of weighted daily sun-minutes (Figs 4–5), in all cases from 7 to 15 h UST, daynumber 170–190. Sun-minutes is the number of minutes that the direct solar beam reaches the detector. When computing a daily weighted sum, the number of sun-minutes for a particular hour was multiplied by the model irradiance of UV-B for that hour, and divided by the average modeled UV-B irradiance for the day (7–15 h UST). This weighting resulted in considerably stronger correlation between sun-minutes per day

and daily radiation exposure (we use the term exposure here as a synonym for time-integrated irradiance) than when unweighted sun-minutes were used. There was no correlation between the residuals and interpolated

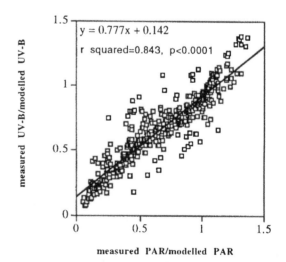

Fig. 3. The ratio between measured and modelled UV-B (cloud transmission factor) plotted versus the corresponding ratio for PAR. Both the PAR and the UV-B data were averaged hourly, and only values from UST 7 to 15 were used. The correlation coefficient is $r^2 = 0.843$ with $p < 0.0001$.

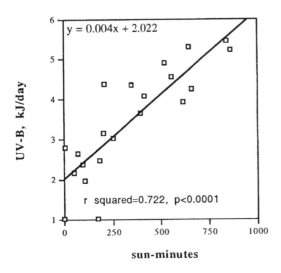

Fig. 4. Dependence of daily UV-B on sunshine minutes per day. As for Fig. 5 only 7–17 h UST and daynumber 170–190 were included.

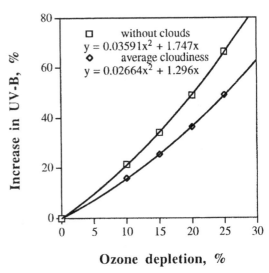

Fig. 6. The relation between ozone depletion and increase in plant-weighted UV-B for Abisko summer conditions. The upper graph is for clear skies, the lower one skies with normal cloud cover.

ozone column. The purpose of these comparisons was to see whether it is possible to estimate daily UV-B exposure from measurements of PAR and sun-minutes per day, which are monitored in a larger number of locations than is UV-B.

Tentatively it can be concluded from the above, that at locations where UV-B is not monitored, the best estimate is obtained by a measurement of PAR in conjunction with use of PAR and UV-B models. The regression equation in Fig. 3 must, of course, be compared to what will be obtained at other locations before a more definite statement can be made. At present it

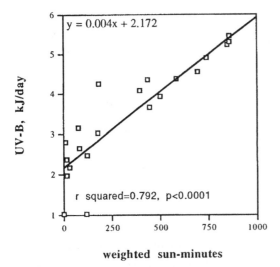

Fig. 5. Dependence of daily UV-B on sunshine minutes per day weighted by the values of modelled UV-B. Thus greater weight is given to the effect of clouds during times when modelled UV-B is high.

seems that UV-B estimation from number of sun-minutes gives a less reliable measure.

All the values of measured UV-B during the period 19 June to 14 August for hours 7–15 were summed, and the corresponding done for modelled UV-B. The ratio of these sums was 0.742. The artificial UV-B administered to the plants in the field experiments (e.g. Johanson et al. 1995a, b), calculated to correspond to 15% ozone depletion under cloudless skies, in fact corresponds to a higher degree of ozone depletion. In Fig. 6 we have plotted the relation between targeted ozone depletion and the corresponding percent increase in plant-weighted UV-B, and also the relation between the corrected ozone depletion and plant-weighted UV-B. For this weighting we used the same function as described above under Models, but the analytical approximation to Caldwell's generalized plant action spectrum designed by Thimijan et al. (1978, see Björn and Teramura 1993 for further explanation) gave practically the same result. Although the true UV-B effect spectrum differs among plants and processes, Caldwell's spectrum is used as a "model" spectrum by many researchers. The two third-order polynomial fits to the data can be used for a computer program that converts targeted ozone depletion to corrected ozone depletion. For instance, a target of 15% depletion (cloud-free conditions) results in 19.0% corrected depletion (for real, i.e. cloudy conditions, and a targeted 20% depletion results in 25.0% corrected depletion.

The interpolated ozone values were compared with the ozone model (Fig. 7). No ozone depletion in relation to the model, based on old measurements, could be seen. As a mean for the whole period, the recent values are 1% above the model values.

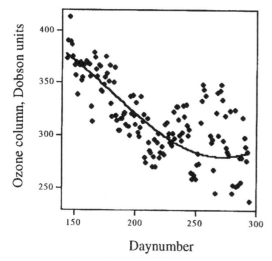

Fig. 7. Ozone column interpolated from measurements (dots) compared to model (solid line).

Acknowledgements – The initiative for this paper comes from the Director of the Abisko Scientific Research Station, M. Sonesson, who also helped us in many ways. Financial support has come from The Wallenberg Foundation and Astra-Draco AB. We are grateful to The Swedish Meteorological and Hydrological Inst., Nordlyslaboratoriet at the University of Tromsø, and Sodankylä Meteorological Observatory of the Finnish Meteorological Inst. for providing us with ozone data. The investigation, although financed from Sweden, was undertaken as a European Union collaboration (EC contract No. EV5V-CT910032).

References

Berger, D. S. 1976. The sunburning ultraviolet meter: Design and performance. – Photochem. Photobiol. 24: 587–593.

Bird, R. E. and Riordan, C. 1986. Simple solar spectral model for direct and diffuse irradiance on horizontal and tilted planes at the earth's surface for cloudless atmospheres. – J. Clim. Appl. Meteorol. 25: 87–97.

Björn, L. O. and Murphy, T. M. 1985. Computer calculation of solar ultraviolet radiation at ground level. – Physiol. Vég. 23: 555–561.

– and Teramura, A. H. 1993. Simulation of daylight ultraviolet radiation and effects of ozone depletion. – In: Young, A. R., Björn, L. O., Moan, J. and Nultsch, W. (eds), Environmental UV photobiology. Plenum Press, pp. 427–471.

– , Callaghan, T. V., Gehrke, C., Holmgren, B., Johanson, U. and Sonesson, M. 1995. Simulating the effects of stratospheric ozone depletion on natural vegetation at Abisko. – In: Callaghan, T. V., Oechel, W. C., Gilmanov, T., Holten, J. I., Maxwell, B., Molau, U., Sveinbjörnsson, B. and Tyson, M. (eds), Ecosystems research report 10: Global change and arctic terrestrial ecosystems. Proc. Int. Conf. 21–26 August 1993, Oppdal, Norway. Directorate-General Sci. Res. Develop., European Union, Brussels, pp. 43–48. (Commission of the European Communities Ecosystems Research Report, Brussels).

– , Callaghan, T. V., Johnsen, I., Lee, J. A., Manetas, Y., Paul, N. D., Sonesson, M., Wellburn, A. R., Coop, D., Heide-Jørgensen, H. S., Gehrke, C., Gwynn-Jones, D., Johanson, U., Kyparissis, A., Levizou, E., Nikolopoulos, D., Petropoulou, Y. and Stephanov, M. 1996. The effects of UV-B radiation on European heathland species. – In: Rozema, J., Gieskes, W. W., van de Geijn, S. C. and de Boois, H. (eds), UV-B and Biosphere. Kluwer Academic Publishers.

Gebhart, R., Bojkov, R. and London, J. 1970. Stratospheric ozone: a comparison of observed and computed models. – Contrib. Atmos. Physics 43: 209–227.

Gehrke, C., Johanson, U., Callaghan, T. V., Chadwick, D. and Robinson, C. H. 1995. The impact of enhanced ultraviolet-B radiation on litter quality and decomposition processes in *Vaccinium* leaves from the subarctic. – Oikos 72: 213–222.

Gwynn-Jones, D., Björn, L. O., Callaghan, T. V., Gehrke, C., Johanson, U., Lee, J. A. and Sonesson, M. 1996. Effects of enhanced UV-B radiation and elevated concentrations of CO_2 on a subarctic heathland. – Physio. Ecol. Ser., in press.

Hilsenrath, E. and Schlesinger, B. M. 1981. Total ozone seasonal and interannual variations derived from the 7 year Nimbus-4 BUV data set. – J. Geophys. Res. 86(C12): 12087–12096.

Johanson, U., Gehrke, C., Björn, L. O., Callaghan, T. V. and Sonesson, M. 1995a. The effects of enhanced UV-B radiation on a subarctic heath ecosystem. — Ambio 24: 108–113.

– , Gehrke, C., Björn, L. O. and Callaghan, T. V. 1995b. The effects of enhanced UV-B radiation on the growth of dwarf shrubs in a subarctic heathland. – Funct. Ecol. 9: 713–719.

Sonesson, M., Callaghan, T. V. and Björn, L. O. 1995. Short term effects of enhanced UV-B and CO_2 on lichens at different latitudes. – Lichenologist, in press.

Thimijan, R. W., Carns, H. R. and Campbell, L. E. 1978. Radiation sources and related environmental control for biological and climatic effects of UV research (BACER) – In Final report. Environmental Protection Agency, BACER Program, Washington, D.C.

Ecological Bulletins 45: 210–219. Copenhagen 1996

Climatic impacts on flowering, growth, and vigour in an arctic-alpine cushion plant, Diapensia lapponica, under different snow cover regimes

Ulf Molau

Molau, U. 1996. Climatic impacts on flowering, growth and vigour in an arctic-alpine cushion plant, *Diapensia lapponica*, under different snow cover regimes. – Ecol. Bull. 45: 210–219.

The impact of climatic variation on growth, flowering, and survival of the arctic-alpine cushion plant, *Diapenia lapponica*, was investigated along a snow-melt gradient at Latnjajaure, N Swedish Lapland, over a 5 yr period. During the study period (1990–94) the live biomass of *D. lapponica* in permanently marked control plots at the study site declined by 22% and their flowering by 55%. Many old (> 50 yr) cushions died off completely during the study period, and little or no recruitment was recorded; these changes are almost irreversible as the species is extremely slow-growing. Snow-cover was the main determinant of performance in the *D. lapponica* community, and the thickness and duration of the snow-pack influence on flower production and survival. Most of the recorded loss in live phytomass and flower production was accounted for by specimens inhabiting the most wind-exposed part of the gradient, with thin winter snow cover and early melt-off. The onset of thawing in the area has been taking place extremely early during the last three seasons, in accordance with circumpolar trends in increasing spring temperatures and progressively earlier snow-melt. The *Diapensia* plants responded by increasingly early virescence, and suffered badly from subsequent June blizzards that killed off vegetative shoot apices and expanding flower buds. The species reaches its lowest level of frost tolerance around midsummer. In more sheltered situations, with much thicker snow-pack and later emergence of the plant cover, little or no damage was observed. The trend of increasing winter precipitation resulted in thick snow-packs in wind-protected sites during the study period, and hence late melt-off in these sites. The *D. lapponica* cushions under these circumstances experienced a very short growing season, and responded by low flower bud production and frequent postponing of final bud expansion to subsequent seasons. This cushion plant species shows superior properties as a climatic indicator, and the investigation provides one of very few present records on the impacts of climate change on the tundra plant cover.

Ulf Molau, Dept of Systematic Botany, Univ. of Göteborg, Carl Skottsbergs Gata 22, S-413 19 Göteborg, Sweden and Inst. of Arctic Biology, Univ. of Alaska Fairbanks, P.O. Box 757000, Fairbanks, AK 99775-7000, USA.

During the past three decades the arctic tundra has experienced a substantial warming, although this effect is not evenly expressed throughout the Arctic (Chapman and Walsh 1993). The strongest warming trend (over $+0.75°C$ per decade in annual means) have occurred in north-central Siberia and in arctic Alaska, whereas a few regions (e.g. west Greenland and adjacent islands of the Canadian High Arctic) have undergone a slight cooling. Northern Scandinavia did not show any detectable trend until 1990, but since then the

climate has been changing there as well, mainly manifested as a prolongation of the snow-free season (Molau 1995, unpubl.), a trend common to most of the Northern hemisphere (Groisman et al. 1994).

The changing climate in polar areas during the past few decades has created significant responses in the tundra plant cover, but reports are extremely sparse in the literature to date. In a ten year study of tundra at Toolik Lake, northern Alaska, Chapin et al. (1995) report a 50% decrease in biomass of the evergreen sedge

210

Eriophorum vaginatum in unmanipulated control plots, correlated with a 30% biomass increase in the dwarf birch, *Betula nana*. In the Alaska range, *Betula glandulosa* has been spreading over vast areas of previously low-stature alpine heathland during the past few decades (Bryant, pers. comm.). From the Antarctic, Fowbert and Lewis Smith (1994) report a substantial increase in the number of individuals of the two native Antarctic vascular plant species, *Colobanthus quitensis* (Caryophyllaceae) and *Deschampsia antarctica* (Poaceae) at a thoroughly mapped area in the Argentine Islands from 1960 to 1990; the increase in numbers was almost five-fold for *Colobanthus* and > 25-fold for *Deschampsia*.

I have studied the changes in a population of the evergreen arctic-alpine cushion plant, *Diapensia lapponica*, for five years (1990–94) at the Latnjajaure Field Station in northernmost Sweden. This evergreen perennial has exceptional properties in terms of the correlation between age and developmental events: its normal life span exceeds 100 yr and the almost hemispherical clones can be aged from their diameter (Molau unpubl.). The primary aim of this long-term study was to investigate the annual growth increment in *Diapensia* cushions at sites with exposure to various microclimates (Molau unpubl.). However, during the course of the study (in August 1992), it became evident that mature clones (particularly in exposed situations) responded negatively to the ongoing changes of climate in Scandinavia, and began to show obvious signs of damage and reduced vitality. Therefore, I extended the study to focus on the effects of climatic forcing on the performance and viability of a larger set of individual clones of this long-lived perennial. The proportion of live and dead parts of each clone in the monitoring plots was already recorded annually from the start of the project in 1990, even though my intention at that time was to study population dynamics and demography in a stable mature population.

When the negative trend in terms of flowering and vitality in mature, reproductive *Diapensia* clones (see below) persisted throughout the five year period, additional sampling at subsites of various exposure to microclimatic factors (time of thawing of the snow-pack, exposure to snow drift, etc.) was undertaken at the end of the 1994 season. The aim was to document the impact of microclimate on the performance of this species, because there are few published reports on how climate change might cause directional changes in composition of tundra plant communities. To date, only two other such reports are available (Fowbert and Lewis Smith 1994, Chapin et al. 1995; see above). Despite *D. lapponica* being one of the most typically developed cushion plants of arctic and alpine environments, ecological studies including this widespread species are astonishingly few. In the present paper, I

demonstrate the potential of long-lived tundra plants as tools and indicators for monitoring climate change in polar regions and add another piece of evidence to the record of the present warming trend.

Material and methods

The plant species

Diapensia lapponica L. (Diapensiaceae) is a long-lived, arctic-alpine, evergreen perennial. The cushions are more or less hemispherical in shape, radiating from a single, stout, woody tap-root. It grows most abundantly in exposed sites, such as windswept ridge crests, and is indifferent to substrate acidity. In sites with active soil processes, e.g. solifluction, it may also be found in microhabitats with longer duration of the annual snowpack. It is obviously a weak competitor, but is able to grow in adverse habitats. The mean annual radial growth rate of *D. lapponica* cushions equal c. 0.6 mm, but the rate is sigmoid in relation to cushion age or size; young established cushions grow faster than seedlings and old (>50 yr) clones (Molau unpubl.).

The leaves are glabrous, and frost-resistance of the cushion is brought about by extremely dense leaf aggregation. Flowers are terminal, initiated from apical shoots of the ramets, preformed at the latest during August the year before flowering. In sites with long duration of the snow pack, flower buds may be postponed for several subsequent growing seasons depending on the snow situation (see below). The flowers emerge on leafless pedicels ranging 2–4 cm above the surface of the cushion, flowering mainly in the month of June (vernal phenoclass; Molau 1993a). The corolla is white, and there are five yellow stamens opening by terminal pores. The single pistil contains on average 150 ovules. The pollinator fauna comprises arctic bumblebees, wasps, and flies (Petersen 1908, Bergman et al. unpubl., Molau unpubl.). Previous investigations of the mating system indicate that the species is slightly self-pollinating, but entirely dependent on insect visitation for seed set and reproductive success (Molau 1993a). The fruit is a 3-locular, tough capsule, born on an erect pedicel, emergent above the snow for much of the winter. The capsules are still undehisced at the onset of winter in late September or early October, but always open and empty at snowmelt in April and May. Abortion rates are high, and capsules with >60% filled seed are rare. The relative reproductive success (RRS), the product of the fruit:flower and seed:ovule ratios (Wiens 1984), was calculated as 0.315 for the Latnjajaure population (Molau 1993a), with whole-fruit abortion being the minor component.

The study site

Field work was carried out at the Latnjajaure Field Station (68°22′N, 18°29′E) west of Abisko in northern-most Swedish Lapland during the growing seasons of 1990–94. The field station is located at 1 000 m a.s.l. at the bottom of a high-alpine valley inhabited by a representative set of low-arctic plant communities. Latnjajaure is the Swedish site within the International Tundra Experiment (ITEX), and is equipped with a standardized climate station, operating year-round since 1992. The annual mean temperature during the years of study ranged from −2.7°C (1993) to c. −4°C (1991). The field work was carried out within 300 m of the station, just SE of Lake Latnjajaure, in a relatively closed lichen-rich heath community with *Cassiope tetragona, Diapensia lapponica, Empetrum hermaphroditum, Loiseleuria procumbens*, and *Salix herbacea* as the dominant species. The study area is situated off the reindeer migration routes, and grazing and trampling pressure is extremely low. Backpackers have not been passing through the site, and the area is not affected by winter sports; the snow surface of the study site remained undisturbed during the winters. Since the station is staffed continuously during the growing season, the study area is under constant observation. The total record of larger mammals crossing through the study site in the summertime comprises a few reindeer and one wolverine.

Sampling methods

During the summer of 1990, ten permanent 1 × 1 m plots were established at Latnjajaure, five in exposed situations on the crest of a nearby ridge (no. 1, 6–9), four somewhat more sheltered further down its gentle north-facing slope (no. 2–5, along a downhill transect), and one in a sheltered west-facing site (no. 10). The exposed plots on the windswept ridge were snowfree most of the year and never experienced snow depths >10 cm; the sheltered plots were covered by a winter snowpack of 0.5–1.5 m and thawed 1–4 weeks later, depending on snow depth, position in transect, and spring temperatures. For the transect along the snow-pack gradient I used a permanently staked transect for snow-melt monitoring, running from the lake shore and ending at the summit of the ridge, 90 m south of the shore and 15 m above lake level. The plots no. 1–5 were positioned at ±equal intervals downhill from the summit, with plot no. 1 at the crest and no. 5 c. 25 m from the shore-line. During the field seasons of 1991–1994, snow depth was measured at probing stations every 5 m along the transect until final snow-melt (every 3 days in 1991–93, once a week in 1994). The position of the snow-line in the transect was recorded daily.

For each plot a detailed map of all *Diapensia* clones was prepared in August 1990. At the end of each growing season 1990–94 (between August 25 and 30) all mapped clones were measured, and the number of flowers produced during the season was counted. The size of each clone was measured as the average of the maximum diameter and the diameter score perpendicular to the first. At the same time, the proportion of clones damaged, dead, or dying from various reasons was recorded. For logistic reasons the monitoring of growth and survival was restricted to adult, reproductive clones except in the plots no. 6, 7, and 10 where juvenile clones recruitment from seed were followed as well. Altogether, a total of 185 genets were included in the analysis, most of them for the entire five year period, but nine were added as they either became established (plots with all genets mapped) or entered the reproductive stage (all other plots). Twenty-two genets died off completely during the study period (Molau unpubl.). In total, the present study is based on 851 records of mean annual cushion diameter.

During the flowering season, the ten plots were inspected daily for number of open flowers. In 1990, the year the plots were marked, this monitoring did not commence until 1 July (day number 182); in the subsequent four summers monitoring started at the time of thawing.

In 1994, three larger rectangular plots (with the longest side perpendicular to the direction of the slope) were sampled once in late August for cushion size, flowering, and damage. These plots were positioned at three points along the snow-cover gradient of the ridge slope, in the summit area (15 m above lake level, 85 m from the shore), and at two locations further downslope (10 m above lake level and 55 m from shore-line, and 6 m above lake level and 25 m from shore-line, respectively). The transect was established parallel to and 100 m east of the transect of 1 × 1 m plots no. 1–5. Because of local differences in microtopography and plant density, the sizes of the three plots were different (6 × 1, 3 × 1, and 10 × 2 m, respectively). All *Diapensia* individuals (clones) in each of the plots were mapped, measured, the number of flowers, and the degree of damaged aboveground tissue assessed (when possible, with notes on cause of damage). The number of clones encountered in the three plots were 65, 36, and 50, respectively.

Data analysis

The age of individual *Diapensia* cushions was calculated using the formula derived from the long-term demographic study (Molau unpubl.)

Estimated age (yr) = $8.736d - 0.263d^2 + 0.026d^3$

where d is the cushion diameter in cm (one decimal position). This equation was found to apply to clones in

Table 1. Climate at Latnjajaure, Sweden, during the summers of 1990–1994. Data from standardized ITEX climate station (Molau unpubl.). May data are lacking for the first two years. Summer mean temperature and total summer precipitation refer to June–August.

Variable	1990	1991	1992	1993	1994	Mean 1990–1994
May mean temperature (°C)			1.91	−0.06	−2.19	
May precipitation (mm)			53.6	80.0	32.1	
June mean temperature (°C)	6.09	4.94	5.69	1.60	2.73	4.21
June precipitation (mm)	34.9	54.2	27.8	60.1	30.1	41.4
July mean temperature (°C)	9.10	8.15	5.43	8.46	7.61	7.75
July precipitation (mm)	55.0	44.2	133.0	63.6	56.7	70.5
August mean temperature (°C)	7.88	8.50	5.60	6.57	7.56	7.22
August precipitation (mm)	57.3	46.3	52.0	63.3	38.9	51.6
Mean summer temperature (°C)	7.69	7.20	5.57	5.54	5.97	6.39
Summer precipitation (mm)	147.2	144.7	212.8	187.0	125.7	163.5

the summit plateau of the study ridge; for sites with longer-lasting snow cover, the values were adjusted for the shorter growing season relative to that of the summit. Statistical analyses were performed on a Macintosh using the StatView 4.02 and SuperANOVA 1.11 packages.

Results

Climate and snow cover

Weather conditions at Latnjajaure varied considerably among the five seasons of monitoring. As seen from Table 1, the summer of 1990 was generally warmer than average, and 1991 was about average in both temperature and precipitation. In 1992, the summer temperatures remained very cold in July and August, despite a long warm spell in late May and early June, early thawing, and early onset of flowering in most species. This summer was also unusually cloudy and rainy, an effect of lingering stratospheric dust from the 1991 Mt. Pinatubo eruption (Minnis et al. 1993). The last two years have shown May temperatures much above average, followed by cold spells in June, particularly in 1993. The remaining part of the summer (July and August) was warmer than normal in 1994.

The snow-pack thawing process depends largely on ambient air temperature during May. However, ambient air temperature *per se* is of minor importance to plant performance since field surface temperature, when thawed, depends much more on global radiation influx than ambient air temperature (Marsden 1992, Molau 1993a, 1995). The Latnjajaure data set for May mean temperature during 1990–94 is incomplete (Table 1). However, the unique, unbroken, long-term record from the nearby subalpine Abisko Scientific Research Station (c. 350 m alt., 68°21′N 18°49′E) shows a highly significant increase in May temperatures from 1913 to 1994 (Fig. 1). Onset of winter (defined as continuous snow cover in the valley, lasting uninterrupted into the next season) occurred extremely early in 1991 (September 4), very late in 1992 (October 15), and about normal for the decade in the other three years.

Thawing in the Latnja valley bottom, with its hilly topography, is a continuous process throughout the vegetation period of the area. Hill-tops and rock ledges have a thin snow cover due to steady redistribution by wind. Snow drift is one of the main determinants of plant community differentiation; hill-tops are always early-thawing, whereas snow beds of variable duration fill up most depressions (Molau 1993a). Even in snow-rich years with an average snow depth >2 m, such as late winter and spring of 1990 and 1993, the snow-pack at exposed sites remains thin. At the *Diapensia* study site, the thickest snow-pack on record at the summit was 18 cm (17 April [day no. 107], 1992) and at the lake shore 280 cm (21 May [day no. 141], 1993). Snow-melt in the Latnja valley in general occurred relatively late in 1990, a little earlier than normal in 1991, and extremely early in 1992. In 1993 and 1994 snow-melt had started by the first week of May, almost a month earlier than normal, but the thawing process was halted through much of June due to cold temperatures both these years (Table 1). The three last summers (1992–1994) have been characterized by recurrent blizzards in June when snow cover had retreated to about the middle of the study

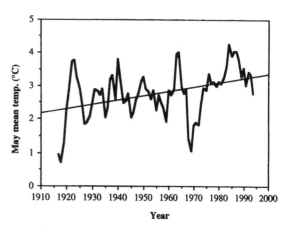

Fig. 1. May temperatures (5 yr running means) at Abisko, N Sweden, 1913–94. The increasing temperature trend is significant in linear regression (adjusted $r^2 = 0.160$, $n = 78$, $p = 0.0002$). Data from Abisko Scientific Research Station.

Table 2. Thawing dates (day no.) and vegetation period (d) of the ten permanent *Diapensia* monitoring plots at Latnjajaure, Sweden, in 1991–1994. Plots no. 1–5 are located along a snow depth gradient (90 m transect) from hill–top to foot (lake shore) of a 15 m high north-facing slope, plot no. 6–9 on the summit plateau, and no. 10 in a sheltered but relatively early-thawing W-facing slope. Days with intermittent snow cover after summer blizzards are subtracted from the vegetation period duration. Thawing dates for 1991 (except plot no. 5) are extrapolations from temperature records, April 1991 snow distribution along the transect, and the snow situation at the onset of monitoring (June 12, day no. 163). TD = Thawing Date, VP = Vegetative Period.

Year		Thawing date (day number) Vegetative period (d)									
Plot no.		1	2	3	4	5	6	7	8	9	10
1991	TD	146	151	157	164	171	146	146	146	146	146
	VP	98	93	87	80	73	98	98	98	98	98
1992	TD	143	145	152	155	167	144	144	144	144	143
	VP	133	131	124	121	109	132	132	132	132	133
1993	TD	130	136	139	153	182	131	131	131	131	142
	VP	129	123	120	106	77	128	128	128	128	117
1994	TD	130	141	158	160	161	130	130	130	130	135
	VP	125	114	97	95	94	125	125	125	125	120
Mean TD 1991–1994		137	143	152	158	170	138	138	138	138	142
Mean VP 1991–1994		121	115	107	101	88	121	121	121	121	117

slope. For plants at exposed sites, such as the summit area of the study site, the four last years have entailed extremely long growing seasons and early virescence (greening) in evergreen perennials, such as *D. lapponica* (Molau unpubl.).

Thawing dates and vegetative period for the ten permanent plots in 1991–1994 are provided in Table 2. By comparing the figures for the last three years with those of 1991, which was a rather "normal" year with regard to climate, it is noticeable that the lowermost plot in the transect (no. 5) has a growing season that is on average 27% shorter than at the summit. Due to the deep snow cover in the winter of 1992–93, snow-melt at the bottom of the transect was late also in 1993, despite early onset of thawing of the area.

Flowering phenology and frequency

The west-facing plot outside the main study area (plot no. 10) turned out to be a useful reference for comparisons (among years and among levels of exposure and thawing time) in the study, as it showed no decreasing trend in flower production over the five consecutive summers (Table 3). The flowering phenology curves are smooth and almost symmetrical (Fig. 2, upper graph), typical of K-strategists (Molau 1993b), and show no signs of disturbance by recurrent early summer blizzards. Furthermore, the graph shows extreme phenological plasticity in *Diapensia*, where climatic differences among years displaced the flowering peak as much as 22 d (1992 vs 1993; Fig. 2, above). Depending on the progression of summer warming, the curves are variously compressed in time. In 1992, flowering progessed extremely rapidly due to a long warm spell early in the season, while below-average temperatures in 1993 and 1994 retarded flowering considerably. Simultaneous flowering of all flowers over small areas as in the site of

plot no. 10 in 1992 induces "majoring" behavior in the pollinating bumblebee queens, resulting in almost instant pollination of all flowers as they open up. (The term "majoring" is used for short-term within-season species specialization in pollinating insects [while "minoring" on other plant species]). The corollas are shed at full turgor as soon as sufficient pollination is achieved, a post-pollination floral change *sensu* Gori (1983); this phenomenon is very pronounced in the 1992 flowering phenology curve. The mean prefloration time of 26.5 d scored in plot 10 in 1991–1994 (Table 3) is close to the mean of 32 d reported by Marsden (1992) from the Latnjajaure area.

If one now compares the curves for plot no. 10 with the flowering phenology in the five exposed plots in the summit plateau of the study ridge (plots no. 1 and 6–9; Fig. 2, lower graph), a similar among-year displacement of flowering peaks is seen. But there are also marked differences, particularly for the 1993 an 1994 seasons, characterized by recurrent blizzards throughout the month of June. Each sharp dip in the annual phenology curves represent a blizzard with below-freezing temperatures and exposure to snow drift, which killed off numerous expanding flower buds and destroying the flowers already open (Fig. 2, below). The graph also shows the marked decreasing trend in total flower production per unit area over the study period (see below). Seed set in *D. lapponica* in exposed situations was reduced by c. 90% in 1993, and c. 50% in 1994 due to blizzard damage. Corollas damaged by frost are not shed, but remain attached, dry and wrinkled; similar effects were seen in *Cassiope tetragona* in the study area these years.

In the snow cover and thawing gradient transect (plots 1–5) monitored from 1990 to 1994, similar responses are seen (Fig. 3). As stated above, the summer of 1991, normal in all respects with no extremely warm nor cold periods, is manifested in the almost symmetri-

Table 3. Prefloration time (days) and total numbers of flowers in the ten permanent *Diapensia* monitoring plots at Latnjajaure, Sweden, in 1990–1994. There are no data on prefloration time from 1990. Plot exposure: E = east-facing; N = north-facing slope (transect); Su = summit, ±horizontal; W = west-facing (not exposed). Means for thawing date (TD) calculated from data in Table 2. Plots no. 1–5 are located along the down-hill snow-melt gradient.

Year	Plot no.										
	1	2	3	4	5	6	7	8	9	10	
Exposure	Su	N	N	N	N	Su	Su	Su	Su(E)	W	
Mean TD	137	143	152	158	170	138	138	138	138	142	
Prefloration time (days)											Mean 1–10
1991	24	34	26	32	29	20	21	23	30	18	25.7
1992	20	33	28	35	39	20	21	22	22	21	26.1
1993	45	45	42	39	30	44	47	50	47	36	42.5
1994	42	50	35	41	40	41	43	48	50	31	42.1
Mean 1991–1994	32.8	40.5	32.8	36.8	34.5	31.3	33.0	35.8	37.3	26.5	34.1
Flower number											Sum 1–10
1990	101	17	13	1	1	105	85	74	113	15	525
1991	59	14	47	4	15	75	47	40	133	27	461
1992	76	14	89	3	14	81	46	36	85	47	491
1993	51	9	26	6	2	44	40	34	7	24	243
1994	43	11	16	1	2	53	51	7	12	38	234

cal curves for all plots that year. Over the years there was a mean difference of 33 d in thawing time between uppermost and lowermost plots of the gradient (Table 2). A corresponding displacement of flowering peak between plots no. 1 and 5 was observed; prefloration time does not differ much among plots (Table 3). Note that prefloration scores for 1993 and 1994, particularly in exposed plots, are abnormal, brought about by extremely early thawing followed by much colder spells in June. The continuous among-year reduction in annual flower production is evident in the upper part of the transect, but the effect decreases gradually with increasing snow cover and increasing thawing date of the habitat.

The flowering in the lowermost plot (no. 5) was always low, but extremely variable among years. All buds produced in August 1990 reached anthesis in 1991, but since then most flower buds have been postponed to subsequent, potentially more favourable seasons. More than half of the buds formed in late summer 1991 remained in the bud stage throughout the 1992 and 1993 seasons, and some of them, though looking quite fresh, were still not coming into flower in 1994. All but one of these finally developed into flowers in 1995, after the termination of the 5 yr monitoring programme.

Decrease in live biomass

Since the first census in late August 1990, there has been a steady decline in the total live area of *Diapen-sia* clones within the permanently marked plots (Fig. 4). Many clones died off entirely (Molau unpubl.). At the same time, the total annual flower production (Fig. 5) decreased by 55.4% (65.3% in exposed plots; Fig. 5). As stated elsewhere (Molau unpubl.), the species faces high risk of clone mortality 1) very early in its life span, i.e. during the recruitment phase, and 2) after reaching the reproductive phase at an average age of 18 yr. There has been almost no recruitment to the population since the monitoring started in 1990, but this does not account for much of the observed decline. Instead, the overall loss of 21.5% of live area in the plots over the 5 yr period is most restricted to mature, reproducing cushions, many of them 50–100 yr old or more (Molau unpubl.). This loss is mainly accounted for by the five plots on the exposed ridge summit (no. 1 and 6–9), whereas the five remaining, less exposed plots showed little or no change (Fig. 5). Damage was usually first expressed as desiccation of the central, topmost part of the cushions. This was initiated as frost damage to young shoots and flower buds during June blizzards. In many of the 50–100 yr old cushions (that were intact in 1990) in the summit area, only a sickle-shaped live area remains along the south-facing, wind-protected half. Large cushions that used to have flowers spread all over the surface are today reduced to ring- or sickle-like formations of flowers surrounding dead tissue. When dying off, the dense *Diapensia* cushions start to crack open, and are frequently invaded by more aggressive competitors, particularly the dwarf willow *Salix herbacea* and the moss *Rhacomitrium lanuginosum*.

Variation with exposure

In addition to the variation along the transect (plots 1–5) described above, the three larger plots sampled in 1994, at approximately the levels of the transect plots no. 1, 3, and 5 respectively, gave some additional data with regards to demography, flowering frequency, and level of damage. Leaves that have died during the last couple of years are steel gray and still smooth in texture; after three or four years such damaged parts turn more yellow and start to decay. Thus, it is possible to discern recent damages from events further back in time, even in a momentaneous census. The mean *Diapensia* cover in these plots was 61, 150, and 56 cm^2 m^{-2}, and the number of mapped clones 65, 36, and 50, respectively. There was also a significant decrease in survival of *Diapensia* clones with increasing exposure and earlier thawing of the habitat (Kruskal-Wallis non-parametric ANOVA, df = 2, H = 5.999, p = 0.0498). Nearly all of this reduction of the live phytomass in the

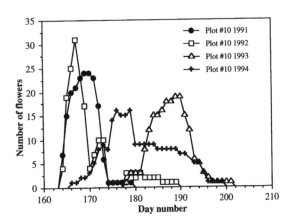

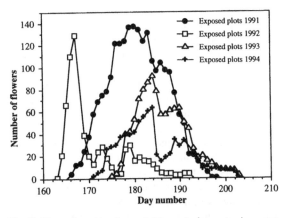

Fig. 2. Flowering phenology of *Diapensia lapponica* in permanent square meter plots at Latnjajaure, N Sweden 1991–1994. The upper graph shows the total daily number of open flowers in plot no. 10, a west-facing, relatively early-thawing plot that is not exposed to snow drift, which serves as a control for comparison with all other plots. The lower graph shows the total daily number of open flowers in the five exposed plots on the summit plateau of the study ridge (plot no. 1 and 6–9).

exposed plots had taken place during the past 2–3 yr. Recruitment was very low in all these larger plots as well, and almost absent in the late-thawing one. The mean computed ages for *D. lapponica* clones in the exposed, intermediate, and sheltered plots were 23.9, 38.5, and 60.0 yr, respectively. The computed age for the largest clone in each of the three plots was 278, 161, and 183 yr, respectively. The high mean clone age at the bottom end of the gradient is a consequence of the rarity of recent recruits.

Discussion

This long-term monitoring has shown that *Diapensia lapponica* is highly plastic in its phenological responses to climatic variations, within as well as among growing seasons. Flowering peaks of the study population were displaced by as much as three weeks between consecutive seasons, and the flowering phenology curves varied from extremely compressed (e.g. 1992) to more expanded over time. The normal prefloration time of *D. lapponica* of about four weeks is unusually long for an arctic-alpine outbreeder (cf. Molau 1993a), but this species occupies the harshest and most exposed of habitats, such as gravelly ridge-tops and windswept solifluction areas. These habitats are the first to thaw in late spring, and at the time virescence and frost-dehardening (Pihakaski 1988, Pihakaski and Junnila 1988) are completed, the flowering normally commences at the same time as most other species of the vernal phenoclass (Molau 1993a). As in *Cassiope tetragona*, flowering in *D. lapponica* never started before foliage virescence was completed, contrary to some other evergreen species, such as *Empetrum hermaphroditum*, inhabiting somewhat less exposed sites.

In later-thawing habitats, where *D. lapponica* is less dominant, the time of snow-melt and its variation among years had strong impacts on flowering. Flower buds, normally preformed in late summer the year before (Sørensen 1941, Molau 1993a), were readily postponed, their development cancelled awaiting more favorable seasons (i.e. earlier thawing). In this study, some flower buds, in clones growing in areas with the longest duration of snow cover within the species' tolerance, were postponed through three consecutive, unfavorable seasons. Such an extreme plasticity is also assumed to be important in the reproduction of *Eriophorum vaginatum* (G. Shaver and U. Molau unpubl.), but has so far never been documented in nature. Thus, *D. lapponica* is well-adapted to the harshest arctic-alpine conditions and has the potential to survive long periods of much more adverse conditions than experienced at the current climatic regime, perhaps even being capable of glacial survival more or less in situ (in ice-free refugia such as nunataks). This is supported

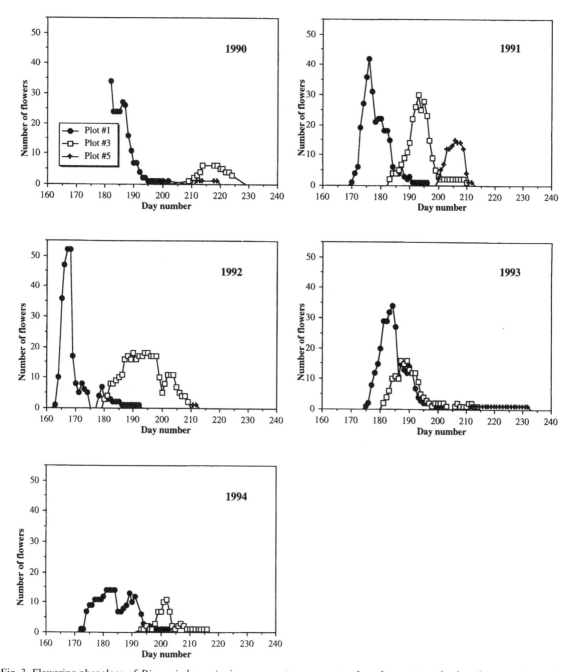

Fig. 3. Flowering phenology of *Diapensia lapponica* in permanent square meter plots along a snow-depth and snow-melt gradient (north–facing slope) at Latnjajaure, N Sweden, during five consecutive seasons, 1990–1994. Data from three plots are displayed in each graph, representing summit with thin cover and early thawing (plot no. 1), midway down the hill with intermediate snow conditions (no. 3) and the lowermost part of the *D. lapponica* community, close to the lake shore, experiencing a thick snow pack melting out late in the season (no. 5). Day no. 160 = 9 June, 200 = 19 July (except in 1991; leap year). Data from plots 2 and 4 omitted to avoid data point crowding.

also by the finding that recruitment of new plant individuals to the population during the century has been taking place mainly during colder periods (highly significant negative correlation between recruitment rate and annual mean temperature; Molau unpubl.).

The current circumpolar warming trend (see e.g.

Chapman and Walsh 1993), by contrast, exposes *D. lapponica* plants to conditions they are not adapted to. During the present century there has been a steady and significant increase in spring temperatures (Fig. 1), and snow-melt in the last few years has occurred much earlier than normal for the last 80 yr period (Groisman

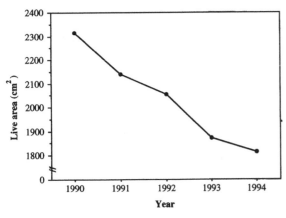

Fig. 4. The declining trend in inclusive live area of the *Diapensia lapponica* cushions monitored at Latnjajaure, N Sweden 1990–1994.

et al. 1994, Molau 1995, this study). As a consequence, the cushions in exposed habitats have undergone virescence earlier than normal, making them more susceptible to frost damage. Recurrent June blizzards, an increasingly common event (judging from my own experience from all summers since 1967 in the mountains of northern Lapland), cause severe damage to the *Diapensia* cushions, killing the expanding flower buds and vegetative shoot apices at a time when the species' annual cold-hardening cycle has reached minimum (cf. Pihakaski 1988, Pihakaski and Junnila 1988). As soon as the firm cushion surface is ruptured, the clone's environmental protection is lost, and secondary damage from drought may spread rapidly throughout the cushion. Invasion by more aggressive competitor species may also proceed very rapidly from this point.

Since the start of this long-term monitoring in 1990 the study population has suffered a loss of nearly a fourth of the live biomass and more than half of its annual flower production. Many old mature clones have died entirely (Molau unpubl.). These changes have

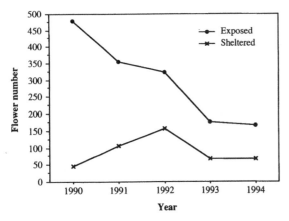

Fig. 5. The declining trend in flower production of exposed *Diapensia lapponica* cushions monitored at Latnjajaure, N Sweden 1990–1994.

all taken place under circumstances where factors other than climate (e.g. trampling, grazing, pollution) are negligible. It is important to note that all age classes except juvenile, non-reproductive clones have been experiencing this rapid decrease in live phytomass at similar levels, i.e., not only old, 'over-mature', and potentially moribund clones (cf. Crawford et al. 1993). Taking the extremely slow turn-over rate of the population into account, this decrease is a virtually irreversible process in the sites that the species inhabits at present. Survival in a changing environment depends on continual adaptation (Crawford 1989), which is not possible in most tundra plants where the life-span of the individual genets by far exceeds the time frame of contemporary climate change. Thus, the only chance of escape is in space through migration and by immediate responses through phenotypic plasticity; the slow population turnover in species like *D. lapponica* gives no opportunity for recombination and adaptation. In other words, massive flowering of *D. lapponica* in the Latnja valley, such as in 1990 and 1991, will never be seen again in our lifetimes.

The forecast for the future of the species in this area is not improved by its reproductive success. Contemporary colonization of other habitats has not occurred, and is not to be expected in the near future, considering the low competitive ability and the low reproductive success in the species at present. The frequent June blizzards have destroyed most of the flowers in the last two seasons (1993 and 1994), and if this climatic regime continues, the role of *D. lapponica* in the local seed pool will diminish substantially.

At the same time, as the part of the local *D. lapponica* population inhabiting plateaux and ridge-tops is suffering from substantial loss in live biomass and flower production due to an earlier thawing, increased winter precipitation and snow-pack thickness has delayed thawing of less exposed sites. The *Diapensia* plants have responded to this shortening of the available growing season by cancelling many of the preformed flower buds, postponing their further development to subsequent seasons. Thus, we should expect a contraction of the ecological amplitude of *D. lapponica* (at least in terms of flowering) to less early- (but not late-) thawing habitats in the near future, even though cushions present in late-thawing sites today may persist for decades or even centuries from now.

This is one of the very few reports of changes in polar terrestrial ecosystems induced by climate change. It parallels the observations by Chapin et al. (1995) from the Alaskan North Slope, showing a significant decrease in evergreen *Eriophorum vaginatum* and a concurrent increase in deciduous *Betula nana* in unmanipulated control plots. Models using experimental data from the Abisko area suggest increasing population growth rate in the sedge *Carex bigelowii* (Carlsson and Callaghan 1994). The observed changes in the plant cover of polar areas, together with results from temperature manipulations in the International Tundra Experiment, ITEX (Henry and Molau unpubl., Welker et al. unpubl.), indicate that tundra plant species react very differently to climatic warming. Walker et al. (1994) reached the conclusion that changes in precipitation and growing season length have significant impact on tundra plant growth and performance, differing among communities and species. Thus, we should expect disintegration of entire plant communities, accompanied by a decrease in biodoversity, if the warming trend continues according to the forecast.

Acknowledgements – This research was supported by grants from the Swedish Natural Science Research Council (NFR), the Kempe Foundation, and the Swedish Society for Nature Conservation. I thank the Abisko Scientific Research Station and its staff for help and hospitality, and J. Alatalo, P. Emanuelsson, M. Engelhard, K. Hagmann, A. Lindskog, E. Nilsson, U. Nordenhäll, H. Pärn, P. Pöyhtäri, M. Sommerkorn, M. Stenström, and M. Svensson for assistance in the field at Latnjajaure during 1990–1994. I am grateful to S. Armbruster, B. Eriksen, D. Murray and P. Wookey for comments and critical reading of earlier drafts of this paper.

References

Carlsson, B. Å. and Callaghan, T. V. 1994. Impact of climate change factors on the clonal sedge *Carex bigelowii*: implications for population growth and vegetative spread. – Ecography 17: 321–330.

Chapin, F. S. III, Shaver, G. R., Giblin, A. E., Nadelhoffer, K. G. and Laundre, J. A. 1995. Responses of arctic tundra to experimental and observed changes in climate. – Ecology 76: 694–711.

Chapman, W. C., and Walsh, J. E. 1993. Recent variations of sea ice and air temperature in high latitudes. – Bull. Am. Meteorol. Soc. 74: 33–47.

Crawford, R. M. M. 1989. Studies in plant survival. Ecological case histories of plant adaptation to adversity. – Stud. Ecol. 11: 1–296.

– , Chapman, H. M., Abbott, R. J. and Balfour, J. 1993. Potential impact of climatic warming on Arctic vegetation. – Flora 188: 367–381.

Fowbert, J. A. and Lewis Smith, R. I. 1994. Rapid population increases in native vascular plants in the Argentine Islands, Antarctic Peninsula. – Arct. Alp. Res. 26: 290–296.

Gori, D. F. 1983. Post-pollination phenomena and adaptive floral changes. – In: Jones, C. E. and Little, R. J. (eds), Handbook of experimental pollination biology. Van Nostrand Reinhold, New York, pp. 31–49.

Groisman, P. Y., Karl, T. R. and Knight, R. W. 1994. Observed impact of snow cover on the heat balance and the rise of continental spring temperatures. – Science 263: 198–200.

Marsden, R. 1992. Snow melt, flowering phenology and reproductive success in tundra plants. – M. Sc. thesis, Univ. of Wales, Bangor.

Minnis, P., Harrison, E. F., Stowe, L. L., Gibson, G. G., Denn, F. M., Doelling, D. R. and Smith, W. R., Jr. 1993. Radiative climate forcing by the Mount Pinatubo eruption. – Science 259: 1411–1414.

Molau, U. 1993a. Relationships between flowering phenology and life history strategies in tundra plants. – Arct. Alp. Res. 25: 391–402.

– 1993b. Phenology and reproductive ecology in six subalpine species of Rhinanthoideae (Scrophulariaceae). – Opera Bot. 121: 7–17.

– 1995. Climate change, plant reproductive ecology, and population dynamics. – In: Guisan, A., Holten, J. I., Spichiger, R. and Tessier, L. (eds), Potential ecological impacts of climate change in the Alps and Fennoscandian mountains. Ed. Conserv. Jard. Bot. Genève, pp. 67–71.

Peterson, H. E. 1908. 2. Diapensiaceae. – In: Warming, E. (ed.), The structure and biology of arctic flowering plants, I. – Meddelelser om Grønland, Vol. 36.

Pihakaski, K. 1988. Seasonal changes in the chloroplast ultrastructure of *Diapensia lapponica*. – Nord. J. Bot. 8: 361–367.

– and Junnila, S. 1988. Cold acclimation of subarctic *Diapensia lapponica* L. – Funct. Ecol. 2: 221–228.

Sørensen, T. 1941. Temperature relations and phenology of the northeast Greenland flowering plants. – Meddelelser om Grønland 125: 1–305 + Pl. 1–15.

Walker, M. D., Webber, P. J., Arnold, E. H. and Ebert-May, D. 1994. Effects of interannual climate variation on aboveground phytomass in alpine vegetation. – Ecology 75: 393–408.

Wiens, D. 1984. Ovule survivorship, brood size, life history, breeding systems, and reproductive success in plants. – Oecologia 64: 47–53.

Ecological Bulletins 45: 220–227. Copenhagen 1996

Plant ecology in subarctic Swedish Lapland: summary and conclusions

T. V. Callaghan and P. S. Karlsson

As the earth's population increases, the areas of wilderness with natural landscapes, biodiversity and ecological processes shrink. Concurrently, increased energy, materials and food consumption by the earth's population create waste products at unprecedented rates. Some of these are transboundary pollutants which have the potential to impact those wilderness areas which have, so far, escaped from intensive land use.

The potential and real anthropogenic environmental threats to our wilderness areas are focusing the ecological community's attention on these areas for several reasons. In order to understand the relationships between the structure and function of ecosystems, their component biota and the complex of numerous abiotic factors which together constitute their environment, it is necessary to investigate pristine areas. Such research has its own considerable intrinsic value and seeks to satisfy our curiosity about the living world and environment surrounding us (Press and Seel 1996, Prock and Körner 1996, Schipperges and Gehrke 1996, Callaghan et al. 1996, Jonsdottir et al. 1996, Karlsson et al. 1996, Sveinbjörnsson et al. 1996, Thorén et al. 1996). Only by understanding and documenting these anthropogenically unperturbed systems and comparing them with those impacted by man can we assess the extent to which man has affected those areas where his activities are concentrated. Also, by understanding the natural ecological and environmental processes occurring in our wilderness areas, we can begin to predict the impacts of environmental change, whether natural (Briffa et al. 1992, Rapp 1996, Holmgren and Tjus 1996, Malmer and Wallén 1996, Berglund et al. 1996) or anthropogenic (Molau 1996, Björn and Holmgren 1996, Gehrke et al. 1996, Jonasson et al. 1996). Such predictions will enable society to develop conservation strategies or remedial measures at the source of anthropogenic perturbation to protect our wilderness areas. An example of such measures at a global scale is the Rio Convention with its specific concern to address problems of decreasing biodiversity.

The contributions to this volume have focused on one important wilderness area on the outskirts of the heavily populated and industrialised countries of Europe: this is the subarctic region of Fennoscandia. This area is an ecotone between taiga and tundra (Andersson et al. 1996) characterised by a low altitudinal treeline of deciduous birch trees (Tenow 1996, Ovhed and Holmgren 1996, Karlsson et al. 1996, Sveinbjörnsson et al. 1996) and extensive heath (Jonasson et al. 1996) and mire communities (Malmer and Wallén 1996). It has a relatively young (Sonesson 1968, 1974, Sonesson and Lundberg 1974, Berglund et al. 1996) but rich and varied landscape (Rapp 1996, Tenow 1996) in which small differences in geology, topography, exposure, aspect and hydrology result in a marked differentiation in plant growth and reproduction (Molau 1996, Callaghan et al. 1996), biodiversity, plant community structure and ecosystem processes while the often harsh subarctic climate results in severe resource constraints (Diemer 1996, Jonasson and Michelsen 1996, Karlsson and Weih 1996, Thorén et al. 1996, Jonsdottir et al. 1996) and the limitation of growth and distribution of biota. Within this area, land use – particularly by the Same over the past several hundred years – is not intensive although locally important (Emanuelsson 1987) and transboundary pollution is relatively low.

Within subarctic Fennoscandia, the Swedish Royal Academy of Science's Abisko Scientific Research Station (ANS) in the Torneträsk region of northern Swedish Lapland represents a focal point for natural sciences including plant ecology, which this volume has addressed to honour the retirement of the ANS Director, Professor Mats Sonesson. In this concluding chapter, we draw together common threads from the various contributions, summarise their findings and we finally seek to identify major challenges to this area of research in the future.

Historical and recent changes in the landscape, climate and ecology of the Torenträsk region

Details of the location, general environment and ecology of the Torneträsk region are presented in Rapp (1996), Tenow (1996), Andersson et al. (1996) and

Berglund et al. (1996). Briefly, The mountain areas of the Abisko area belong to the alpine zone above treeline which is formed by the mountain birch *Betula pubescens* ssp. *tortuosa* between 550–650 m a.s.l. in the west and 700–800 m a.s.l. in the east. *Pinus sylvestris* has a continuous distribution only in the more continental east but occurs sporadically in the Abisko valley, *Picea abies* already reaches its western and northern limit to the east of Lake Torneträsk. Both heath and meadow forest ecosystems occur in the Abisko area (Sonesson and Lundberg 1974) while subalpine heaths occur locally within the subalpine forest belt and above it. In the alpine regions, diverse communities occur according to local environmental conditions such as geology and topography.

The Abisko area was deglaciated at c. 8500 ^{14}C yr BP (Berglund et al. 1996). Following this, vegetation history based on pollen analyses in formative research by Sonesson (1968, 1974) and now confirmed by Berglund et al. (1996) using recent sedimentological, mineral magnetic, oxygen isotope and plant macrofossil records from a lake sequence at tree-line, shows that the Abisko area has been dominated by a "subalpine birch woodland tundra" except for the period between 5500 and 3500 ^{14}C yr BP when the boreal pine-birch forest zone reached the area and the treeline extended up to 100 m above the present limit. During this warmer period the climate was temperate-continental rather than subarctic-oceanic. The climatic deterioration at the end of the temperate-continental climatic period was associated with an increase in soil erosion and the reformation of the Kårsa glacier in the area.

During the early birch period (c. 7700–8500 ^{14}C yr BP) *Hippophaë* was abundant following deglaciation to be followed by herbs and *Lycopodium* (Berglund et al. 1996). This period was followed by the birch-alder period (5500–7700 ^{14}C yr BP) during which these dominants were associated with *Juniperus* and herbs. Following the pine period, between the present and 3400 ^{14}C yr BP, the late birch period has been associated with the abundance of *Salix*, *Juniperus*, Ericales and herbs.

Within the last 1400 yr, summer temperatures determined from pine treerings collected at the eastern part of the Torneträsk region have varied from modern normals (1951–70) by +1.99 to −2.06 for individual years, +0.68 to −0.90 for 20 yr means and +0.54 to −0.67 for 50 yr means (Briffa et al. 1992). Cool periods include the years AD 500–700, 790–870, 1110–1150, 1190–1360, 1570–1750 whereas relatively warm periods include 720–790, 870–1110 and 1360–1570. The cool period between c. 1190 and 1360 seems to be associated with the onset of peat formation on the ombrotrophic mires around Abisko (Malmer and Wallén 1996), although the current relatively warm period seems to be associated with a switch on the bogs from carbon sink to source.

In 1859, a network of Swedish climate stations was established and some were located in the eastern low mountains. Within the Torneträsk region per se, climate records started c. 1900 AD and records have been kept at the Abisko observatory from 1912. Within this instrumental period, there have been only minor changes in summer air temperatures from the 1870's to c. 1910 when an increase of c. 1°C occurred (Holmgren and Tjus 1996). This warmer period was maintained until c. 1940 when there was a decrease of a few tenths of a degree although the average summer temperature at the end of the present century is still c. 1°C higher than at the end of the last century. Spring temperatures in particular have risen; since 1870 there has been a steady increase reaching 1.5°C at present and this is associated with an increase in length of growing season of 1–3 weeks (1913–1994) (Holmgren and Tjus 1996). During this period, models calculate that summer evapotranspiration of the birch forest near Abisko (130 mm) has been higher than precipitation (126 mm) (Ovhed and Holmgren 1996). Of the evapotranspiration of the forest, the ground vegetation is an important component contributing 37%.

Recent temperature trends are associated with the recession of several glaciers, e.g. the Kårsa, Ekman and Kärkereppe glaciers, in the western Abisko mountains during this century (Rapp 1996). They are also associated to some extent with the dynamics of the birch treeline but the relationships are obscured in some areas due to traditional land use impacts such as grazing by reindeer, cows, goats and horses and the modern construction of the railway (Tenow 1996, Holmgren and Tjus 1996, Sveinbjörnsson et al. 1996). Holmgren and Tjus (1996) suggest that the birch treeline has been relatively stable since 500 AD although there has been a recent increase in altitude of 25–50 m (Sonesson and Hoogesteger 1983) in response to the warm period in the 1930's. Nevertheless, this increase is small compared with that of 200 m expected from the increase in temperature over the past 100 yr and this has been interpreted such that the present treeline is a relict of past temperatures and did not respond to the lower temperatures at the end of the last century (Holmgren and Tjus 1996).

While the birch treeline may respond to climatic variations, other environmental conditions are also locally important such as land use and insect herbivory (Tenow 1996) and these may obscure the relationships between climate and treeline. Within the birch forest, the growth of the trees and sometimes the ground flora is affected periodically, i.e. every 9–10 yr, by defoliation by the moth *Epirrita autumnata*. Population peaks of the caterpillars reach extreme levels (e.g. 100–200 caterpillars per 100 birch shoots which is equivalent to 1000 caterpillars m[−2] of forest floor) between longer intervals (Tenow 1996). In 1955, the caterpillars completely defoliated the forest of the Abisko valley and there were also changes in the ground vegetation from dwarf shrubs to grasses through grazing of the dwarf shrubs by the caterpillars, wilting of the dwarf shrubs after defoliation of the covering birch canopy and increase in nutrient availability following the processing of birch leaves by the caterpillars. Estimated duration of the recovery period varies from 25 to 70 yr (Tenow 1996) and it could be inferred that cyclicity in birch forest dynamics could occur over these time scales.

Periodicity in vegetation dynamics in the subarctic through lemming cycles is well known with population cycles of 4–5 yr (Laine and Henttonen 1983) but population peaks have not been recorded in the Abisko area since the early 1980's (Andersson pers. comm.). Over even shorter time periods of 3–4 yr, flowering and shoot population dynamics of the sedge *Carex bigelowii* have been demonstrated over the period 1985–1986 (Callaghan et al. 1996).

Current relationships between the ecosystems, biota and environments of the Torneträsk region

The contributions to this volume summarised above show that the vegetation of the Abisko area is dynamic: there have been changes since the last glacial period associated with natural climatic change events, and there are changes which still occur at scales from decades to 3–4 yr. These vegetation dynamics have been identified by retrospective analyses of historical vegetation and plant growth records and by observation and monitoring over relatively long periods. Against this dynamic background, many ecological studies based at the ANS have explored the current relationships between organisms, ecosystems and the environment in detail and often experimentally, rather than through observation alone. This approach seeks, among other issues, to determine the mechanisms which result in the plant behaviour and distributions we see currently. Once these mechanisms are understood, we then have the potential to predict future behaviour and responses to changing environmental and biotic conditions.

It is well known that plant performance and diversity decrease along a latitudinal gradient from temperate latitudes to the Arctic but the causes are debatable. Severe abiotic stress and physical disturbance from the harsh arctic weather, short growing seasons and low resource availability all contribute to low performance but the exact causes are species – and can be even genet – specific.

Cryptogams are a dominant component of many arctic ecosystems and have important impacts on local temperature and moisture (mosses) or are an important component of the diet of semi-domesticated reindeer (lichens) (Schipperges and Gehrke 1996). Some cryptogams exploit particularly narrow niches, for example some epiphytic lichens are restricted to growth above (*Parmelia olivacea*) or above and below (*Parmeliopsis ambigua*) the height of winter snow on birch trunks (Sonesson 1989) whereas some mosses can photosynthesise using low light conditions and high concentrations of CO_2 accumulated under snow cover in spring (Schipperges and Gehrke 1996). The ability to exploit narrow niches can be achieved through genecological differentiation and different genetically determined patterns of gas exchange have been described for alpine and subalpine populations of the lichen *Nephroma arcticum* (Sonesson et al. 1992).

The most dramatic distributional limit of a species in the Subarctic is that of the mountain birch. The patterns of treeline dynamics described above have been addressed mechanistically by Sveinbjörnsson et al. (1996). Differences in carbon balance between trees at treeline and below showed no great differences although the period during which treeline individuals maintained their leaf canopy was less in treeline individuals than in those growing below. However, application of fertiliser and measurements of soil and tissue N (nitrogen) and P (phosphorus) showed that nitrogen availability to the trees decreased with altitude. Also, the growth of treeline individuals was more limited by nitrogen balance, through differences in N availability, uptake and loss, than that of trees growing below treeline (Sveinbjörnsson et al. 1996).

Nutrient limitation to plant growth is a general phenomenon in arctic ecosystems (Jonasson and Michelsen 1996). In the Torneträsk region, the living phytomass in various ecosystems contained between 5 and 30% of the ecosystem pools of biologically fixed N and P. Between 4 and 14% of the ecosystem N occurred in plant tissues with a turnover time of < 10 yr. About 6% of the soil pool of N and 35% of the soil pool of P were immobilised in soil microbial biomass. The supply of N from atmospheric deposition and fixation appear to be important only on mires; elsewhere, vascular plants received > 50% of the N and P for annual growth from stored reserves (Jonasson and Michelsen 1996). The remainder was acquired through mineralisation of N and P in organic matter, which was a slow process, and leaching of litter and soil organic matter. The determination of natural abundance of 15N in various plant species with and without different types of mycorrhizae suggest that the dwarf shrubs of the Torneträsk area have access to organically bound N in the organic soils there through their mycorrhizal symbionts (Jonasson and Michelsen 1996).

The internal storage and recycling of nutrients within perennial subarctic plants together with uptake of organic N through mycorrhizal symbionts are two forms of adaptation to nutrient limitations in cold, organic soils. Other adaptations are also important, including the foraging for nutrients in habitats spatially heterogeneous for nutrients by rhizomatous/stoloniferous clonal plants (Jonsdottir et al. 1996), the dependence of hemiparasites on host plants (Press and Seel 1996) and plant carnivory (Karlsson et al. 1996).

Clonal perennial plants dominate the vegetation of the Torneträsk region and a higher proportion of them is rhizomatous or stoloniferous compared with more southerly locations (Jonsdottir et al. 1996). Rhizomatous arctic plants such as *Carex bigelowii* have a modular construction in which proliferation depends on module production rather than risky seed production and seedling recruitment. The production of rhizomes by modules leads to a foraging behaviour by long series of modules which are both physically and physiologically connected, allowing translocation of nutrients, water, carbon and hormones within a clone (Jonsdottir et al. 1996). Young modules initially receive resources from the ancestral modules and have growth which is subsidised in unfavourable patches i.e. nutrient poor, where competition is intense or when herbivory occurs. As young modules develop, they provide ancestral modules with carbon and receive nutrients and water from them thereby exhibiting a division of labour in time within the clone. There is also a physiological, morphological and reproductive differentiation of sibling modules within some species which also results in a division of labour in space (Jonsdottir et al. 1996). The dominance of clonal plants with these abilities to buffer environmental adversity in time and space contributes to the resilience of subarctic ecosystems to environmental change.

Hemiparasitism is another important strategy whereby some angiosperms in the Subarctic acquire their nutrients supply (Press and Seel 1996). In the Abisko area, ten species (5 of which are common) from 5 genera of hemiparasitic Scrophulariaceae have been recorded. Some species are facultative parasites and can grow in the absence of a host plant but when doing so their growth is limited by their ability to both absorb and assimilate inorganic solutes. Attachment to a host greatly stimulates their growth, particularly if the host is a nitrogen-fixing legume, suggesting that heterotrophic nitrogen supply may play a leading role in determining the success of the hemiparasite (Press and Seel 1996). Although less is known about the effect of the hemiparasite on the host, a study of the association between the hemiparasite *Rhinanthus minor* and its host *Poa alpina* showed that the host accumulated less biomass and had lower shoot:root ratios than uninfected controls but that rates of photosynthesis were unaffected by infection (Press and Seel 1996).

Plant carnivory, i.e. the ability of some plants to capture prey, digest them and assimilate nutrients from them, is another plant adaptation to low soil nutrient availability (Karlsson et al. 1996). In subarctic parts of northern Scandinavia, three species in the carnivorous

genus *Pinguicula* are found: *P. alpina* (characteristic of nutrient rich calcareous soils, *P. villosa* (characteristic of nutrient poor *Sphagnum* bogs) and *P. vulgaris* (tolerant of a wide ecological amplitude). Experiments evaluating the effects of trapping rate on growth and demographic parameters contributing to the fitness of the three plant species showed that prey capture allowed *P. vulgaris* to increase its seed output, through an increase in the amount of seeds produced per reproductive plant and an increase in the number of reproductive events. This species also exhibited higher survival rates as a result of prey capture (Karlsson et al. 1996). The other two species benefited only marginally from carnivory and this benefit was related to the lower rate of prey capture than that of *P. vulgaris* (Karlsson et al. 1996).

The increased sexual reproductive effort (RE) of *P. vulgaris* in response to prey capture and enhanced nutrition (Karlsson et al. 1996) shows one example of allocation of resources whereas the foraging strategy of *Carex bigelowii* shows an alternative great investment of resources in vegetatively produced modules (Jonsdottir et al. 1996). Such functions as reproduction, proliferation and growth in a plant compete for resources and there may be a relative somatic cost (RSC) associated with reproductive effort (Thorén et al. 1996). The relationship between RSC and RE varied among a range of subarctic plant species and also varied between different habitats within a species and even between resources (Thorén et al. 1996). The number of species for which detailed calculations of RE and RSC are available are relatively few. So far, no pattern has emerged as to whether plants from cold environments such as the Subarctic differ from temperate plants with respect to their reproductive investments. Neither is it clear if a specific investment in reproduction leads to similar costs in different environments. It also remains to be seen under what conditions plants can compensate for their reproductive resource investments and thus avoid reproductive costs (Thorén et al. 1996).

The likely responses of subarctic ecosystems and their plants to future environmental changes

The fundamental understanding of many of the interactions between plants, soils and the subarctic environment presented in this volume have formed a focus and framework for exploring the likely responses of subarctic ecosystems to various aspects of future environmental changes. In addition, our knowledge of vegetation responses to varying climate in the past provides an appropriate context for interpreting the magnitude of future changes. Although we can compare predictions of future ecosystem responses to climatic change with past responses, we have much more limited potential to use analogous contexts for the depletion of stratospheric ozone and the eutrophication of ecosystems through atmospheric nitrogen deposition which are recent, anthropogenic perturbations.

Depletion of stratospheric ozone caused by anthropogenically released gases was first highlighted in the Antarctic where depletion episodes have reached 70% (Madronich et al. 1995). Damage to the ozone layer is enhanced by low and stable stratospheric temperatures and there has been a recent (1996) depletion of stratospheric ozone in the Arctic of 50% (WMO Bulletin, 12 March 1996). It is estimated that for about each 1% reduction in stratospheric ozone, there could be an increase by 2% in biologically harmful UV-B radiation reaching the earth's surface. However, the actual quantity of UV-B radiation penetrating to the earth's surface is lower because of attenuation by cloud and atmospheric aerosol pollutants (Björn and Holmgren 1996). At Abisko, models of UV-B and PAR (photosynthetically active radiation) radiation show that clouds decrease UV-B receipt by 25.8% and that over 3–4 decades prior to 1994, the ozone column at Abisko showed no depletion (Björn and Holmgren 1996). Although no long term increases in UV-B radiation have been identified yet, there is a general increase in UV-B radiation in the north (Madronich et al. 1995) and the recent dramatic ozone depletion episode demonstrates the need to predict possible future impacts of enhanced UV-B radiation on arctic ecosystems which, hitherto, have experienced low levels because of low solar angles and increased atmospheric attenuation of the radiation. At Abisko, whole ecosystem experiments have been carried out in which birch forest heath vegetation and an ombrotrophic bog have been irradiated with enhanced UV-B radiation simulating a 15% stratospheric ozone reduction (equivalent to 19% when cloud cover is included in calculations). Such field perturbations have included interactions with other climate change variables particularly enhanced atmospheric CO_2 concentrations (600 ppm) and increased summer precipitation while controlled environment experiments have addressed the detailed and short term plant physiological and microbial decomposition responses to enhanced UV-B radiation (Gehrke et al. 1996).

Responses of subarctic mosses, lichens, and deciduous and evergreen dwarf shrubs were subtle rather than dramatic: the four dwarf shrubs showed reduced shoot growth and differences in leaf thickness while the mosses *Hylocomium splendens* and *Sphagnum fuscum* showed a more rapid response and greater reduction in growth (Gehrke et al. 1996). Some counter-intuitive responses were however, also observed: *H. splendens* growth increased when subjected to enhanced UV-B and 50% additional summer precipitation. Exposure to combined increases of UV-B and CO_2 resulted in reduced growth of the deciduous dwarf shrub *Vaccinium*

myrtillus but increased growth of the moss *H. splendens*. Effects of enhanced UV-B on the tissue chemistry of *V. myrtillus* and *V. uliginosum* resulted in reductions of decomposition of leaf litter while decomposition of leaf litter under enhanced UV-B was further reduced by its impact on microbial decomposer communities (Gehrke et al. 1996). These trends could affect nutrient cycling in the strongly nutrient limited ecosystems of the Subarctic (Jonasson and Michelsen 1996) while affecting the storage of carbon in soils there.

Impacts of likely climatic change resulting from anthropogenically increased greenhouse gases are of particular concern in the Arctic because climatic change is predicted to be greatest there, particularly during winter (mean annual temperature increases of 2–4°C in summer, Cattle and Crossley 1995 compared with an increase of only 1°C over the past 100 yr, Holmgren and Tjus 1996) and even small increase in temperature are likely to result in disproportionately large responses in the biota and ecosystems there (Chapin et al. 1992, Callaghan et al. 1992, Callaghan and Jonasson 1995). As the Arctic has important feedback mechanisms on global atmospheric processes through its high albedo (reflectivity) and long history of sequestration of atmospheric CO_2 (an important greenhouse gas) in its soils (Jonasson et al. 1996), there is also concern that climatic warming could change historical negative feedbacks from arctic ecosystems to positive feedbacks. Such changes would occur if boreal forests with low albedo (e.g. the birch trees near Abisko absorb 51% of the net radiation: Ovhed and Holmgren 1996) displaced tundra with high albedo and if warming arctic soils stimulated a net loss of carbon as a result of increased microbial decomposer activity (Malmer and Wallén 1996, Jonasson et al. 1996).

To predict the responses of dominant subarctic ecosystems to various aspects of environmental change, particularly climatic warming during summer, a series of whole ecosystem manipulation experiments were established by international groups along an environmental severity gradient from the subarctic birch forest, through a treeline heath to an alpine fellfield community which formed part of a wider gradient including a temperate upland community and a high arctic heath and polar semi-desert (Jonasson et al. 1996). In the Abisko area, manipulations consisted of temperature enhancements using passive greenhouses (2–4°C warming over the summer period), nutrient additions to simulate both predicted increased microbial decomposer activity related increase in available plant nutrients and increased deposition of atmospheric nitrogen, shade treatments to simulate increased cloudiness and increased competition for light, and increased summer precipitation.

Soil nutrient mineralisation was not affected by increased temperatures inside passive greenhouses as soil temperatures changed little compared to those of the air (Jonasson et al. 1996). Fertiliser application increased the N and P content in soil microbial biomass indicating their potential to immobilise any additional nutrients. Plant growth and reproductive output responded to increases in temperature, particularly at those sites where temperatures were lowest or where the species were close to their distribution limits, and particularly to nutrient addition. Increased temperature promoted demographic responses in meristem dynamics more than the growth of individual leaves and shoots (Jonasson et al. 1996). As in the experimental manipulations of UV-B and CO_2 radiation, many of the plant responses were rather subtle and there were no dramatic implications for early changes in biodiversity. However, shading clearly disadvantaged some species and there was a pronounced reduction in the cover of the moss *Hylocomium splendens* at the forest site in response to the manipulations and probably increased growth and shade of the overstorey (Potter et al. 1995).

Monitoring of plant performance and relating this to interannual variations in climate has been carried out on the cushion plant *Diapensia lapponica* for the period 1990–1994 (Molau 1996) and for the sedge *Carex bigelowii* for the 12 yr period (1984–1995) (Callaghan et al. 1996) at alpine fellfield sites near Abisko. In addition, retrospective analyses of the growth and growing point dynamics of the vascular cryptogam *Lycopodium annotinum* in the forest floor community, and the dwarf shrub *Cassiope tetragona* from alpine and subalpine heaths have provided climate related growth records for at least two decades (Callaghan et al. 1996). Such analyses show individualistic responses of the plants to climate variability. *Diapensia lapponica* performance was retarded both by increased precipitation as snow and shorter growing seasons in sheltered areas over the 5 yr of monitoring and by earlier snow melt and warmer spring temperatures in more exposed sites (Molau 1996). During the monitoring period, biomass decreased by 22%, flowering decreased by 55% and many old cushions died while recruitment was not observed. At a nearby alpine site, the flowering of *C. bigelowii* showed a cyclical pattern with a cyclicity of c. 3–4 yr and an apparent decrease over the 12 yr period. In contrast, *L. annotinum* showed a strong buffering ability against climatic variations which led to stable clonal growth. As many of the plants of the Arctic and Subarctic are long-lived clonal perennials, and recruitment may be infrequent or sporadic, it is necessary to interpret such demographic responses within a long term context (Callaghan et al. 1996).

Over even longer periods, climatic change is expected to result in the displacement of subarctic plant populations by temperate species and populations. However, a study of photosynthetic performance and leaf carbon gain of temperate and subarctic genotypes of *Geum rivale* and *Ranunculus acris* in which the populations were grown together in common garden experiments suggest that it is unlikely that southern provenances attain greater plant carbon balances in the Subarctic than native genotypes and it is therefore, unlikely that southern provenances will displace existing subarctic populations in the future (Diemer 1996). Measurements of plant performance and phenology in a number of species at temperate lowland and alpine, subarctic lowland and alpine, and high arctic lowland locations show that there are few latitudinal differences in parameters such as leaf weight ratio and leaf life span in species of the genera *Ranunculus* and *Geum*. However, differences in specific leaf area and fine root weight ratio were pronounced (Prock and Körner 1996). Reciprocal transplants within the lowland or alpine life zone revealed strong but species-specific phenorhythmic disorder related to differences in photoperiod. These experiments imply that during climatic warming and species migration, photoperiod will be a constraint on those species with strong photoperiod controls (Prock and Körner 1996) additional to the limitations on the displacement of subarctic species by temperate alpine species through inappropriate carbon balance (Diemer 1996). However, opportunistic species will benefit in terms of biomass production, reproduction and, potentially, abundance (Prock and Körner 1996).

While many of the responses of plants and vegetation to predicted climatic warming are expected to be directly related to climate as discussed above, indirect effects are also likely to be important. In particular, the future response of mountain birch to predicted increases in temperature may be mediated through herbivory by moth caterpillars. Currently, the large scale defoliation of birch forest by caterpillars such as *Epirrita autumnata* is controlled to some extent by the mortality of overwintering eggs in particularly cold winters ($< -36°C$, Tenow 1996). Increases in winter temperatures could, therefore, increase the incidence of pest outbreaks. In addition, other herbivore species such as *Operophtera brumata* from warmer areas adjacent to Abisko may become more active in the Abisko area and increase the defoliation impact on the trees. To some (unknown) extent, this process might be moderated by the birch trees' defence to herbivory and changes in chemistry due to warmer temperatures, increased CO_2 concentrations and enhanced UV-B radiation.

Herbivory is a major cause of loss of nitrogen in mountain birch trees (Karlsson and Weih 1996). Leaf abscission and fine root turnover are other major mechanisms of N loss which is important for the performance of the subarctic birch as leaf nitrogen concentrations in situ are less (1.24 mmol N g^{-1}) than the optimum (1.5 mmol N g^{-1}) for leaf productivity. In addition to soil nutrient availability, soil temperature is an important determinant of N uptake rate and thus of plant nitrogen status and the growth of birch seedlings (Karlsson and Weih 1996). N uptake is particularly responsive to soil temperatures between 5 and 15°C which suggests that mountain birch performance could be greatly affected by climatic warming (Karlsson and Weih 1996).

While impacts on subarctic ecosystems have obvious local and regional concerns, the interaction between northern ecosystems and the atmosphere have the potential for global implications as discussed above. The apparent short term litter formation rate in the dominating *Sphagnum* communities which characterise the extensive ombrotrophic parts of mires around Abisko have been investigated over a 14 yr period by following the fate of plant matter labelled with ^{14}C. These studies have been interpreted within the context of longer term dating of rates of peat accumulation which show that the formation of ombrotrophic peat started c. 800 yr ago and that litter annual decay has been only approximately 20–25% of the annual production rate until recently. However, currently the litter production rate is only c. 35 g m^{-2} yr^{-1} which is not adequate to compensate for decay losses. Thus, the carbon accumulation patterns suggest that these bogs have recently changed from sink to sources of CO_2 (Malmer and Wallén 1996), a pattern that has also been suggested for the tundra of Alaska (Oechel et al. 1993).

Future challenges for plant ecological studies in the Subarctic

This volume has shown how a wide range of approaches have been successfully employed to investigate past and present relationships between subarctic ecosystems and their environments and to predict future responses of the ecosystems to likely environmental change. The linking of the various approaches, all focused on the range of ecosystems and ecotones surrounding Abisko, have provided an important framework for future research and an extensive database for interrogation and interpretation of ecological and environmental problems of current concern.

Although the contributions to this volume have covered a wide range of biological foci, from organs and tissues within plants to ecosystems, and an equally large range of environmental variables, our capacity to address interactions among organisms and/or taxa, and among co- or independently varying environmental factors is limited. Multiple interactions among environmental variables which are known to be changing, e.g. temperature, precipitation, atmospheric CO_2, UV-B radiation, atmospheric nitrogen deposition, and increased reindeer grazing, need to be addressed in large scale multifactorial experiments. Because of their scale, width of required expertise and cost, such experiments would need to be collaborative ventures.

Interactions between biological units also need to be specifically targeted for research: the work presented here often exemplifies in depth understandings of processes within individual plants and monospecific populations, between a plant host and its parasite or mycorrhiza, between a plant species and its insect prey or herbivore and – at a very different level – between vegetation zones over time. However, we do not yet have the ability or tools to synthesise the dynamics of ecosystems from the dynamics of the component species. This is a daunting challenge but again, the baseline information available through the focusing of research at the ANS, is a particularly useful starting point.

The ecological work presented here has spanned temporal variability from 8500 ^{14}C yr BP to instantaneous gas exchange. However, the studies are relatively constrained to a small range of spatial variability centred around the recording quadrat or the experimental plot. A future challenge will be to increase our understanding of the scales at which ecological processes operate at both ends of the spectrum. Thus, high level resolution molecular techniques are required to understand the basis of changes in biodiversity and the activity of micro-organisms at one end of the spectrum while landscape processes addressed within this volume (Rapp 1996) at the other end of the spectrum need to be related to processes operating at smaller spatial scales. It is noticeable that the modelling of ecological processes is limited within this volume (but see the welcome exception by Ovhed and Holmgren 1996): one challenge for such modelling is to provide conceptual and mathematical frameworks for the linking together of multispecies responses to multifactor experiments at a range of temporal and spatial scales.

The width and depth of knowledge surrounding the outstanding and varied wilderness of the Fennoscandian Subarctic, together with the gaps in our research which need to be filled, should be used to ensure that this heritage is not destroyed by local or global anthropogenic perturbation: this application of our knowledge is, perhaps, the greatest challenge for the future.

Acknowledgements – We thank the Director of the Abisko Scientific Research Station, M. Sonesson, to whom this volume is dedicated, for his collaboration, encouragement and hospitality and facilities provided at the ANS. We also thank the staff of the ANS for many types of help, N. Malmer for advice and encouragement during this project and L. Svensson, Managing Editor of Ecography, for support and considerable patience during the editing process. Finally, we are greatly indebted to our friends and colleagues for their most welcome contributions to this Festschrift.

References

Andersson, N. Å., Callaghan, T. V. and Karlsson, P. S. 1996. The Abisko Scientific Research Station. – Ecol. Bull. 45: 11–14.

Berglund, B. E., Barnekow, L., Hammarlund, D., Sandgren, P. and Snowball, I. F. 1996. Holocene forest dynamics and climate changes in the Abisko area, northern Sweden – the Sonesson model of vegetation history reconsidered and confirmed. – Ecol. Bull. 45: 15–30.

Björn, L. O. and Holmgren, B. 1996. Monitoring and modelling of the radiation climate at Abisko – Ecol. Bull. 45: 204–209.

Briffa, K. R., Jones, J. D., Plicher, J. R., Karlén, W., Schweigruber, F. H. and Zetterberg, P. 1990. A 1400 year tree-ring record of summer temperatures in Fennoscandia. – Nature 434–439.

Callaghan, T. V., Sonesson, M. and Sømme, L. 1992. Responses of terrestrial plants and invertebrates to environmental change at high latitudes. – Phil. Trans. R. Soc. Lond. B. 338: 279–288.

– and Jonasson, S. 1995. Arctic terrestrial ecosystems and environmental change. – Phil Trans. R. Soc. Lond. A. 352: 259–276.

–, Carlsson, B. Å. and Svensson, B. M. 1996. Some apparently paradoxical aspects of the life cycles, demography and population dynamics of plants from the subarctic Abisko area. – Ecol. Bull. 45: 133–143.

Chapin, F. S. III, Jefferies, R. L., Reynolds, J. F., Shaver, G. R. and Svoboda, J. 1992. Arctic ecosystems in a changing climate: an ecological perspective. – Academic Press, pp. 11–34.

Cattle, H. and Crossley, J. 1995. Modelling arctic climate change. – Phil. Trans. R. Soc. Lond. A. 352: 201–213.

Diemer, M. 1996. A comparison of photosynthetic performance and leaf carbon gain of temperate and subarctic genotypes of *Geum rivale* and *Ranunculus acris* in northern Sweden. – Ecol. Bull. 45: 144–150.

Emanuelsson, U. 1987. Human influence on vegetation in the Torneträsk area during the last three centuries. – Ecol. Bull. (Stockholm) 38: 95–111.

Gehrke, C., Johanson, U., Gwynn-Jones, D., Björn, L. O., Callaghan, T. V. and Lee, J. A. 1996. Effects of ultraviolet-B radiation on terrestrial subarctic ecosystems and implications for interactions with increased atmospheric CO_2. Ecol. Bull. 45: 192–203.

Holmgren, B. and Tjus, M. 1996. Summer air temperatures and tree line dynamics at Abisko – Ecol. Bull. 45: 159–169.

Jonsdottir, I. S., Callaghan, T. V. and Headley, A. D. 1996. Resource dynamics within arctic clonal plants. – Ecol. Bull. 45: 53–64.

Jonasson, S. and Michelsen, A. 1996. Nutrient cycling in subarctic and arctic ecosystems, with special reference to the Abisko and Torneträsk region. – Ecol. Bull. 45: 45–52.

–, Lee, J. A., Callaghan, T. V., Havström, M. and Parsons, A. N. 1996. Direct and indirect effects of increasing temperatures on subarctic ecosystems. – Ecol. Bull. 45: 180–191.

Karlsson, P. S. and Weih, M. 1996. Relationships between nitrogen economy and performance in the mountain birch *Betula pubescens* spp. *tortuosa*. – Ecol. Bull. 45: 71–78.

–, Svensson, B. M. and Carlsson, B. Å. 1996. The significance of carnivory for three *Pinguicula* species in a subarctic environment. – Ecol. Bull. 45: 115–120.

Laine, K. and Henttonen, H. 1983. The role of plant production in microtine cycles in northern Fennoscandia. – Oikos 40: 407–418.

Madronich, S., McKenzie, R. L., Caldwell, M. M. and Björn, L. O. 1995. Changes in ultraviolet rdiation reaching the earth's surface. – Ambio 24: 143–152.

Malmer, N. and Wallén, B. 1996. Peat formation and mass balance in subarctic ombrotrophic peatlands around Abisko, northern Scandinavia. – Ecol. Bull. 45: 79–92.

Molau, U. 1996. Climatic impacts on flowering, growth and vigour in an arctic-alpine cushion plant, *Diapensia lapponica*, under different snow cover regimes. – Ecol. Bull. 45: 209–219.

Oechel, W. C., Hastings, S. J., Jenkins, M., Reichers, G., Grulke, N. and Vorlitis, G. 1993. Recent change of Arctic tundra ecosystems from a carbon sink to a source. – Nature 361: 520–526.

Ovhed, M. and Holmgren, B. 1996. Modelling and measuring evapotranspiration in a mountain birch forest. – Ecol. Bull. 45: 31–44.

Potter, J. A., Press, M. C., Callaghan, T. V. and Lee, J. A. 1995. Growth responses of *Polytrichum commune* Hedw. and *Hylocomium splendens* (Hedw.) Br. Eur. to simulated environmental change. – New Phytol. 131: 533–541.

Press, M. C. and Seel, W. E. 1996. Interactions between hemiparasitic angiosperms and their hosts in the subarctic. – Ecol. Bull. 45: 151–158.

Prock, S. and Körner, C. 1996. A cross-continental comparison of phenology, leaf dynamics and dry matter allocation in arctic and temperate zone herbaceous plants from contrasting altitudes. – Ecol. Bull. 45: 93–103.

Rapp, A. 1996. Photodocumentation of landscape change in northern Swedish mountains. – Ecol. Bull. 45: 170–179.

Schipperges B. and Gehrke. C. 1996. Photosynethic characteristics of subarctic mosses and lichens. – Ecol. Bull. 45: 121–126.

Sonesson, M. 1968. Pollen zones at Abisko, Torne Lappmark, Sweden. – Bot. Notiser 121: 491–500.

– Late Quaternary development of the Torneträsk area, northern Sweden: 2. Pollen analytical eveidence. – Oikos 25: 288–307.

– Water, light and temperature relations of the epiphytic lichens *Parmelia olivacea* and *Parmeliopsis ambigua* in northern Swedish Lapland. – Oikos 56: 402–415.

– and Lundberg, B. 1974. Late quaternary forest development of the Torneträsk area, north Sweden. 1. Structure of modern forest ecosystems. – Oikos 25: 121–133.

– and Hoogesteger, J. 1983. Recent tree-line dynamics (*Betula pubescens* Ehrh. ssp *tortuosa* (Ledeb.) Nyman in northern Sweden. – Nordicana 47: 47–54.

–, Schipperges. B. and Carlsson, B. Å. 1992. Seasonal patterns of photosynthesis in alpine and subalpine populations of the lichen *Nephroma arcticum*. – Oikos 65: 3–12.

Sveinbjörnsson, B., Kauhanen, H. and Nordell, O. 1996. Treeline ecology of mountain birch in the Torneträsk area. – Ecol. Bull. 45: 65–70.

Tenow, O. 1996. Hazards to a mountain birch forest – Abisko in perspective. – Ecol. Bull. 45: 104–114.

Thorén, L. M., Hemborg, A. and Karlsson, P. S. 1996. Resource investment in reproduction and its consequences for subarctic plants – Ecol. Bull. 45: 127–132.

Printed and bound by CPI Group (UK) Ltd, Croydon, CR0 4YY

27/10/2024

14580391-0004